Grundlehren der mathematischen Wissenschaften 239

A Series of Comprehensive Studies in Mathematics

P. Turán (1910–1976)

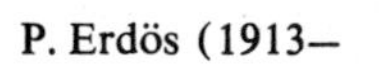

P. Erdös (1913–   )

M. Kac (1914–　　)

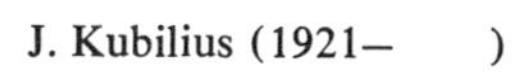

J. Kubilius (1921–　　)

P.D.T.A. Elliott

Probabilistic Number Theory I

Mean-Value Theorems

Springer-Verlag
New York Heidelberg Berlin

P.D.T.A. Elliott
Department of Mathematics
University of Colorado
Boulder, Colorado 80309
USA

AMS Subject Classifications (1980): 10KXX, 60B99, 60F99

Library of Congress Cataloging in Publication Data

Elliott, Peter D
Probabilistic number theory.

(Grundlehren der mathematischen Wissenschaften;
239-240)
Bibliography: p.
Includes index.
CONTENTS: v. 1. Mean-value theorems.—v. 2. Central
limit theorems.
1. Probabilistic number theory. I. Title.
II. Series.
QA241.7.E55 512′.7 79-20824

Printed in the United States of America.

9 8 7 6 5 4 3 2 1

ISBN 0-387-90437-9 Springer-Verlag New York
ISBN 3-540-90437-9 Springer-Verlag Berlin Heidelberg

Grundlehren der mathematischen Wissenschaften 239

Probabilistic Number Theory I: Mean-Value Theorems
by P.D.T.A. Elliott

Errata for Volume I

Page 2, line 16 should read: function $g(n)$ non-decreasing for all sufficiently large integers n, with the

Line 23 should read:

$$g(n) = \begin{cases} \log\log n & \text{if } n \geq 3, \\ 0 & \text{if } 1 \leq n < 3. \end{cases}$$

Page 13, line 5. In the second summation condition read: $p^k \| m$

Page 82, line 20 and page 88, line 7. The sum should be over the range $i = 1$ *to* $i = s - 1$.

Page 167, line 22. For D read: **D**.

Page 175, line 17. For which are distinct *read*: with distinct coordinates;

Page 177, line 16 should read: Define further, for $m \geq 2$,

Line 17. The final formula should read:

$$k_{m-1}(k - k_{m-1}) \prod_{i=1}^{m-2} (k_{m-1} - k_i).$$

Line 19. For $\Lambda\Gamma - 1$ *read:* $\Lambda\Gamma^{-1}$

Page 178, line 3 should read:

$$a_1 \ldots a_m \leq a_1^m + \cdots + a_m^m$$

Line 5. The first range of summation should be $i = 0$ *to* m, *with* $i \neq m - 1$. *The numerator* $m^2 - 1$ *should be replaced by* $m^2 + m$.

Lines 7, 9, 10. Replace m^2 *at each occurrence by* m^3, *and delete* $\rho(d)$ *from line 10.*

Page 211, line 8. The first summation condition should read: $u > t^{1/10}$.

Page 223, line 17 should read: $|\sin^2(u_1 + \cdots + u_l)|$

Line 19 should read: $\sin^2 ur(p)/2$

Page 235, line 10. The first summand should read: $g(p)p^{-s}\log p$
Page 242, line 1. In place of $m \geqslant n_1$ *read:* $m \leqslant n_j$

Lines 6 and 7 should read: Considering m to be a fixed positive integer and letting $j \to \infty$, we deduce that

Page 254, line 25. The prize should read: $\$10^{10!}$.
Page 277, line 6. For zero *read:* one

Line 11. The summand in the exponential is:

$$p^{-1}\left|\left|\log\left|g(p)\right|p^{-c}\right|\right|.$$

Page 280, line 15. For weak, *read:* proper weak

Line 18 should read: converge, and that g(n) not be identically one.

Page. 300, line 5. Replace A by: $Ai\beta$

Line 7. Replace the coefficient $i\beta$ *by:* $-i\beta$

Line 11 should read:

$$J = \left[\frac{-y^{-i\beta}}{i\beta\log y}\right]_2^x - \frac{1}{i\beta}\int_2^x \frac{y^{-i\beta-1}}{(\log y)^2}\,dy$$

Line 13. For $A + \frac{i\beta}{2}$ *read:* $Ai\beta$

Page 306, line 1. In place of $q = k$ *read:* $q = k$ or $k/2$
Page 316, line 14. The first summand should be:

$$\Lambda(n)^{1/2} n^{-(\sigma/2)-i\tau_k} c_k$$

Line 18. The formula should read:

$$\sum_{k=1}^{N}\sum_{l=1}^{N} c_k\bar{c}_l \sum_{n=1}^{\infty} \Lambda(n)n^{-\sigma-i(\tau_k-\tau_l)}$$

Page 332, line 9. For $\sim\eta\log p$ *read:* $\sim\eta\log x$
Page 348, line 14. The summation condition should read: $p^k \leqslant \alpha^{-1}$.
References, p. II. Cauchy's first name should read: Augustin.

See also the comments made in the Preface to Volume II.

This book is dedicated to:
Arthur George Elliott,
my father,
Martha Chalk Elliott, née Ralph,
my mother,
to the dark melody of life,
with its firefly flashes,
to the journey along the mathematical rainbow,
and to the memory of E. T. A. Hoffmann,
who would have appreciated the irony.

Acknowledgments

The author would like to thank Thomas Y. Crowell for permission to reproduce an extract from G. Marek's book *Gentle Genius,* and Albert Blanchard for permission to reproduce (in translation) a passage from *Quelques aspects de la pensée d'un mathématicien,* by P. Lévy. I would also like to thank The Pacific Journal of Mathematics for allowing me to reproduce part of the paper *On the distribution of numbers of the form* $\sigma(n)/n$ *and some related questions,* by P. Erdös, which appeared in volume 52 of their 1974 issue. Detailed references are given at the appropriate places.

It has been my very good fortune to correspond with P. Erdös. M. Kac, J. Kubilius, and the late P. Turán, who played important rôles in the foundation and development of the Probabilistic Theory of Numbers. I would like to thank them for their permission to reproduce some of this correspondence which is of historical as well as mathematical interest. I would also like to thank P. Erdös and C. Ryavec for reading and commenting upon the manuscript.

I thank Janice Wilson who, in the summer of 1977, expertly typed the bulk of the manuscript.

My thanks also go to the National Science Foundation of the United States for support on contracts numbers GP33026X, MCS75-08233 and MCS78-04374.

Last, but not least, I would like to thank my wife Jean for her continued interest and support, and for sailing serenely through the sea of yellow pages with which I for several years covered the floor of our apartment.

Boulder, Colorado
August 1979

P.D.T.A. Elliott

Contents Volume I

Notation xix

Introduction 1

About This Book 14

Chapter 1. Necessary Results from Measure Theory 16

Steinhaus' Lemma 16
Cauchy's Functional Equation 17
Slowly Oscillating Functions 18
Halasz' Lemma 21
Fourier Analysis on the Line: Plancherel's Theory 22
The Theory of Probability 24
Weak Convergence 24
Lévy's Metric 24
Characteristic Functions 27
Random Variables 29
Concentration Functions 31
Infinite Convolutions 37
Kolmogorov's Inequality 44; Lévy's Continuity Criterion 46; Purity of Type 46; Wiener's Continuity Criterion 48
Infinitely Divisible Laws 49
Convergence of Infinitely Divisible Laws 53
Limit Theorems for Sums of Independent Infinitesimal Random Variables 54
Analytic Characteristic Functions 57
The Method of Moments 59
Mellin–Stieltjes Transforms 61
Distribution Functions (mod 1) 65
Quantitative Fourier Inversion 69
Berry-Esseen Theorem 74
Concluding Remarks 76

Chapter 2. Arithmetical Results, Dirichlet Series 79

Selberg's Sieve Method; a Fundamental Lemma 79
Upper Bound 84
Lower Bound 87
Distribution of Prime Numbers 89

Dirichlet Series 94
Euler Products 95
Riemann Zeta Function 96
Wiener–Ikehara Tauberian Theorem 100; Hardy–Littlewood Tauberian Theorem 102; Quadratic Class Number, Dirichlet's Identity 110
Concluding Remarks 111

Chapter 3. Finite Probability Spaces 115

The Model of Kubilius 119
Large Deviation Inequality 127
A General Model 129
Multiplicative Functions 140
Concluding Remarks 144

Chapter 4. The Turán–Kubilius Inequality and Its Dual 147

A Principle of Duality 150
The Least Pair of Quadratic Non-Residues (mod p) 153
Further Inequalities 158
More on the Duality Principle 162
The Large Sieve 165
An Application of the Large Sieve 170
Concluding Remarks 179

Chapter 5. The Erdös–Wintner Theorem 187

The Erdös–Wintner Theorem 187; Examples $\varphi(n)$, $\sigma(n)$ 188; Limiting Distributions with Finite Mean and Variance 196
The Function σ(n) 203
Modulus of Continuity, an Example of an Erdös Proof 207; Commentary on Erdös' Proof 210
Concluding Remarks 213
Alternative Proof of the Continuity of the Limit Law 220

Chapter 6. Theorems of Delange, Wirsing, and Halász 225

Statement of the Main Theorems 225
Application of Parseval's Formula 228
Montgomery's Lemma 229; Product Representation of Dirichlet Series (Lemma 6.6) 230; Quantitative form of Halász' Theorem for Mean-Value Zero 252
Concluding Remarks 254

Chapter 7. Translates of Additive and Multiplicative Functions 257

Translates of Additive Functions 257; Finitely Distributed Additive Functions 258; The Surrealistic Continuity Theorem (Theorem 7.3) 265; Additive Functions with Finite First and Second Means 269

Distribution of Multiplicative Functions 272
Criterion for Essential Vanishing 272; Modified-weak Convergence 273; Main Theorems for Multiplicative Functions 274; Examples 282
Concluding Remarks 283

Chapter 8. Distribution of Additive Functions (mod 1) 284

Existence of Limiting Distributions 284; Erdös' Conjecture 285
The Nature of the Limit Law 291
The Application of Schnirelmann Density 293; Falsity of Erdös' Conjecture 302; Translation of Additive Functions (mod 1), Existence of Limiting Distribution 302
Concluding Remarks 305

Chapter 9. Mean Values of Multiplicative Functions, Halász' Method 308

Halász' Main Theorem (Theorem (9.1)) 308; Halász' Lemma (Lemma (9.4)) 311; Connections with the Large Sieve 317; Halász's Second Lemma (Lemma (9.5)) 318; Quantitative Form of Perron's Theorem (Lemma (9.6)) 322; Proof of Theorem (9.1) 326; Remarks 330

Chapter 10. Multiplicative Functions with First and Second Means 333

Statement of the Main Result (Theorem 10.1) 333; Outline of the Argument 335
Application of the Dual of the Turán–Kubilius Inequality 335
Study of Dirichlet Series 337
Removal of the Condition $p > p_0$ 340
Application of a Method of Halász 343
Application of the Hardy–Littlewood Tauberian Theorem 348
Application of a Theorem of Halász 351
Conclusion of Proof 354
Concluding Remarks 356

References (Roman) I

References (Cyrillic) XIX

Author Index XXIII

Subject Index XXIX

Contents Volume II

Notation xv

Chapter 11. Unbounded Renormalisations: Preliminary Results 1

Chapter 12. The Erdös–Kac Theorem. Kubilius Models 12

Definition of Class H 12; Statement of Kubilius' Main Theorem 12; Archetypal Application of a Kubilius Model 14; Analogue of the Feller–Lindeberg Condition 17; The Erdös–Kac Theorem 18; Turán's Letter 18; Remarks Upon Turán's letter; LeVeque's Conjecture 20; Erdös at Kac' Lecture 24; Kac' Letter 24; Remarks upon Kac' Letter 24; Further Examples 26; Analogues on Shifted Primes 27; Examples 30; Further Analogues on Shifted Primes, Application of Lévy's Distance Function 31; Examples 36; Additive Functions on the Sequence N-p, p Prime 37; Barban's Theorem on the Normal Order of $f(p + 1)$ 41; Additive Functions on Polynomials 44; Additive Functions on Polynomials with Prime Arguments 44; Further Theorems and Examples 45; Quantitative Form of the Application of a Kubilius Model 48

Concluding Remarks 50

Chapter 13. The Weak Law of Large Numbers. I 52

Theorem Concerning the Approximation of Additive Functions by Sums of Independent Random Variables 52; Essential Lemma (Lemma 13.2) 53

Concluding Remark 57

Chapter 14. The Weak Law of Large Numbers. II 58

Statement of the Main Results 58

The Approximate Functional Equation for $\alpha(x)$ 61

Introduction of Haar Measures 63

Introduction of Dirichlet Series, Fourier Analysis on $\mathbb{R}$ 71

Study of the Integrals J 75

Approximate Differential Equation 75

A Compactness Lemma 76

Solution of the Differential Equation 80

Further Study of Dirichlet Series 83

The Decomposition of $\alpha(x)$ 86
Proof of Theorem (14.1): Necessity 93; Proof of Theorem (14.1): Sufficiency 96; Proof of Theorem (14.2) 97
Concluding Remark 97

Chapter 15. A Problem of Hardy and Ramanujan 98

Theorems of Birch and Erdös 99; The Hardy–Ramanujan Problem. Statement of Theorem 101
Commentary on a Method of Turán 112
Examples 118
Concluding Remarks 119

Chapter 16. General Laws for Additive Functions I: Including the Stable Laws 122

Statement of Isomorphism Theorem 122; Stable Laws 134; Convergence to Normal Law 136; Convergence to Cauchy Law 136; Fractional Part of $p\sqrt{2}$, p prime 137
Construction of Stable Laws 140
The Cauchy Law 143
Concluding Remarks 145

Chapter 17. The Limit Laws and the Renormalising Functions 147

Growth of $\beta(x)$, (Theorem (17.1)) 147; Class M Laws 148; Continuity of Limit Laws (Theorem (17.2)) 148; Laws of Class L are Absolutely Continuous (Lemma (17.11), Zolotarev) 167
Laws Which Cannot Occur 169
The Poisson Law 172
Further Continuity Properties 173
Conjectures 177
Conjectures (Summing Up) 182

Chapter 18. General Laws for Additive Functions II: Logarithmic Renormalisation 184

Statement of the Main Theorems 184; Example of Erdös 202; Non-infinitely Divisible Law 204
Concluding Remarks 207

Chapter 19. Quantitative Mean-Value Theorems 211

Statement of the Main Results 211; Reduction to Application of Parseval's Theorem (Lemma (19.5)) 219; Upper Bounds for Dirichlet Series (Lemma (19.6)) 221
The Prime Number Theorem 238
Axer's Lemma (Lemma (19.8)) 239
Primes in Arithmetic Progression; Character Sums 241
L-Series Estimates (Theorem (19.9)) 245

The Position of the Elementary Proof of the Prime Number Theorem in the Theory of Arithmetic Functions 248
Hardy's Copenhagen Remarks 248; Bohr's Address at the International Mathematics Congress 248; Elementary Proof of Prime Number Theorem 248; Method of Delange 255; Method of Wirsing 256; Theorem of Wirsing 259; Historical Remark on the Application of Parseval's Identity 260; Ingham's Review 260
Concluding Remarks 261

Chapter 20. Rate of Convergence to the Normal Law 262

Theorem of Kubilius and Improvements (Theorem (20.1)) 262; Examples 266; Additive Functions on Polynomials 267; Additive Functions on Polynomials with Prime Arguments 268; Examples 269; Conjugate Problem, (Theorem (20.4)) 271; Example 272
Improved Error Term for a Single Additive Function 273
Statement of the Main Theorem, (Theorem (20.5)) 273; Examples 285
Concluding Remarks 286

Chapter 21. Local Theorems for Additive Functions 290

Existence of Densities 290; Example of Rényi 294; Hardy–Ramanujan Estimate 296; Local Behaviour of Additive Functions Which Assume Values 0 and 1 296; Remarks and Examples 301; Connections with Hardy and Ramanujan Inequality 302; Uniform Local Upper Bound (Theorem (21.5)) 303
Concluding Remarks 311

Chapter 22. The Distribution of the Quadratic Class Number 313

Statement of the Theorem 314
Approximation by Finite Euler Products 314
An Application of Duality 315
Construction of the Finite Probability Spaces 323
Approximation by Sums of Independent Random Variables 324
Concluding Remarks 328

Chapter 23 Problems 330

References (Roman) I

References (Cyrillic) XIX

Author Index XXIII

Subject Index XXIX

Notation

We list here some of the more important symbols/definitions which occur in this book.

$\mathbb{R}$	The real numbers.
$\mathbb{C}$	The complex numbers.
$s, = \sigma + i\tau$	complex variable.
$\mathbb{Z}$	The ring of rational integers.
n	will generally denote a positive (natural) integer.
p	will generally denote a positive (natural) prime.
$[a, b]$	The least common multiple of the integers a and b. It also denotes the closed interval of real numbers x, $a \leqslant x \leqslant b$.
(a, b)	The highest common factor of the integers a and b. It also denotes the open interval of real numbers x, $a < x < b$.
$(a, b]$	The interval of real numbers x, $a < x \leqslant b$.
positive interval (bounded)	(72, Vol. II)
An arithmetic function	is a function which is defined on the positive natural integers.
An additive function	which will generally be denoted by $f(n)$, is an arithmetic function which satisfies $f(ab) = f(a) + f(b)$ whenever a and b are coprime integers.
A strongly additive function	is an additive function which also satisfies $f(p^m) = f(p)$ for every prime-power p^m, $m \geqslant 1$.
A completely additive function	satisfies $f(ab) = f(a) + f(b)$ for every pair of (positive) integers a and b.
A multiplicative function	which will generally be denoted by $g(n)$, is an arithmetic function which satisfies $g(ab) = g(a)g(b)$ whenever a and b are coprime integers.
A strongly multiplicative function	is a multiplicative function which also satisfies $g(p^m) = g(p)$ for every prime-power p^m, $m \geqslant 1$.

A completely multiplicative function	satisfies $g(ab) = g(a)g(b)$ for every pair of (positive) integers a and b.
class H	of strongly additive functions. (12, Vol. II).
$\pi_k(x)$	The number of (positive) integers, not exceeding x, which are the product of k distinct primes; (1).
$\pi(x)$	The number of primes not exceeding x; (90).
$\omega(n)$	The number of distinct prime divisors of the integer n; (2).
$\Omega(n)$	The number of prime divisors of the integer n, counted with multiplicity; (2).
$[x]$	The largest integer not exceeding x; (115). Thus $[3/2] = 1$ and $[-3/2] = -2$.
$\nu_x(n; \ldots)$	Let $N_x(n; \ldots)$ denote the number of positive integers not exceeding x which have the property $\ldots$; then $\nu_x(n; \ldots) = \text{the frequency } [x]^{-1} N_x(n; \ldots)$. Similarly $\nu_x(p; \ldots) = \pi(x)^{-1} \sum_{p \leqslant x}' 1$ where $'$ indicates that summation is confined to those prime numbers p for which the property $\ldots$ is valid.
weak convergence	The distribution functions $F_y(z)$, parametrized by an increasing set of real numbers y, converge weakly to a distribution function $G(z)$ if $\lim_{y\to\infty} F_y(z) = G(z)$ for every point z at which $G(z)$ is continuous. We write $F_y(z) \Rightarrow G(z),\ y \to \infty$.
modified-weak convergence	(273, 274).
$\tau(n)$, $d_2(n)$, $d(n)$	The number of positive integers which divide the integer n. See page (108).
$d_k(n)$	The number of representations of n as the product of k positive integers.
$\sigma(n)$	The sum of the positive divisors of n.

$$A(n) = \sum_{p \leqslant n} \frac{f(p)}{p},$$

$$B(n) = \left(\sum_{p \leqslant n} \frac{f(p)^2}{p} \right)^{1/2} \geqslant 0,$$

$$D(n) = \left(\sum_{p^k \leqslant n} \frac{|f(p^k)|^2}{p^k} \right)^{1/2} \geqslant 0.$$

$p^k \| n$	The prime-power p^k divides n, p^{k+1} does not.
$\pi(x, D, l)$	The number of primes p not exceeding x which satisfy $p \equiv l \pmod{D}$; (90).
$\Lambda(n)$	von Mangoldt's function; $= \log p$ if n is a power of a prime p, $= 0$ otherwise; (97).
$\mu(n)$	Möbius' function; zero if n is not squarefree, $(-1)^{\omega(n)}$ if it is; (85).
$\lambda(n)$	Liouville's function; $= (-1)^{\Omega(n)}$; (238, Vol. II).
$\rho(r)$	for a polynomial $f(x)$ with integer coefficients, is the number of distinct solutions (mod r) to the equation $f(x) \equiv 0 \pmod{r}$.
Kronecker symbol	(110).
$A(x)$	If $A : a_1 < a_2 < \dots$ is a sequence of positive integers, $A(x)$ will often be used to denote the number of these which do not exceed x.
Asymptotic density	A sequence of positive integers A is said to have an asymptotic density d if $$d = \lim_{x\to\infty} x^{-1}A(x)$$ exists.
$d(A)$	Lower asymptotic density. For a sequence of positive integers A $$d(A) = \liminf_{x\to\infty} x^{-1}A(x).$$ See page (295).
$\sigma(A)$	Schnirelmann density. For a sequence of positive integers A $$\sigma(A) = \inf_{n\geq 1} n^{-1}A(n).$$ See page (293).
basis	(294).
Δ	for the Turán–Kubilius inequality, (147).
Δ	for the Large Sieve, (166).
dual (conjugate) operator	(181, 182).
adjoint operator	(181, 182).
component	of (characteristic function, distribution function), (113).
class L	of Khinchine, (125, Vol. II).
class M	laws, (148, Vol. II).

For a random variable X, with associated distribution function $F(z)$,

$\bar{X}$, *mean* $$= \int_{-\infty}^{\infty} z\, dF(z),$$

σ^2, D^2, *variance* $$= \int_{-\infty}^{\infty} (z - \bar{X})^2\, dF(z).$$

$\rho(F, G)$, *Lévy metric.*	For any two distribution functions $F(= F(z))$, $G(= G(z))$, we define $\rho(F, G)$ to be the greatest lower bound of those numbers h which have the property that the inequality $$G(z-h) - h \leqslant F(z) \leqslant G(z+h) + h$$ holds for all real values of z.
$c_1, c_2, \ldots$	will denote constants. These are renumbered from the beginning of each chapter, and on occasion, when no confusion thereby arises, from the beginning of a new section.
$f(x) = O(g(x))$	for a range of x-values, means that there is a constant A so that the inequality $$\lvert f(x) \rvert \leqslant A g(x)$$ holds over the range.
$f(x) = o(g(x))$	as $x \to \infty$, means $$\lim_{x \to \infty} \frac{f(x)}{g(x)} = 0.$$
$f(x) \sim g(x)$	as $x \to \infty$, means $$\lim_{x \to \infty} \frac{f(x)}{g(x)} = 1.$$

These last two will only be used with functions $g(x)$ which do not vanish when x is sufficiently large.

Introduction

In 1791 Gauss made the following assertions (collected works, Vol. 10, p. 11, Teubner, Leipzig 1917):

Primzahlen unter a $(=\infty)$

$$\frac{a}{la}$$

Zahlen aus zwei Factoren

$$\frac{lla \cdot a}{la}$$

(warsch.) aus 3 Factoren

$$\frac{1}{2}\frac{(lla)^2 a}{la}$$

et sic in inf.

In more modern notation, let $\pi_k(x)$ denote the number of integers not exceeding x which are made up of k distinct prime factors, $k = 1, 2, \ldots$. Then his assertions amount to the asymptotic estimate

$$\pi_k(x) \sim \frac{x}{\log x}\frac{(\log\log x)^{k-1}}{(k-1)!} \qquad (x \to \infty).$$

The case $k = 1$, known as the Prime Number Theorem, was independently established by Hadamard and de la Vallée Poussin in 1896, just over a hundred years later. The general case was deduced by Landau in 1900; it needs only an integration by parts.

Nevertheless, one can scarcely say that Probabilistic Number Theory began with Gauss.

In 1914 the Indian original mathematician Srinivasa Ramanujan arrived in England. Six years of his short life remained to him during which he wrote, amongst other things, five papers and two notes jointly with G. H. Hardy. Two of these five papers gave new directions to the theory of numbers.

The first of these two particular papers appeared in 1917 [1], under the title "The normal number of prime factors of a number n". Hardy and Ramanujan considered the functions $\Omega(n)$ and $\omega(n)$ which count the number of prime divisors of an integer n, with and without multiplicity respectively. They proved inductively that there are not more than

$$\frac{c_1 x}{\log x} \frac{(\log \log x + c_2)^{k-1}}{(k-1)!}$$

integers n not exceeding $x(\geq 2)$ for which $\omega(n) = k$. With certain positive constants c_1 and c_2 this result holds uniformly for all positive integers k.

As a consequence they deduced that whenever the function $\psi(x)$ becomes unbounded with x, the frequency

$$\nu_x(n; |\omega(n) - \log \log x| > \psi(x)(\log \log x)^{1/2}) \tag{1}$$

of those integers n in the interval $1 \leq n \leq x$ for which the inequality

$$|\omega(n) - \log \log x| > \psi(x)(\log \log x)^{1/2}$$

is satisfied, approaches zero as $x \to \infty$.

More generally, if for a given arithmetic function $f(n)$ we can find a further function $g(n)$ be "elementary" and increasing, Hardy and Ramanujan proved property that whenever $\varepsilon > 0$ those integers n for which

$$|f(n) - g(n)| \geq \varepsilon g(n)$$

have asymptotic density zero, then we say that $f(n)$ has the normal order $g(n)$. Introducing this concept in their paper, actually they require that the function $g(n)$ be "elementary" and increasing, Hardy and Ramanujan proved that $\omega(n)$ and $\Omega(n)$ both have the normal order "$\log \log n$". Here we may set

$$g(n) = \begin{cases} \log \log n & \text{if } n \geq 3, \\ 0 & \text{if } 1 \leq n \leq 2. \end{cases}$$

They asked whether other well-known arithmetic functions such as $\tau(n)$, the divisor function, possessed (simple) normal orders, (§V, p. 92 of their paper).

It seems that this paper did not make a splash.

The second of their two papers in which we are presently interested appeared in the following year, 1918 [2]. It gave an asymptotic estimate for the number of partitions of a number n, and in it may be found the genesis of the Hardy–Littlewood, or Circle method. The importance of the circle method was immediately clear, and as elaborated by Hardy and Littlewood, and Vinogradov, in particular, enabled many hitherto difficult problems concerning the additive representation of integers to be solved. It is not our purpose to pursue this topic here, suffice it to say that Vinogradov in 1937 proved by means of the circle method together with ingenious ideas of his own, that every sufficiently large odd integer N may be expressed as the sum of three prime numbers,

$$N = p_1 + p_2 + p_3. \tag{2}$$

We shall consider the circle method again presently.

In 1934 Turán [2] gave a new proof of Hardy and Ramanujan's result concerning the frequency (1). It depended upon the readily obtained estimate

$$\sum_{n \le x} (\omega(n) - \log\log x)^2 \le c_3 x \log\log x.$$

Moreover, Turán's method of proof was susceptible to generalization. As Turán wrote to me, seventeen years had passed without the slightest sign of anyone realizing such general theorems might exist at all.

Although his method is similar in appearance to an argument of Tchebycheff in the theory of probability, at that time Turán knew no probability theory (see Chapter 12 of this monograph). Indeed, the first (to-be) widely accepted axiom system for the theory of probability, due to Kolmogorov, had only appeared in 1933. An interesting short account of the history of early attempts at axiomatization may be found in Rényi [7], pp. 53–57. To put the matter into better context, Lévy [3] pp. 72–74 remarks that in 1919 he was unaware of Tchebycheff's argument, it not having been mentioned by Bertrand, Poincaré or Borel amongst the pioneers in the theory of probability, and he rediscovered it for himself. Although published in France, from about 1867, and in French, Tchebycheff's argument was apparently obscured by the lack of a succinct notation. I am reminded of the Fejér–Erdös dictum: "Everybody writes, nobody reads."

It seems that Hardy was not overly impressed with Turán's proof. Erdös remembers very clearly a conversation with Wintner, who told him of a letter in which Hardy had written that although clever Turán's proof was not the most illuminating. It was Wintner's opinion that on the contrary it *was* the most illuminating. Erdös had thought this letter of Hardy had been sent to the English journal "Nature". However, three searches, two by myself and one by the pair of us, failed to turn it up. This anecdote must at present remain based upon oral tradition. There is a letter in Nature [2] 1941, sent by Wintner, pointing out the parallel between the behaviour of the function $\omega(n)$ and Brownian motion from Einstein's statistical point of view.

Be that as it may, the paper of Turán had other effects. In a book on Number Theory Bessel–Hagen had remarked that it was not known whether abundant numbers, those positive integers n for which

$$\sigma(n) = \sum_{d|n} d > 2n,$$

had an asymptotic density. Interest in the function $\sigma(n)$ goes back at least as far as Euclid, who could generate the even *perfect numbers*, that is to say those even integers n for which $\sigma(n) = 2n$. It is still not known if an odd perfect number exists. The question of Bessel–Hagen was answered in the affirmative by Chowla [1], Davenport [1], and Erdös [1], independently. The same result was apparently also obtained by Behrend, who in [1] pp. 146–149 reviewed the papers of these three authors. Widening his scope somewhat Erdös, who very soon completely adopted Turán's method of proof into his personal armory, showed by 1938 [7] that whenever the three series

$$\sum_{|f(p)|>1} \frac{1}{p} \qquad \sum_{|f(p)|\le 1} \frac{f(p)}{p} \qquad \sum_{|f(p)|\le 1} \frac{f^2(p)}{p}$$

converge then the strongly additive function $f(n)$ possesses a limiting distribution, that is to say the frequencies

$$\nu_n(m; f(m) \le z)$$

converge as $n \to \infty$ to a distribution function. It turned out that the convergence of these three series was in fact necessary (Erdös and Wintner [1]).

Mention should be made here of the papers of Schönberg [1] from 1928, and [3].

A second effect of Turán's paper was to set Kac thinking about the rôle of independence in the application of probability to number theory. A more detailed discussion of this and related topics is given in Chapter 12. Suffice it here to say that in 1939 [1], 1940 [2], Erdös and Kac proved that for strongly-additive functions which satisfy $|f(p)| \le 1$, and for which

$$A(n) = \sum_{p\le n} \frac{f(p)}{p}$$

(3)

$$B(n) = \sum_{p\le n} \left(\frac{f(p)^2}{p}\right)^{1/2} \to \infty \qquad (n \to \infty),$$

we have

$$\text{(4)} \qquad \nu_n\left(m; \frac{f(m) - A(n)}{B(n)} \le z\right) \Rightarrow \frac{1}{\sqrt{2\pi}} \int_{-\infty}^{z} e^{-w^2/2}\, dw \qquad (n \to \infty).$$

This result, of immediate appeal, was the archetype of many results to follow. It firmly established the application of the theory of probability to the study of a fairly wide class of additive and multiplicative arithmetic functions. In particular Erdös and Kac made essential use of the notation of independent random variables, the central limit theorem of probability, and the sieve method of Brun.

In the case that $f(n) = \omega(n)$, the function considered by Hardy and Ramanujan, Erdös and Kac thus obtained that

$$(5)\quad \nu_n\left(m; \frac{\omega(m) - \log\log n}{\sqrt{\log\log n}} \le z\right) \Rightarrow \frac{1}{\sqrt{2\pi}} \int_{-\infty}^{z} e^{-w^2/2}\, dw \qquad (n \to \infty).$$

As we shall further show in Chapter 12, by 1936 Erdös had marshalled enough results (of a combinatorial nature) to establish this example (5), (but not the more general result (4)), were it not for the fact that his knowledge of the theory of probability was as slight as that of his Hungarian compatriot Turán. A remark to this effect concerning Erdös was made by Kac in a 1949 address to the American Mathematical Society, [2], p. 658.

A further understanding of the method of Erdös and Kac was obtained by Kubilius (1954–1955). He constructed a finite probability space on which independent random variables could be defined so as to mimic the (in-frequency) behaviour of truncated additive functions

$$\sum_{p|n,\ p \le r} f(p).$$

As we shall indicate in Chapters 3 and 12, this approach is effective if the ratio

$$\frac{\log r}{\log n}$$

essentially approaches zero as n becomes large. With this formal basis, and assuming that $f(n)$ belonged to a certain fairly wide class of additive functions, Kubilius was able to give necessary and sufficient conditions in order that the frequencies

$$\nu_n(m; f(m) - A(n) \le zB(n))$$

converge weakly as $n \to \infty$. In particular he showed that although the limiting distributions so obtained necessarily possessed mean and variance, laws other than the normal law could occur.

Kubilius summarized many of the applications, then extant, of probability to the study of additive arithmetic functions, in a monograph. Of its several editions we note here that which appeared in 1962 as number 11 in the American Mathematical Society series of Translations of Mathematical Monographs [5].

Besides obtaining results like the Erdös–Kac theorem, in his book Kubilius applies finite probability models to the study of the joint distribution of several additive functions, thus extending earlier work of Erdös [4],

Theorem VII, p. 4, LeVeque [1], and Halberstam [3]. Moreover, he discusses the finer behaviour, as $n \to \infty$, of the frequency

$$\nu_n(m; \omega(m) - \log\log n \le z\sqrt{\log\log n}).$$

This includes the theorem of Rényi and Turan [1], conjectured by LeVeque [1], that this frequency may be uniformly approximated by

$$\frac{1}{\sqrt{2\pi}} \int_{-\infty}^{z} e^{-w^2/2}\, dw + O\left(\frac{1}{\sqrt{\log\log n}}\right) \qquad n \ge 3.$$

Many of the results in Kubilius' monograph will be contained amongst the theorems of the present book. Some, such as the study of truncated additive functions for their own sake, will not. Besides lack of space, we have at present little to add to such a topic. Thus Kubilius' book remains a useful reference. In our present treatment we shall however give a more extensive account of the early history of the subject.

We make here a particular note of the following result, which appears in Chapter III of Kubilius' book [5], and which extends the argument of Turán's paper [3] to arbitrary complex-valued additive functions.

Define

$$D(n) = \left(\sum_{p^k \le n} \frac{|f(p^k)|^2}{p^k} \right)^{1/2} \ge 0.$$

Then there is a positive constant c_1 so that the inequality

$$\sum_{m=1}^{n} |f(m) - A(n)|^2 \le c_1 n D(n)^2 \tag{6}$$

holds uniformly for all complex-valued additive functions $f(n)$, for all positive integers n. Results of this kind have become collectively known as "The Turán–Kubilius inequality". It turns out that in this form (6) and related inequalities suit naturally the study of additive and multiplicative arithmetic functions. We shall discuss this and related matters later in this introduction, and in more detail in Chapter 4.

Let us now go back and see how the circle method had progressed.

Let $n_1 < n_2 < \ldots$ be a sequence of positive integers. For complex numbers z, $|z| < 1$, define the function

$$w(z) = \sum_{j=1}^{\infty} z^{n_j}.$$

According to Cauchy's theorem the number of representations of the integer N in the form

$$N = n_{j_1} + n_{j_2} + n_{j_3},$$

which we shall denote by $R(N)$, is given by

$$R(N) = \frac{1}{2\pi i}\int_{|z|=\rho} w(z)^3 z^{-N-1}\,dz, \tag{7}$$

the integral being taken over any circle $|z| = \rho$, $0 < \rho < 1$. For many sequences n_j, $j = 1, 2, \ldots$, the function $w(z)$ has singularities at the points $\exp(2\pi iaq^{-1})$ for rational integers a and q. In the circle method the contribution towards the integral (7) which arises from short arcs containing those singular points with q not too large compared with N is carefully estimated, in the hope that it will provide most towards the value of $R(N)$. Suitably chosen these arcs about the points $\exp(2\pi iaq^{-1})$ are designated *The Major Arcs*. On what remains of the contour $|z| = \rho$, called *The Minor Arcs*, one looks for some method to show that the function $w(z)$ is "sufficiently small". One may often set $\rho = 1 - N^{-1}$.

Although it disguises somewhat the underlying method, the following presentation of Vinogradov is technically convenient to apply.

For real θ define the sum

$$S(\theta) = \sum_{n_j \le N} e^{2\pi i n_j \theta},$$

then

$$R(N) = \int_0^1 S(\theta)^3 e^{-2\pi i\theta N}\,d\theta.$$

The rôle of the major arcs is now played by intervals

$$I_{a,q} = \left\{\theta; \frac{a}{q} - \eta \le \theta \le \frac{a}{q} + \eta (\text{mod } 1)\right\},$$

where η is chosen so that these intervals do not overlap.

After an integration by parts the behaviour of the function $S(\theta)$ on such an interval may be related to the sum $S(aq^{-1})$ which may be in turn represented in the form

$$S\left(\frac{a}{q}\right) = \sum_{k=1}^{q} \exp\left(\frac{2\pi ika}{q}\right) \sum_{\substack{n_j \le N \\ n_j \equiv k (\text{mod } q)}} 1.$$

For further progress information concerning the distribution of the sequence n_j, $j = 1, 2, \ldots$, in the various residue classes (mod q) is clearly helpful.

The contribution of the minor-arc intervals, which we shall denote by m, will not exceed

$$\max_{m} |S(\theta)| \int_{m} |S(\theta)|^2 \, d\theta \leq \max_{m} |S(\theta) S(0),$$

this last step since

$$\int_{m} |S(\theta)|^2 \, d\theta \leq \int_0^1 |S(\theta)|^2 \, d\theta = \sum_{n_j \leq N} 1 = S(0).$$

The application of Parseval's relation in this step is quite powerful.

Thus, in order to obtain the representation (2) Vinogradov applied characteristically ingenious ideas to prove that an estimate

$$\sum_{p \leq N} e^{2\pi i \theta p} = O(N(\log N)^{-2-\delta}),$$

for some $\delta > 0$, could be obtained on the minor-arc intervals. (See, for example, I. M. Vinogradov [1], Chapter IX, theorems 1, 2a and 2b, and Chapter X, p. 168).

On the face of it the circle method is not applicable, at any rate in this form, if one is interested in binary representations such as

$$N = n_{j_1} + n_{j_2}.$$

In 1941, in a small note [1], Linnik showed that information could be obtained concerning the distribution of an arbitrary sequence $n_1 < n_2 < \ldots$ in arithmetic progressions with differences q quite large compared to N. His idea was to put the circle method into reverse. Thus, the study of the quantities

$$\sum_{n_j \equiv k (\bmod q)} 1$$

is related to sums

$$\left| S\left(\frac{a}{q} + w\right) \right|^2 \qquad |w| \leq \eta,$$

and these, in turn, to a sum of the form

$$\sum_{q} \sum_{a} \int_{I_{a,q}} |S(\theta)|^2 \, d\theta. \tag{8}$$

This is an expression concerning the minor-arc intervals when the circle method is applied to the representation of zero,

$$0 = n_{j_1} - n_{j_2}.$$

Choosing η so that the intervals $I_{a,q}$ do not overlap, this restricting the variable q, and arguing as before, we obtain for the sum in (8) the upper bound

$$\int_0^1 |S(\theta)|^2 \, d\theta = S(0).$$

Non-trivial results are then often obtained.

Linnik called this procedure "The Large Sieve" since it gave good results even when the sequence $n_j, j = 1, 2, \ldots$, missed many residue classes (mod q) for many q, a situation in which Brun's sieve method became unwieldy, and (apparently) less efficient. Linnik made only a limited use of this idea (see [2] and [3]), but in 1948 his student Rényi [2] applied it to the study of character sums, and proved that every sufficiently large even integer $2N$ may be represented in the form

$$2N = p + P_k.$$

Here p is a prime, and the integer P_k has at most k prime factors, where k is a fixed number for which a value can be given.

With a further development of these ideas Barban [1] proved that for a certain positive constant $\delta > 0$, and every fixed A,

$$\sum_{d \le x^\delta} \mu^2(d) \max_{(l,d)=1} \left| \pi(x, d, l) - \frac{Li(x)}{\varphi(d)} \right| = O(x(\log x)^{-A}).$$

This enabled him (Barban [2]) to transfer the method of Erdös–Kac and Kubilius to the study of additive functions on sequences such as the "shifted primes" $p + 1$, p prime. Such an application, together with earlier and related results concerning the behaviour of arithmetic functions $f(p + a)$, p prime, $a \neq 0$, is considered in Chapter 12.

For additive functions $f(m)$ one can study the frequency

$$\nu_n(m; f(m) \le z) \tag{9}$$

by means of its characteristic function

$$n^{-1} \sum_{m=1}^{n} e^{itf(m)} \qquad t \text{ real}.$$

The function

$$g(m) = e^{itf(m)}$$

is multiplicative and satisfies $|g(m)| \leq 1$ for every integer m. The average values of arithmetic functions have long been an object of study in the theory of numbers. As is well known, some of the questions to which their study leads are still unsolved. Such is the case with Dirichlet's divisor function. (See, for example, Titchmarsh [4], Chapter XII).

Delange [4] characterized those multiplicative functions $g(m)$ which satisfy $|g(m)| \leq 1$ and for which a non-zero mean-value

$$A = \lim_{n \to \infty} n^{-1} \sum_{m=1}^{n} g(m)$$

exists. This enabled him to give an alternative treatment of the theorem of Erdös and Erdös and Wintner concerning the limiting behaviour of the frequencies (9).

To characterize those multiplicative functions $g(m)$, $|g(m)| \leq 1$, for which such a mean-value A exists and is zero turned out to be more difficult. It was done in essentially two steps; for real-valued functions by Wirsing [1], [4], this included an old conjecture, variously ascribed to Erdös and Wintner, to the effect that a mean-value A always exists whenever $g(m)$ assumes only the values ± 1; and for complex-valued functions by Halász [1]. Each of these authors made important steps, and although related, their methods were quite different.

Since we give an extended treatment of these results in Chapter 6 little more will be said here. We mention only that Halász' paper contained an argument which gave further impetus to the study of the Large Sieve. We discuss this a little in the concluding remarks of Chapter 4. Towards the end of Chapter 19 we consider some of these matters in the light of the 1948 Erdös and Selberg elementary proof of the Prime Number Theorem.

More generally, for additive functions $f(n)$ it is natural to consider when functions $\alpha(x)$ and $\beta(x) > 0$ may be found such that the frequencies

$$\nu_x\left(n; \frac{f(n) - \alpha(x)}{\beta(x)} \leq z\right) \tag{10}$$

possess a limiting distribution as $x \to \infty$.

The theorem of Erdös and Wintner corresponds to the choice $\beta(x) = 1$, $\alpha(x) = 0$. The next case, so to speak, was $\beta(x) = 1$ but with $\alpha(x)$ unrestricted at the outset. It was shown independently by the author and Ryavec [1], and by Levin and Timofeev [1], that the frequencies

$$\nu_x(n; f(n) - \alpha(x) \leq z)$$

with a suitably chosen $\alpha(x)$, converge weakly if and only if there is a constant c so that the function $h(n) = f(n) - c \log n$ satisfies

$$(11) \qquad \sum_{|h(p)|>1} \frac{1}{p} < \infty \qquad \sum_{|h(p)|\le 1} \frac{h^2(p)}{p} < \infty.$$

Apparently this result was also obtained, but not published, by Delange, and by Kubilius.

Although perhaps unusual from a probabilistic point of view, the form of condition (11) was not a surprise to number-theorists. In a paper published in 1946 [10], to which we have already alluded, Erdös introduced the notion of "finitely distributed additive function". He had established a characterization of such functions, in fact an additive function is finitely distributed if and only if, for some c, $h(n) = f(n) - c \log n$ satisfies the condition (11). This notion of Erdös, which is often convenient to apply to the study of additive and multiplicative arithmetic functions, is considered in Chapter 7.

It turned out that if in (10) no satisfactory choice of $\alpha(x)$ exists when $\beta(x) = 1$ for all x, then the only cases remaining to be considered are those when $\beta(x) \to \infty$ as $x \to \infty$.

In Wirsing's treatment of multiplicative functions $g(n)$, a central rôle is played by approximate integral equations, such as the asymptotic relation

$$m(x)\log x \sim (1 + \tau) \int_1^x \frac{m(w)}{w}\, dw$$

which appears in [1], Hilfssatz 4, p. 97. In this particular paper somewhat strong assumptions are made concerning the average behaviour of the $g(p)$, and in order to eliminate the need for such assumptions Wirsing developed a more delicate method in his second paper [4]. However, starting from Wirsing's original point of view Levin and Faĭnleĭb in a series of papers [1], [2], [3], [4], developed a method which in certain circumstances leads to good quantitative results. An application concerning those integers which are representable as the sum of two squares may be found in Postnikov [1], Section 4.11, pp. 379–393. Of interest here is that in their fourth paper [4], Levin and Faĭnleĭb give, apparently for the first time, an additive function $f(n)$ for which the frequencies (10) converge to a proper stable law that is not the normal law. We shall discuss this and related examples in Chapter 16. Note that apart from the normal law no proper stable law has a (finite) variance.

A further advance in the study of the frequencies (10) when $\beta(x)$ is unbounded with x was made by Levin and Timofeev [1]. Assuming essentially that

$$(12) \qquad \lim_{x\to\infty} \frac{\beta(x^y)}{\beta(x)} = 1 \qquad \lim_{x\to\infty} \frac{\alpha(x^y) - \alpha(x)}{\beta(x)} = \psi(y) < \infty$$

for each fixed $y > 0$ they gave necessary and sufficient conditions in order that the frequencies (10) should possess a limiting distribution as $x \to \infty$. (Actually their notation is slightly different, the variable x is confined to integer values). They use the theory of Dirichlet series, and for the sufficiency part of the argument apply a method of Halász [1]. For an alternative treatment see the author's paper (Elliott [16]).

In fact a satisfactory analysis of the limiting behaviour of the frequencies (10) may be carried out assuming only that the limits in (12) exist; the first need not have the value 1. This was shown independently, with different methods, by Levin and Timofeev [3], and the author, Elliott [19]. However, the argument of Halász mentioned above now plays an essential part. Indeed, it seems likely that whenever a set of conditions upon the values $f(p)$ of an additive function $f(n)$ ensures the weak convergence of the frequencies (10), the argument of Halász [1], or a suitable modification of it, may be used to construct a proof. What the necessary conditions upon the $f(p)$ should be is not so clear.

In this introduction we do not extensively pursue the ramifications of the Large Sieve method in the analytic theory of numbers. This would lead us rather far from our present topic. However, we do discuss in Chapter 2 following lemma (2.10), and in the concluding remarks of Chapter 4, contributions of Rényi, Barban, Roth, Bombieri, Davenport and Halberstam, Gallagher and others to this subject. Some connections between the Large Sieve and the topics pursued in the present monograph are also considered.

It was pointed out by the author, Elliott [7], that to obtain an inequality of Large Sieve type amounts to determining the spectral radius of an Hermitian operator, and that such inequalities always come in pairs, the inequality together with its conjugate (or dual). The derivation of a Large Sieve inequality by considering its dual (or what amounts to the same thing) was also independently carried out, explicitly by Matthews [1], [2], [3], and implicitly by Kobayashi [1]. Moreover, Bombieri pointed out (see Forti and Viola [1], p. 32) that one need not confine oneself to the L^2-norm.

At the meeting on Analytic Number Theory, held in St. Louis, Missouri, in 1972, the author (Elliott [12]) suggested that there is sometimes a correspondence

$$\begin{aligned} &\text{operator} \to \text{sufficiency} \\ &\text{dual of operator} \to \text{necessity} \end{aligned} \tag{13}$$

and gave some applications of this idea to the study of additive and multiplicative arithmetic functions.

The application of the correspondence (13) is certainly useful in the study of Plane Projective Geometry, where the duality is between point and line. For example, the dual of Desargues' Theorem is in fact its converse. As a rule, however, one expects to supplement the use of a dual proposition with further argument.

The Turán–Kubilius inequality, which we here write in the form

$$\sum_{m=1}^{n} \left| f(m) - \sum_{p^k \le n} f(p^k) p^{-k} \right|^2 \le c_1 n \sum_{p^k \le n} |f(p^k)|^2 p^{-k}, \tag{14}$$

is a considerable aid when establishing the limiting behaviour of the frequencies (10), as we have already remarked. Its dual is

$$\sum_{p^k \le n} \left| \sum_{\substack{m=1,\, p^k \| m}}^{n} a_m - p^{-k} \sum_{m=1}^{n} a_m \right|^2 \le c_1 n \sum_{m=1}^{n} |a_m|^2, \tag{15}$$

an inequality of Large Sieve type. Here the range of summation $p^k \le n$ differs a little from that which is more usual, but is suited to the study of arithmetic functions (see, for example, Chapter 10). One might regard the inequalities (14) and (15) as indirect descendents of the two original Hardy and Ramanujan papers considered at the beginning of this introduction. It seems fitting that they are dual to one another.

By adopting the point of view in (13) many of the results of the probabilistic theory of numbers may be given a unified treatment.

A satisfactory treatment of the convergence of the frequencies (10) to the improper law may be obtained if

$$\limsup_{x \to \infty} \frac{\beta(x^y)}{\beta(x)} \tag{16}$$

is finite for each fixed $y > 0$. This particular condition is not unreasonable; when the frequencies (10) converge to a proper (limit) law it is indeed satisfied. These results, due to the present author, are featured in Chapters 13, 14 and 17, and are obtained by constructing a non-linear approximate differential equation of a type usually associated with the mathematical theory of control. It transpires that the renormalizing function $\alpha(x)$ must have a decomposition

$$\alpha(x) = \alpha_1(x) + \alpha_2(x)$$

where, for each fixed $y > 0$,

$$\frac{\alpha_1(x^y) - y\alpha_1(x)}{\beta(x)} \to 0 \qquad \frac{\alpha_2(x^y) - \alpha_2(x)}{\beta(x)} \to 0,$$

as $x \to \infty$. This might be viewed as the first step in classifying the asymptotic behaviour of the renormalizing functions $\alpha(x)$ and $\beta(x)$ with respect to the group of substitutions $x \mapsto x^y$.

In Chapter 15 we characterize those additive functions which possess non-decreasing normal orders. Since Birch [1] showed that the only multiplicative functions which have such normal orders are the fixed powers n^c, this completes the solution of Hardy and Ramanujan's original problem for the first two widest classes of arithmetic functions presently known.

By strengthening the condition (16) to

$$\beta(x^y) \sim \beta(x) \qquad x \to \infty, \tag{17}$$

we obtain in Chapter 16 necessary and sufficient conditions in order that the frequencies (10) possess a limiting distribution, proper or improper. It proves convenient to introduce there the Lévy metric from the theory of probability. Considering convergence to general laws, without assuming the condition (17), much remains to be done, as we shall indicate.

Lest it be thought that the example (5) of Erdös and Kac exhausts those results in the probabilistic theory of numbers which are simple to state we give the following:

Denote the fractional part of the number v by $\{v\}$. Define the additive function

$$h(n) = \sum_{\substack{p|n \\ p \geq 3}} \frac{(-1)^{(p-1)/2}}{\{p\sqrt{2}\}}.$$

Then

$$\nu_x\left(n; \frac{h(n)}{\log \log x} \leq z\right) \Rightarrow \int_{-\infty}^{z} \frac{dw}{\pi^2 + w^2} \qquad (x \to \infty),$$

where the limiting distribution is the Cauchy law, see Chapter 16, and has the alternative representation

$$\frac{1}{2} + \frac{1}{\pi} \tan^{-1}\left(\frac{z}{\pi}\right).$$

Our volumes essentially end with the application of independent random variables to the study of Dirichlet L-series and hence to the in-frequency behaviour of the quadratic class numbers. There is, however, a final chapter containing unsolved problems.

About This Book

Permit me to make a few remarks.

It seems to me that the way a man or woman thinks and the way that they act are inextricably linked together. This monograph may therefore be viewed as, to some extent, a portrait of various mathematicians through a selection of the results which they have proved. It has occurred to many of us, on the occasion of going into a darkened room, to press the switch and illuminate the interior. In this way we neither discover nor invent electricity. Accordingly I have considered it appropriate, and besides a pleasure, to discuss the historical antecedents of particular results, and some of the authors who contributed much towards them. This is sometimes done in the body of the text, and sometimes in the concluding remarks. These considerations are meant to be an integral part of the book, for I believe in the man and not the method.

The book is written as if the central problem were to find necessary and sufficient conditions in order that the frequencies

$$\nu_x(n;\ f(n) - \alpha(x) \leq z\beta(x))$$

possess a limiting distribution, as $x \to \infty$, when the functions $\alpha(x)$ and $\beta(x) > 0$ are chosen suitably. Nevertheless there are many detours.

The first four chapters contain the necessary introductory material. The study of the application of probability to the theory of numbers begins with Chapter 5. The classical results are contained in Chapters 5 and 12. For the remainder some familiarity with the application of Dirichlet series is assumed.

Not having had the opportunity to give a course in the Probabilistic Theory of Numbers I can only hazard what such a course might reasonably contain. One could begin with a selection of results from Chapters 5 and 12, say theorems (5.1), (12.1), (12.2) and (12.3), picking up various results from the first four chapters when needed. At a leisurely rate this might already take up a large part of one semester. One could complete the course by establishing the sufficiency part of theorems (16.1) and (16.2), going on to generate all the stable laws as limit laws. In a more extensive course after theorem (12.3) one might proceed via Chapters 6, 7, 9, 13, 14 and 16.

Chapter 1

Necessary Results from Measure Theory

In this chapter we collect together a number of results in measure theory and the theory of probability. We shall make use of these in subsequent chapters.

Unless defined otherwise, *measure* will denote *Lebesgue measure.*

For the purposes of this chapter, the differences generated by a set E of real numbers will be those numbers of the form $x - y$, where x and y are members of E.

Lemma 1.1 (Steinhaus). *The differences generated by a set of real numbers of positive measure, cover an open interval about the origin.*

Proof. Let S be a set of real numbers, of positive measure μS. Then there exists a countable collection of open intervals, whose union contains S, and whose total measure does not exceed $\frac{4}{3}\mu S$. For at least one of these intervals, say I, we shall have $\mu(S_\cap I) > \frac{2}{3}\mu I$. Otherwise, we can label the intervals I_j, $(j = 1, 2, \ldots)$, and then have

$$\mu S \leq \sum_{j=1}^{\infty} \mu(S_\cap I_j) \leq \frac{2}{3} \sum_{j=1}^{\infty} \mu I_j \leq \frac{2}{3} \cdot \frac{4}{3} \mu S < \mu S,$$

which is impossible. We maintain that the differences generated by the set $E = S_\cap I$ already contains an open interval about the origin.

Suppose that the number α is not amongst the differences generated by E. Then E and the translated set $\{E - \alpha\}$ do not intersect. Hence

$$\mu I \geq \mu E + \mu(\{E - \alpha\}_\cap I) \geq 2\mu E - |\alpha| > 2 \cdot \tfrac{2}{3}\mu I - |\alpha|,$$

which is a contradiction if $|\alpha| \leq \frac{1}{3}\mu I$.

This completes the proof of lemma (1.1).

Besides making direct use of this lemma in Chapter 7, it is convenient to apply it in the following lemma to the study of Cauchy's functional equation.

Cauchy's Functional Equation

Lemma 1.2. *Let the real-valued, measurable function $f(x)$ satisfy the functional equation*

$$f(x + y) = f(x) + f(y)$$

for all real numbers x and y. Then there is a constant A, so that for all real numbers x it has the form $f(x) = Ax$.

Proof. We first note that $f(0) = f(0 + 0) = 2f(0)$, so that $f(0) = 0$.

Let ε be a positive real number. The union of the countable collection of sets

$$\{x; x \in \mathbb{R}, |f(x) - n\varepsilon| < \varepsilon\} \qquad (n = 0, \pm 1, \pm 2, \ldots),$$

contains the real line $\mathbb{R}$. Therefore, there is a real number w so that the set

$$E = \{x; |f(x) - w| < \varepsilon\}$$

has positive measure.

Apply lemma (1.1) to the set E, and let $(-\delta, \delta)$ be an open interval contained in the difference set of E. If z is a real number which is contained in this interval, then there is a representation $z = x - y$, with x and y belonging to E, and so

$$\begin{aligned} |f(z) - f(0)| &= |f(z)| = |f(x - y)| = |f(x) - f(y)| \\ &\leq |f(x) - w| + |w - f(y)| < 2\varepsilon. \end{aligned}$$

Therefore $f(x)$ is continuous at $x = 0$, and so by translation continuous at every point x.

We can now continue with the proof, on the classical lines due to Cauchy himself.

A simple induction argument shows that for each integer n and real number y the relation $f(ny) = nf(y)$ is satisfied. It follows from this that if m is a further, positive integer, then $mf(m^{-1}y) = f(y)$, so that $f(nm^{-1}y) = nm^{-1}f(y)$. Thus for every rational number r we have $f(r) = rf(1)$.

Each real number x is the limit of a sequence of rational numbers r, and since $f(x)$ is an everywhere continuous function

$$f(x) = \lim_{r \to x} f(r) = \lim_{r \to x} rf(1) = xf(1).$$

We set $A = f(1)$, and the proof of lemma (1.2) is complete.

Slowly Oscillating Functions

If a function $L(x)$ is defined and non-zero for all sufficiently large positive real numbers x, and if for each fixed positive number u it satisfies the condition

$$\lim_{x\to\infty} \frac{L(ux)}{L(x)} = 1$$

then we shall say that it is *slowly oscillating*. It is convenient to allow such functions to assume complex values.

A form of slow oscillation, involving discrete variables, was introduced and studied by R. Schmidt [1]. Functions $L(x)$ of the above type, but with the extra conditions that they assume real values, and be positive for $x > 0$, were studied by Karamata, ([1], [2]).

The most commonly used property of slowly oscillating functions is contained in the following result, which in its present form is due to van Aardenne-Ehrenfest, de Bruijn and Korevaar [1].

Lemma 1.3. *Let $L(x)$ be a measurable slowly oscillating function. Then the relation*

$$\frac{L(ux)}{L(x)} \to 1 \qquad (x \to \infty)$$

holds uniformly on any interval of the form $a \le u \le b$, where $0 < a < b < \infty$.

Remark. Although $L(x)$ may only be defined for $x > x_0$ say, by setting $L(x) = 1$ when $x \le x_0$ we can apply the standard definition of measurability.

Proof. Suppose first that the assertion of lemma (1.3) fails to be true if $a = e^{-1}$ and $b = e$. In other words, there is an unbounded sequence of numbers $x_1 < x_2 < \cdots < x_k < \ldots$, and further numbers $w_1, w_2, \ldots, w_k, \ldots$ where each $|w_k| \le 1$, so that the ratio

$$\frac{L(x_k e^{w_k})}{L(x_k)}$$

does *not* converge to 1 as $k \to \infty$.

Since the function $L(x)$ is measurable, by a theorem of Egoroff (see Munroe [1] Theorem (21.3), pp. 108–109) there is a subset E of the interval $[-1, 1]$, of measure at least $\frac{7}{4}$, so that

$$\frac{L(x_k e^{y})}{L(x_k)} \to 1$$

uniformly for y in E.

Similarly there is a subset F of the interval $[-1, 1]$, of measure at least $\frac{7}{4}$, so that

$$\frac{L(x_k e^{w_k+z})}{L(x_k e^{w_k})} \to 1$$

uniformly for z in F.

The sets $w_k + F$ and E must contain at least one common point. Otherwise (for example) if $w_k \geq 0$ then they are both contained in the interval $[-1, 1 + w_k]$, so that (in terms of Lebesgue measure)

$$2 + w_k \geq \mu(w_k + F) + \mu E \geq \tfrac{7}{4} + \tfrac{7}{4},$$

which is impossible.

If $w_k + z_k = y_k$ then

$$\frac{L(x_k e^{w_k})}{L(x_k)} = \frac{L(x_k e^{w_k})}{L(x_k e^{w_k+z_k})} \cdot \frac{L(x_k e^{y_k})}{L(x_k)} \to 1$$

as $k \to \infty$, contradicting our initial assumption.

We have therefore shown that the convergence in the statement of lemma (1.3) is uniform over the interval $e^{-1} \leq u \leq e$. By making use of this result with x replaced by $e^m x$, setting $m = \pm 1, \pm 2, \ldots$, in turn, this uniformity may be extended to any interval of the form $e^{-m} \leq u \leq e^m$, where m is a fixed positive integer.

This completes the proof of lemma (1.3).

We shall need a number of other results that are related to lemma (1.3).

Lemma 1.4. *Let $\beta(x)$ be a real-valued, measurable function of x, which is defined and positive for all sufficiently large values of x. Assume that for each positive real number y the limit*

$$g(y) = \lim_{x \to \infty} \frac{\beta(x^y)}{\beta(x)}$$

exists, and is finite. Then there is a constant c so that the identity $g(y) = y^c$ holds for all $y > 0$.

Proof. The function $g(y)$ is clearly measurable. Moreover, for any pair of positive numbers y and z,

$$g(yz) = \lim_{x \to \infty} \frac{\beta(x^{yz})}{\beta(x^z)} \cdot \frac{\beta(x^z)}{\beta(x)} = g(y)g(z).$$

In particular

$$g(y)g(y^{-1}) = g(1) = 1,$$

so that $g(y)$ is positive for all $y > 0$.

Taking logarithms we see that $\log g(y)$ satisfies Cauchy's functional equation. The result of the present lemma now follows from that of lemma (1.2).

Remark. By considering the slowly oscillating function $L(x) = \beta(x)(\log x)^{-c}$, we can prove that the convergence to $g(y)$ in the hypothesis of lemma (1.4) is necessarily uniform over any interval of the form $a \le y \le b$, where $0 < a < b < \infty$.

We need a further result in this circle of ideas.

Lemma 1.5. *Let $\beta(x)$ be a positive, real-valued function of x, defined for all sufficiently large values of x. Let there be a positive constant C so that for all sufficiently large values of x*

$$\sup_{x^{1/2} \le w \le x^2} \beta(w) \le C\beta(x).$$

Let $\alpha(x)$ be a measurable, real-valued function of x, also defined for all large values of x. Suppose that for each positive real number y the asymptotic relation

$$\alpha(x^y) = \alpha(x) + o(\beta(x)) \qquad (x \to \infty),$$

is satisfied.

Then this relation holds uniformly on any interval of the form $a \le y \le b$, where $0 < a < b < \infty$.

Moreover, a similar statement can be made if the relation to be satisfied by $\alpha(x)$ is replaced by

$$\alpha(x^y) = y\alpha(x) + o(\beta(x)) \qquad (x \to \infty).$$

Remark. In the hypotheses of this lemma we do not need $\beta(x)$ to be a measurable function. In fact, the condition upon the growth of $\beta(x)$ ensures that the above asymptotic relations can hold if, and only if, they also hold when $\beta(x)$ is replaced by the function $\gamma(x)$ which is defined by

$$\gamma(n) = \beta(n) \qquad (n = 1, 2, \ldots),$$

and by linear interpolation otherwise

$$\gamma(x) = (n + 1 - x)\beta(n) + (x - n)\beta(x) \qquad (n < x < n + 1).$$

Since $\gamma(x)$ is continuous for $x \geq 1$, it is a measurable function. Without loss of generality, we may therefore assume that $\beta(x)$ is measurable. This ends the remark.

Proof. A straightforward induction argument shows that for each positive real number τ the expression

$$\beta(x)^{-1} \sup_{x^{\tau^{-1}} \leq w \leq x^{\tau}} \beta(x)$$

will be bounded, in terms of τ, for all large enough values of x.

A proof of lemma (1.5) can now be readily constructed along the lines of the proof given for lemma (1.3). We note only the following pertinent identities:

$$\frac{\alpha(x^{yz}) - \alpha(x)}{\beta(x)} = \left\{\frac{\alpha(x^{yz}) - \alpha(x^z)}{\beta(x^z)}\right\} \frac{\beta(x^z)}{\beta(x)} + \frac{\{\alpha(x^z) - \alpha(x)\}}{\beta(x)}$$

and

$$\frac{\alpha(x^{yz}) - yz\alpha(x)}{\beta(x)} = \left\{\frac{\alpha(x^{yz}) - y\alpha(x^z)}{\beta(x^z)}\right\} \frac{\beta(x^z)}{\beta(x)} + \frac{y\{\alpha(x^z) - z\alpha(x)\}}{\beta(x)}.$$

This completes our present considerations concerning slowly oscillating functions.

In the following lemma all sets S will be of real numbers, and will be Lebesgue measurable. $|S|$ will denote the measure of that part of the set which lies in the interval $[-\pi, \pi]$.

Lemma 1.6. (Halász). *Let S be a set of real numbers, which is closed, periodic with period 2π, and symmetric with respect to the origin. For each positive integer k let S_k denote the set of real numbers which are representable as the sum of k elements taken from S.*

Then either S_k is the whole real line, or $|S_k| \geq k|S|$.

Proof. We may assume that $|S| > 0$, otherwise the lemma is trivially valid.

We next note that for each integer k the set S_k is closed, periodic with period 2π, and symmetric about the origin. It is easy to check the validity of the second and third of these assertions. Concerning the first, we make use of the fact that S is periodic, and that $u_1 + u_2 = (u_1 - 2\pi) + (u_1 + 2\pi)$, and so on, to prove that each number u which is contained in S_k has a representation of the form $u = u_1 + \cdots + u_k$, with $|u_i| \leq |u| + 2\pi(k - 1)$, $(i = 1, 2, \ldots, k)$. That S_k is closed now follows readily.

We establish lemma (1.6) by induction on k.

In the case $k = 1$ there is nothing to prove.

Suppose that for some integer k, $k \geq 2$, S_k is not the whole real line. Consider the set

$$E = S_k \cap [-\pi, \pi].$$

Since S is not empty, and S_k contains at least one translation of S, then S_k, and so E, is non-empty. Let v be a point on the boundary of E, considered in the space of the interval $-\pi \leq x \leq \pi$, with the topology induced by the standard topology on the reals. By the hypothesis on S_k such a point must exist. Let $v_1, v_2, \ldots,$ be a sequence of points, not in S_k, so that as $n \to \infty$, $v_n \to v$. For each value of n the set

$$v_n + S = v_n - S$$

does not have any points in common with S_{k-1}. Hence, letting $n \to \infty$ and applying the dominated convergence theorem, we see that

$$|(v + S) \cap S_{k-1}| = \lim_{n \to \infty} |(v_n + S) \cap S_{k-1}| = 0.$$

But v belongs to the (closed) set S_k. Therefore there are numbers α and β such that $v = \alpha - \beta$, with α contained in S_{k-1}, and β contained in S. Translating by β yields

$$|(\alpha + S) \cap (\beta + S_{k-1})| = |(v + S) \cap S_{k-1}| = 0.$$

Hence, making use of the inductive hypothesis, we deduce that

$$\begin{aligned} |(\alpha + S) \cup (\beta + S_{k-1})| &= |\alpha + S| + |\beta + S_{k-1}| \\ &= |S| + |S_{k-1}| \geq |S| + (k - 1)|S| = k|S|. \end{aligned}$$

Since $\alpha + S$ and $\beta + S_{k-1}$ are disjoint subsets of S_k, we must have $|S_k| \geq k|S|$, which was the second alternative of the lemma.

This completes the proof of lemma (1.6).

Fourier Analysis on the Line: Plancherel's Theory

For a full account of this elegant theory we refer to Titchmarsh [3], Chapter 3. Here, we state those results which we shall need.

Let $f(x)$ be a complex-valued function which belongs to the Lebesgue class $L^2(-\infty, \infty)$. For each positive real number a define the function

$$h(y, a) = \frac{1}{\sqrt{(2\pi)}} \int_{-a}^{a} f(x) e^{ixy} \, dx.$$

Then there is an $L^2(-\infty, \infty)$ function, $h(y)$, to which $h(y, a)$ converges in the topology induced by the L^2-norm. In other words, we have

$$\lim_{a\to\infty} \int_{-\infty}^{\infty} |h(y) - h(y, a)|^2 \, dy = 0.$$

We write

$$h(y) = \underset{a\to\infty}{\text{l.i.m.}}\; h(y, a)$$

Conversely, if we define

$$f(x, a) = \frac{1}{\sqrt{(2\pi)}} \int_{-a}^{a} h(y) e^{-iyx} \, dy,$$

then

$$f(x) = \underset{\alpha\to\infty}{\text{l.i.m.}}\; f(x, a).$$

Moreover, we have Plancherel's (analogue of Parseval's) identity

$$\int_{-\infty}^{\infty} |h(y)|^2 \, dy = \int_{-\infty}^{\infty} |f(x)|^2 \, dx.$$

Almost surely in x the relations

$$h(y) = \frac{1}{\sqrt{(2\pi)}} \frac{d}{dy} \int_{-\infty}^{\infty} f(x) \frac{e^{ixy} - 1}{ix} \, dx,$$

$$f(x) = \frac{1}{\sqrt{(2\pi)}} \frac{d}{dx} \int_{-\infty}^{\infty} h(y) \frac{e^{-iyx} - 1}{-iy} \, dy,$$

are valid, although we shall have no general need of them. However, it will be convenient to note a special case.

Suppose that for some L^2-function $f(x)$ the Fourier transform $h(y)$ belongs to the class $L(-\infty, \infty)$. Then, applying Lebesgue's theorem on dominated convergence, we see that almost surely in x,

$$f(x) = \frac{1}{\sqrt{(2\pi)}} \int_{-\infty}^{\infty} h(y) e^{-iyx} \, dy.$$

In particular, if $f(x)$ is continuous then this relation holds for all real values of x.

The Theory of Probability

A general account of the foundations of probability is given in Rényi [7]. Almost all of the following results which are well-known may be found either in that reference, or in Gnedenko and Kolmogorov [1]. These we shall state without proof. We do, however, give proofs for those results which are perhaps less well known, or of which we shall have a particular need.

For the purposes of this monograph a distribution function (or law) $F(z)$ will be a real-valued, non-decreasing function of the real variable z, defined on the whole real line. It will be right-continuous, that is to say, for each number z

$$F(z) = \lim_{\varepsilon \to 0+} F(z + \varepsilon)$$

where ε is confined to positive values. Moreover, it will be assumed to satisfy the two additional conditions

$$\lim_{z \to -\infty} F(z) = 0 \qquad \lim_{z \to +\infty} F(z) = 1.$$

Weak Convergence

We shall say that a system of distribution functions $F_y(z)$, parametrized by an increasing set of real numbers y, converges weakly as $y \to \infty$ if there is a distribution function, $G(z)$, so that the limiting relation

$$\lim_{y \to \infty} F_y(z) = G(z)$$

holds at every point z at which $G(z)$ is continuous. We shall write

$$F_y(z) \Rightarrow G(z) \qquad (y \to \infty).$$

Sometimes $G(z)$ will be referred to as *the limit law*. (See, also, later remarks concerning random variables.)

Lévy's Metric

A useful metric was introduced into the space of distribution functions on the line, by Lévy [2], p. 47. For any two distribution functions, $F(z)$ and $G(z)$, we define $\rho(F, G)$ to be the greatest lower bound of those numbers h which have the property that the inequality

$$G(z - h) - h \leq F(z) \leq G(z + h) + h$$

holds for all real values of z.

It is easy to establish the following result.

Lemma 1.7. *The distribution functions $F_y(z)$ converge weakly to $G(z)$, as $y \to \infty$, if and only if $\rho(F_y, G) \to 0$, as $y \to \infty$.*

Remark. In fact, in terms of this metric the space of distribution functions is complete.

A special case of this lemma, which often arises in practice, is the following example. Let $u(n, x)$ and $w(n, x)$ be real numbers, defined for all positive integers n and real numbers $x \geq 1$. Assume that for each $\varepsilon > 0$ the condition

$$\nu_x(n; |u(n, x) - w(n, x)| > \varepsilon) \to 0 \qquad (x \to \infty),$$

is satisfied. (See Notation.) Then the distributions

$$F_x(z) = \nu_x(n; u(n, x) \leq z)$$

converge weakly as $x \to \infty$, if and only if the distributions

$$H_x(z) = \nu_x(n; w(n, x) \leq z)$$

converge weakly as $x \to \infty$. The limit laws if they exist, will then coincide.

In fact, in terms of Lévy's metric, the condition stated initially in this remark may be formulated as: $\rho(F_x, H_x) \to 0$ as $x \to \infty$. Let the distribution functions $H_x(z)$ converge weakly to $G(z)$, as $x \to \infty$. Then by the triangle inequality of Lévy's metric

$$\rho(F_x, G) \leq \rho(F_x, H_x) + \rho(H_x, G).$$

Since the right-hand side of this inequality approaches zero as x becomes unbounded, we deduce from lemma (1.7) that the $F_x(z)$ also converge weakly to $G(z)$ as $x \to \infty$.

The next lemma is often useful.

Lemma 1.8 (Compactness lemma). *Assume that for each $\varepsilon > 0$ there is an $N > 0$ so that the sequence of distribution functions $F_n(z)$, $(n = 1, 2, \ldots)$, satisfies*

$$F_n(N) - F_n(-N) > 1 - \varepsilon,$$

for all sufficiently large values of n. Then there is a subsequence n_j, $(j = 1, 2, \ldots)$, so that the corresponding distribution functions $F_{n_j}(z)$ converge weakly as $j \to \infty$.

A distribution function will be said to be *proper* if it does not consist of a jump at one point (a step function), and to be *improper* otherwise.

The next lemma will be useful many times.

Lemma 1.9. *Let the distribution functions $F_y(z)$ be defined for all sufficiently large values of y. The following propositions are valid:*

(i) *If $\beta_y > 0$, $b_y > 0$, α_y and a_y are real numbers such that*

$$F_y(\beta_y z + \alpha_y) \Rightarrow G(z),$$

$$F_y(b_y z + a_y) \Rightarrow K(z) \qquad (y \to \infty)$$

holds for some proper distribution functions $G(z)$ and $K(z)$, then there are numbers $A > 0$ and B so that as $y \to \infty$

$$\frac{\beta_y}{b_y} \to A \qquad \frac{\alpha_y - a_y}{b_y} \to B$$

If $G(z) = K(z)$ then $A = 1$ and $B = 0$.

(ii) *If*

$$F_y(z + \alpha_y) \Rightarrow G(z)$$

$$F_y(z + a_y) \Rightarrow H(z) \qquad (y \to \infty),$$

holds for any two distribution functions $G(z)$ and $H(z)$, which need not be proper, then there is a number C, so that as $y \to \infty$

$$\alpha_y - a_y \to C.$$

Proof. Part (i) of this lemma in the case that $G(z) = K(z)$ appears in its essentials as Theorem 2, p. 42 of Chapter 2 of Gnedenko and Kolmogorov [1].

According to Theorem 1, p. 40 of the same reference, $K(z)$ must anyway be of the same type as $G(z)$. Thus there are constants $A > 0$ and B so that the relation

$$K(z) = G(A^{-1}(z - B))$$

holds identically for all real values of z. The second part of the hypothesis in (i) may be equivalently written as

$$F_y(Ab_y z + a_y + Bb_y) \Rightarrow G(z) \qquad (y \to \infty).$$

We are now reduced to the particular case already considered.

Moreover, according to this same Theorem 1, p. 40, if, in (ii), both $G(z)$ and $H(z)$ are proper, then they must be of the same type. In other words, there are constants $L > 0$ and M so that

$$H(z) = G(Lz + M).$$

But then

$$F_y\left(\frac{1}{L}\{z + a_y - M\}\right) \Rightarrow G(z) \qquad (y \to \infty),$$

and, by the result of part (i), $L = 1$ and

$$\alpha_y - a_y + M \to 0 \qquad (y \to \infty).$$

Suppose, therefore, that $H(z)$ is the improper law with a jump at the point $z = b$. Then, if $\varepsilon > 0$ and $\delta > 0$, the inequalities

$$F(b + \varepsilon + a_y) - F(b - \varepsilon + a_y) > 1 - \delta,$$

and, therefore,

$$G(b - \alpha_y + a_y + \varepsilon) - G(b - \alpha_y + a_y - \varepsilon) > 1 - \delta,$$

hold for all sufficiently large values of y. It is now straightforward to establish, first that the numbers $n - \alpha_y - a_y$ are uniformly bounded for all large y, and then that they converge to some number w, with $G(z)$ being the improper law with jump at $z = w$.

This completes the proof of lemma (1.9).

Characteristic Functions

Associated with each distribution function $F(z)$ is the characteristic function

$$\phi(t) = \int_{-\infty}^{\infty} e^{itz}\, dF(z).$$

This characteristic function is defined for all real values of t. It is uniformly continuous on the whole real line, and satisfies

$$\phi(0) = 1 \qquad |\phi(t)| \leq 1.$$

Lemma 1.10 (Inversion formula). *Let u and v be continuity points of the distribution function $F(z)$ whose associated characteristic function is $\phi(t)$. Then,*

$$F(u) - F(v) = \frac{1}{2\pi} \lim_{N \to \infty} \int_{-N}^{N} \frac{e^{-itv} - e^{-itu}}{it} \cdot \phi(t) dt.$$

Remark. This formula is usually not very practical, and we shall return to consider quantitative Fourier inversion. However, it does show that the characteristic function determines $F(z)$ uniquely.

We can formulate weak convergence in terms of characteristic functions.

Lemma 1.11. *Let $F_y(z)$ have the characteristic function $\phi_y(t)$. Then the following propositions are equivalent:*

(i) *The distributions $F_y(z)$ converge weakly to a distribution function as $y \to \infty$.*
(ii) *There is a function $\psi(t)$, defined for all real values of t, and continuous at $t = 0$, so that as $y \to \infty$*

$$\phi_y(t) \to \psi(t) \qquad (-\infty < t < \infty).$$

(iii) *There is a function $\psi(t)$, defined for all real values of t, such that as $y \to \infty$ the relation*

$$\phi_y(t) \to \psi(t)$$

holds uniformly on any bounded set of t-values.

In cases (ii) *and* (iii) *$\psi(t)$ will be characteristic function, and as $y \to \infty$ the $F_y(z)$ will converge weakly to its corresponding distribution function.*

Proof. For a proof of this important theorem we refer to Gnedenko and Kolmogorov [1], Chapter 2, Theorems 1 and 2, p. 53.

The following theorem proves useful, often in connection with lemma (1.8).

Lemma 1.12. *Let $\phi(t)$ be the characteristic function of $F(z)$. Let τ and X be positive real numbers such that $\tau X > 1$. Let $\pm X$ be continuity points of $F(z)$. Then the inequality*

$$F(X) - F(-X) \geq \frac{|(1/2\tau) \int_{-\tau}^{\tau} \phi(t) dt| - 1/\tau X}{1 - 1/\tau X}$$

is satisfied. Moreover, this inequality remains valid if $\phi(t)$ *is replaced by* Re $\phi(t)$.

Proof. We refer to Gnedenko and Kolmogorov [1] Chapter 2, Theorem 2 and following, pp. 53–54.

Lemma 1.13. *If* $\phi(t)$ *is a characteristic function, then for all real numbers* t

$$1 - \operatorname{Re} \phi(2t) \leq 4(1 - \operatorname{Re} \phi(t))$$

Proof. For each real number θ,

$$3 - 4 \cos \theta + \cos 2\theta = 2(1 - \cos \theta)^2 \geq 0.$$

In another form, we write

$$1 - \cos 2\theta \leq 4(1 - \cos \theta).$$

We set $\theta = tz$, and integrate over the whole line as follows:

$$\begin{aligned} 1 - \operatorname{Re} \phi(2t) &= \int_{-\infty}^{\infty} (1 - \cos 2tz) dF(z) \leq 4 \int_{-\infty}^{\infty} (1 - \cos tz) dF(z) \\ &= 4(1 - \operatorname{Re} \phi(t)). \end{aligned}$$

This completes the proof.

Remark. The inequality given at the beginning of this proof plays a fundamental part in the proof of the prime–number theorem given by Hadamard, and de la Vallée Poussin.

Random Variables

Let $\mathfrak{S}$ be a σ-algebra of sets upon which a probability measure P is defined. We say that the pair $(\mathfrak{S}, P)$ is a probability space. Let Ω denote the member of $\mathfrak{S}$ which represents the total event. Thus Ω is the union of all those sets which are members of $\mathfrak{S}$.

A random variable X, $= X(\alpha)$, is a real-valued function, which is defined for every element α of Ω, with the property that for every pair of real numbers a, b, $(-\infty \leq a \leq b \leq \infty)$, the set

$$\{\alpha; a < X(\alpha) \leq b\}$$

belongs to the algebra $\mathfrak{S}$. A suitable meaning is to be given when $b = \pm\infty$. As is conventional, we shall, in future, omit the dummy variable α.

Associated with each random variable X will be the distribution function $F(z)$ which is defined by

$$F(b) - F(a) = P(a < X \leq b) \qquad (-\infty \leq a \leq b \leq \infty),$$

We sometimes say that the variable X is distributed according to the law $F(z)$.

If X and Y are independent random variables, with corresponding distribution functions $F(z)$ and $G(z)$, then associated with the variable $X + Y$ is the convolution

$$(F * G)(z) = \int_{-\infty}^{\infty} F(z - u)dG(u) = \int_{-\infty}^{\infty} G(z - u)dF(u).$$

This property amounts to a definition of independence. The characteristic function of the sum of independent random variables is therefore the product of their respective characteristic functions. The converse need not be true.

We remark that a set of measures $P_y(E)$, defined on a suitable topological space. are said to be weakly convergent to a measure $\mu(E)$ if

$$P_y(E) \to \mu(E) \qquad (y \to \infty),$$

whenever the frontier $\partial(E)$ of E has μ-measure zero. If the P_y are defined on the Borel subsets of the real line in a manner consistent with

$$P_y((a, b]) = F_y(b) - F_y(a),$$

for some distribution functions $F_y(z)$, then the weak convergence of measures and the weak convergence of distribution functions are in fact equivalent. For further details of this aspect of probability theory we refer to Billingsley [1].

The *mean* and *variance* of a random variable X are respectively defined by

$$\bar{X} = E(X) = \int_{-\infty}^{\infty} z\, dF(z),$$

and

$$D^2(X) = E(X - E(X))^2 = \int_{-\infty}^{\infty} (z - \bar{X})^2\, dF(z),$$

where $F(z) = P(X \leq z)$, whenever the integrals exist. For independent random variables X_j, $(j = 1, \ldots, n)$, we have

$$D^2(X_1 + \cdots + X_n) = \sum_{j=1}^{n} D^2(X_j).$$

Note that some texts use the symbol σ in place of D.

Let the event B have a positive probability. The conditional probability that, given B, the event A will occur, is defined to be

$$\frac{P(AB)}{P(B)},$$

where AB denotes the intersection of the events A and B.

Concentration Functions

It is convenient to follow Lévy [2], p. 44, and to introduce, for each random variable X, the concentration function

$$Q(l) = Q_X(l) = \sup_z P(z < X \leq z + l),$$

where the supremum is taken over all real numbers z. This function is clearly non-decreasing for increasing values of l.

If X and Y are independent random variables, then

$$Q_{X+Y}(l) \leq \min(Q_X(l), Q_Y(l)).$$

To see this, for example we note that

$$P(z < X + Y \leq z + l) = \int_{-\infty}^{\infty} \{F(z + l - u) - F(z - u)\} dG(u)$$
$$\leq Q_X(l) \int_{-\infty}^{\infty} dG(u) = Q_X(l).$$

We now establish the following, sharper theorem.

Let $X_1, \ldots, X_n$ be n independent random variables, $n \geq 1$. Let S_n denote their sum. For non-negative real numbers r and l.

$$Q_i(r) = Q_{X_i}(r) \qquad (i = 1, \ldots, n) \qquad Q(r) = Q_{S_n}(r),$$

and

$$s = s(l) = \sum_{i=1}^{n} \{1 - Q_i(l)\}.$$

Lemma 1.14 (Kolmogorov, Rogozin). *There is an absolute constant c_1 so that if $L \geq l > 0$, then*

$$Q(L) \leq \frac{c_1 L}{l\sqrt{s}}.$$

In order to establish this lemma we shall need a number of preliminary results.

Lemma 1.15 (Sperner). *Let E be a finite set of n elements. Then any collection of subsets of E, no one subset being contained in another, cannot comprise of more than*

$$\binom{n}{\left[\frac{n}{2}\right]}$$

members.

Remark. The symbol $\binom{n}{r}$ denotes the (binomial) coefficient of x^r in the expansion of $(1 + x)^n$.

Proof. Consider the collection of all towers of subsets of E

$$E_1 \subseteq E_2 \subseteq \cdots \subseteq E_n \subseteq E,$$

where each E_i contains exactly i elements, $(i = 1, \ldots, n)$. Clearly there are $n!$ distinct such towers.

Let A be a subset of our special collection, and which contains exactly r elements. Then A is contained in $r!(n - r)!$ of the above towers. Owing to the monotonic properties of the binomial coefficients

$$r!(n - r)! \geq n! \binom{n}{\left[\frac{n}{2}\right]}^{-1}.$$

Let the total number of the various sets A in the collection be denoted by N. It follows from the hypothesis of the lemma that no two distinct sets amongst the A can belong to the same tower. Hence

$$Nn!\binom{n}{\left[\frac{n}{2}\right]}^{-1} \leq n!$$

and lemma (1.15) is proved.

Remark. We shall have more to say concerning the use of this lemma in the remarks at the end of this chapter, and of Chapter 5.

We can now prove a special case of lemma (1.14) and from this special case shall deduce the whole result.

Lemma 1.16. *Let the random variables X_i in lemma* (1.14) *have distributions which satisfy*

$$P(X_i = \alpha_i) = P(X_i = -\alpha_i) = \tfrac{1}{2},$$

where each of the numbers α_i satisfies $\alpha_i > l$, $(i = 1, 2, \ldots, n)$. Then we have

$$Q(2l) \leq 2^{-n}\binom{n}{\left[\frac{n}{2}\right]}.$$

Proof. Let z be a temporarily fixed real number. The probability that $z < S_n \leq z + 2l$ is 2^{-n} times the number of sums of the form

$$\sum_{k=1}^{n} \varepsilon_k \alpha_k \qquad (\text{each } \varepsilon_k = \pm 1)$$

which fall into the interval $(z, z + 2l]$. For each sum consider the subset of the integers between 1 and n which is defined by those k for which $\varepsilon_k = 1$. No one of these subsets can be contained in another. Otherwise there will be choices ε_k and ε_k', $(k = 1, \ldots, n)$, say, so that

$$\sum_{k=1}^{n} \varepsilon_k \alpha_k - \sum_{k=1}^{n} \varepsilon_k' \alpha_k > 2l,$$

which is impossible. An application of lemma (1.15) now shows that the number of such sums is at most

$$\binom{n}{\left[\frac{n}{2}\right]}.$$

Lemma 1.17. *Under the conditions of lemma* (1.16), *if* $L \geq l$ *then*

$$Q(L) \leq \frac{c_2 L}{l\sqrt{n}}.$$

Proof. From the previous lemma

$$Q(l) \leq Q(2l) \leq 2^{-n}\binom{n}{\left[\frac{n}{2}\right]}.$$

By Stirling's approximation there is an absolute constant c_3 so that uniformly for all positive integers n

$$2^{-n}\binom{n}{\left[\frac{n}{2}\right]} < \frac{c_3}{\sqrt{n}}.$$

Thus

$$\begin{aligned} Q(L) &= \sup_z P(z < S_n \leq z + L) \\ &\leq \sum_{j=0}^{[L/l]} P(z + jl < S_n \leq z + (j+1)l) \\ &\leq \left\{\left[\frac{L}{l}\right] + 1\right\} Q(l) < \frac{2c_3 L}{l\sqrt{n}}, \end{aligned}$$

as was asserted.

Remark. If we introduce numbers β_i into the hypothesis of lemma (1.16), so that

$$P(X_i = \alpha_i + \beta_i) = P(X_i = -\alpha_i + \beta_i) = \tfrac{1}{2},$$

then we do not alter any of the concentration functions, nor, therefore, any of the conclusions of the last two lemmas.

Proof of lemma 1.14. We first prove the theorem with the additional assumption that the variables X_i have continuous, strictly increasing distribution functions $F_i(z)$. Then the inverse functions $F_i^{-1}(\xi)$, $-\infty < \xi < \infty$, are well defined. The random variable $\xi_i = F_i(X_i)$ satisfies

$$\begin{aligned} P(\xi_i < y) &= P(F_i(X_i) < y) \\ &= P(X_i < F_i^{-1}(y)) \\ &= F_i(F_i^{-1}(y)) = y, \end{aligned}$$

so that

$$X_i = F_i^{-1}(\xi_i),$$

where $\xi_1, \xi_2, \ldots, \xi_n$ are independent random variables, each being uniformly distributed over the interval (0, 1).

For $i = 1, \ldots, n$, define

$$\varepsilon_i = \tfrac{1}{4}\{1 - Q_i(l)\} \qquad x_i' = F_i^{-1}(\varepsilon_i) \qquad x_i'' = F_i^{-1}(1 - \varepsilon_i),$$

and note that

$$F_i(x_i'') - F_i(x_i') = 1 - 2\varepsilon_i = \tfrac{1}{2}\{1 + Q_i(l)\} > Q_i(l),$$

so that $x_i'' - x_i' > l$. Note, also, that our temporary assumptions concerning $F_i(z)$ ensure that if l is positive then $Q_i(l) < 1$.

Define new random variables by

$$Z_i = \begin{cases} \xi_i & \text{if } \xi_i \leq \varepsilon_i \\ 1 - \xi_i & \text{if } \xi_i > \varepsilon_i \end{cases} \qquad (i = 1, \ldots, n).$$

Consider the random subset $i_1, i_2, \ldots, i_m$ of the integers 1 to n on which the variables ξ_i satisfy $\xi_i \leq \varepsilon_i$ or $\xi_i > 1 - \varepsilon_i$. Let $\bar{P}$ denote probabilities conditional upon the i_j and the corresponding variables Z_{i_j}, $(j = 1, \ldots, m)$, having prescribed values. Let $\bar{Q}$ denote the corresponding concentration function.

Then

$$\bar{P}(X_k = a_k + x_k) = \bar{P}(X_k = a_k - x_k) = \tfrac{1}{2},$$

where

$$x_k = \tfrac{1}{2}\{F_k^{-1}(1 - Z_k) - F_k^{-1}(Z_k)\}$$

$$a_k = \tfrac{1}{2}\{F_k^{-1}(1 - Z_k) + F_k^{-1}(Z_k)\} \qquad (k = i_1, i_2, \ldots, i_m).$$

We note that owing to the conditioning we have $Z_k \leq \varepsilon_k$ and $1 - Z_k > \varepsilon_k$, so that

$$x_k \geq \tfrac{1}{2}\{F_k^{-1}(1-\varepsilon_k) - F_k^{-1}(\varepsilon_k)\} = \tfrac{1}{2}\{x_k'' - x_k'\} > \frac{l}{2}.$$

Therefore $\bar{Q}(L)$ does not exceed the (conditioned) concentration function of the variable

$$X_{i_1} + \cdots + X_{i_m}.$$

In turn, this may be estimated by means of lemma (1.17), taking into account the remark made at the end of that lemma, and does not exceed

$$\frac{2c_2 L}{l\sqrt{m}}.$$

Hence

$$Q(L) \leq \text{Expect } \bar{Q}(L) \leq P(m \leq \tfrac{1}{4}s) + \frac{4c_2 L}{l\sqrt{s}}.$$

In order to estimate $P(m \leq s/4)$ we regard it as the probability of m ($\leq s/4$) successful outcomes in n independent Bernoulli trials. The probability of a successful outcome at the kth trial is $2\varepsilon_k$, so that

$$E(m) = \sum_{k=1}^{n} 2\varepsilon_k = \frac{s}{2}.$$

Moreover,

$$D^2(m) = \operatorname{Var}(m) = \sum_{k=1}^{n} 2\varepsilon_k(1-\varepsilon_k) \leq \frac{s}{2},$$

so that by applying Tchebycheff 's argument

$$P\left(m \leq \frac{s}{4}\right) \leq P\left(|m - E(m)| > \frac{s}{4}\right) \leq 16s^{-2}D^{-2}(m) < 8s^{-1}.$$

Since $L \geq l$, we deduce that

$$Q(L) \leq \frac{8}{s} + \frac{4c_2 L}{l\sqrt{s}} \leq \frac{c_1 L}{l\sqrt{s}}.$$

In order to complete the proof of lemma (1.14) it remains for us to remove the auxiliary condition which we imposed upon the $F_i(z)$. In fact, since each $F_i(z)$ is measurable we can find continuous and strictly increasing distribution functions $G_i(z)$ which approximate $F_i(z)$ uniformly closely over whole real line. In particular, if $\eta > 0$ we can find such distribution functions, one for each value of i, $1 \leq i \leq n$, so that the estimate

$$|Q_i(r) - Q_i'(r)| < \eta$$

holds uniformly for $1 \leq i \leq n$, $r \geq 0$. Here $Q_i'(r)$ denotes the appropriate concentration function, defined in terms of the distribution function $G_i(z)$. We apply the previous result to the convolution of the $G_i(z)$, and deduce that

$$Q(L) \leq c_1 L l^{-1}(s - n\eta)^{-1/2}.$$

Since η may be chosen arbitrarily small, the proof of lemma (1.14) is complete.

Infinite Convolutions

Let $X_1, X_2, \ldots$ be an infinite sequence of independent random variables, with corresponding distribution functions $F_j(z)$, $j = 1, 2, \ldots$. For each positive integer n let $S_n = X_1 + \cdots + X_n$. We shall be interested in the weak limit of the distributions

$$P(S_n \leq z) \qquad (n = 1, 2, \ldots).$$

When this limit exists we shall denote it by $F(z)$.

The most famous result concerning infinite convolutions of random variables is the following.

Lemma 1.18. *The following three propositions are equivalent:*

(i) *The series*

$$X_1 + X_2 + \ldots$$

converges almost surely (i.e. with probability one).

(ii) *The distributions*

$$P(S_n \leq z)$$

converge weakly to a limiting distribution as $n \to \infty$.

(iii) *For any* $\varepsilon > 0$ *the three series*

$$\sum_{j=1}^{\infty} P(|X_j| > \varepsilon)$$

$$\sum_{j=1}^{\infty} \int_{|z| \le \varepsilon} z \, dF_j(z)$$

$$\sum_{j=1}^{\infty} \left(\int_{|z| \le \varepsilon} z^2 \, dF_j(z) - \left(\int_{|z| \le \varepsilon} z \, dF_j(z) \right)^2 \right)$$

all converge.

It is convenient to begin the proof of this lemma by concentrating upon the equivalence of propositions (ii) and (iii). This part of the proof is the most complicated.

Proof that (ii) *implies* (iii). Let $\phi_j(t)$ denote the characteristic function of $F_j(z)$, $(j = 1, 2, \ldots, n)$. According to our hypothesis the limiting distribution $F(z)$ exists. Let its characteristic function be $\phi(t)$. Then by lemma (1.11) the limiting relation

$$\prod_{j=1}^{n} \phi_j(t) \to \phi(t) \qquad (n \to \infty),$$

holds uniformly on any bounded interval of t-values.

Since $\phi(t)$ is continuous at $t = 0$, there is an interval, $|t| \le \tau_1$ say, on which it satisfies $|\phi(t)| \ge \frac{1}{2}$. Therefore

$$\phi_j(t) \to 1 \qquad (j \to \infty)$$

holds uniformly over this interval. By means of the inequality of lemma (1.13) we deduce that as $j \to \infty$,

$$1 - \operatorname{Re} \phi_j(2t) \le 4(1 - \operatorname{Re} \phi_j(t)) = o(1).$$

This allows us to extend the region of uniformity for the asymptotic estimate of $\phi_j(t)$, from $|t| \le \tau_1$ to $|t| \le 2\tau_1$. Arguing in this way we see that $\phi_j(t) \to 1$ uniformly on any bounded interval of t-values.

Let ε be a positive number, for which $\pm\varepsilon$ are continuity points of the distribution function $F_j(z)$. We apply lemma (1.12), with a suitably large value of τ, to deduce that as $j \to \infty$

$$P(|X_j| > \varepsilon) \le 1 - \left\{ \frac{1}{2\tau} \int_{-\tau}^{\tau} (1 + o(1)dt - \frac{1}{\tau\varepsilon} \right\} \left(1 - \frac{1}{\tau\varepsilon} \right)^{-1}.$$

Since τ may be chosen arbitrarily large it is clear that for any fixed value of ε,

$$P(|X_j| > \varepsilon) \to 0 \qquad (j \to \infty).$$

Let m_j be a median for the variable X_j, that is to say a number such that

$$P(X_j \geq m_j) \geq \tfrac{1}{2} \qquad P(X_j \leq m_j) \geq \tfrac{1}{2}.$$

Of course m_j need not be unique. However, we may conclude from the result of the previous paragraph that any choice of the m_j must satisfy $m_j \to 0$ as $j \to \infty$. We shall presently make use of this remark.

For each value of j, $(j = 1, 2, \ldots)$, it is convenient to define a new random variable Y_j, with the same distribution as X_j, and such that the variables X_i, Y_j, $(i, j = 1, 2, \ldots)$ are independent. Let $G_j(z)$ be the distribution function of the variable $X_j - Y_j$. Its characteristic function is $|\phi_j(t)|^2$. In particular, it is real.

Our next step is to prove that the series

$$\sum_{j=1}^{\infty} (1 - |\phi_j(t)|^2)$$

converges, and is, indeed, uniformly bounded on any bounded interval of t-values. In fact the inequality $-\log(1 - w) \geq w$ certainly holds for real numbers w in the range $0 \leq w < 1$, so that if $|t| \leq \tau_1$, (τ_1 being defined earlier in this proof),

$$\sum_{j=1}^{\infty} (1 - |\phi_j(t)|^2) \leq -\sum_{j=1}^{\infty} \log|\phi_j(t)|^2 = -2 \log|\phi(t)| \leq 2 \log 2.$$

This justifies our assertion when t is confined to the range $|t| \leq \tau_1$. By means of lemma (1.13) we see that

$$1 - |\phi_j(2t)|^2 \leq 4(1 - |\phi_j(t)|^2)$$

and we can extend the uniformity to the interval $|t| \leq 2\tau_1$, and so on.

Before proceeding to the next part of the proof we recall the inequalities

$$1 - \frac{\sin z}{z} \geq \begin{cases} \dfrac{1}{10} & \text{if } |z| > 1, \\[2ex] \dfrac{z^2}{8} & \text{if } |z| \leq 1. \end{cases}$$

Applying these for each value of $j \geq 1$ yields

$$\frac{1}{2}\int_{-1}^{1}(1 - |\phi_j(t)|^2)dt = \int_{-\infty}^{\infty}\left(1 - \frac{\sin z}{z}\right)dG_j(z)$$
$$\geq \frac{1}{8}\int_{|z|\geq 1} z^2\, dG_j(z) + \frac{1}{10}\int_{|z|>1} dG_j(z).$$

Summing over $j = 1, 2, \ldots$, and making use of the result of the previous paragraph, we see that

$$\frac{1}{5}\sum_{j=1}^{\infty}\int_{|z|>1} dG_j(z) + \frac{1}{4}\sum_{j=1}^{\infty}\int_{|z|\leq 1} z^2\, dG_j(z) \leq \int_{-1}^{1}\sum_{j=1}^{\infty}(1 - |\phi_j(t)|^2)dt < \infty.$$

We can express this result more succinctly in the form

$$\sum_{j=1}^{\infty}\int_{-\infty}^{\infty}\frac{z^2}{1+z^2}\, dG_j(z) < \infty.$$

In order to relate this result to our original variables X_j we shall need the following simple lemma.

Lemma 1.19. *Let ξ and η be independent and identically distributed random variables. Let m be a median of the variable ξ. Define the function*

$$s(z) = \begin{cases} z^2 & \text{if } |z| \leq 1 \\ 1 & \text{if } |z| > 1 \end{cases}$$

Then the inequality

$$2E\{s(\xi - \eta)\} \geq E\{s(\xi - m)\}$$

is satisfied.

Remark. The function $s(z)$ is bounded above and below by

$$\frac{z^2}{1+z^2} \leq s(z) \leq \frac{2z^2}{1+z^2}.$$

Proof. By replacing the variables ξ and η by $\xi - m$ and $\eta - m$ we see that it will suffice to establish the lemma when $m = 0$. Let $H(z)$ denote the distribution function of the variable ξ. Then, since $s(w - z) \geq s(w)$ when w and z have opposite signs,

$$\begin{aligned} E\{s(\xi - \eta)\} &= \int_{-\infty}^{\infty} \int_{-\infty}^{\infty} s(w - z) dH(w) dH(z) \\ &\geq \int_{w \geq 0} \int_{z \leq 0} + \int_{w < 0} \int_{z \geq 0} s(w - z) dH(w) dH(z) \\ &\geq \int_{z \leq 0} dH(z) \int_{w \geq 0} s(w) dH(w) + \int_{z \geq 0} dH(z) \int_{w < 0} s(w) dH(w) \\ &\geq \frac{1}{2} \int_{-\infty}^{\infty} s(w) dH(w) = \frac{1}{2} E\{s(\xi)\}. \end{aligned}$$

Note that we make use of the fact that $m\ (= 0)$ is a median of the variable ξ in order to obtain the final inequality of the chain.

Applying this lemma to what we have proved so far we see that for any $\varepsilon > 0$ both of the series

$$\sum_{j=1}^{\infty} P(|X_j - m_j| > \varepsilon) \qquad \sum_{j=1}^{\infty} \int_{|z - m_j| \leq \varepsilon} (z - m_j)^2 \, dF_j(z)$$

are convergent. But $m_j \to 0$ as $j \to \infty$, so that for any fixed $\varepsilon > 0$ the series

$$\sum_{j=1}^{\infty} P(|X_j| > \varepsilon)$$

is also convergent. This is the first of the three conditions in (iii) which we wish to establish.

Let $\varepsilon > 0$ and let

$$\alpha_j = \alpha_j(\varepsilon) = \int_{|z| \leq \varepsilon} z \, dF_j(z).$$

Then

$$\alpha_j - m_j = \int_{|z| \leq \varepsilon} (z - m_j) dF_j(z) - m_j \int_{|z| > \varepsilon} dF_j(z)$$

and by virtue of the elementary inequality $(a - b)^2 \leq 2(a^2 + b^2)$ we have

$$(\alpha_j - m_j)^2 \leq 2\left(\int_{|z| \leq \varepsilon} (z - m_j) dF_j(z)\right)^2 + 2m_j^2 \left(\int_{|z| > \varepsilon} dF_j(z)\right)^2.$$

If j is sufficiently large, $|m_j| \leq \varepsilon$ will hold, so that applying the Cauchy–Schwarz inequality twice, the right-hand side of this inequality is seen not to exceed

$$2\varepsilon \int_{|z| \leq \varepsilon} (z - m_j)^2 \, dF_j(z) + 2\varepsilon^2 \int_{|z| > \varepsilon} dF_j(z)$$

$$\leq 2\varepsilon \int_{|z - m_j| \leq 2\varepsilon} |z - m_j|^2 \, dF_j(z) + 2\varepsilon^2 \int_{|z| > \varepsilon} dF_j(z).$$

It follows at once that the series

$$\sum_{j=1}^{\infty} (\alpha_j - m_j)^2$$

is convergent, and therefore that

$$\sum_{j=1}^{\infty} \int_{|z| \leq \varepsilon} (z - \alpha_j)^2 \, dF_j(z) \leq \sum_{j=1}^{\infty} \int_{|z| \leq \varepsilon} \{2(z - m_j)^2 + 2(m_j - \alpha_j)^2\} \, dF_j(z)$$

$$\leq 2 \sum_{j=1}^{\infty} \int_{|z| \leq \varepsilon} (z - m_j)^2 \, dF_j(z)$$

$$+ 2 \sum_{j=1}^{\infty} (m_j - \alpha_j)^2 < \infty.$$

This is the third of the three conditions in (iii).

We cannot expect to derive any further information by arguing in the above manner, for we have only made use of the facts that $\phi_j(t) \to 1$ and that

$$\prod_{j=1}^{n} |\phi_j(t)| \to |\phi(t)| \qquad (n \to \infty).$$

In order to complete the proof that (ii) implies (iii) we need the following result, which is of independent interest.

Lemma 1.20. *Let* $\varepsilon > 0$, *and define, as earlier:*

$$\alpha_j = \alpha_j(\varepsilon) = \int_{|z| \leq \varepsilon} z \, dF_j(z).$$

Assume that the series

$$\sum_{j=1}^{\infty} P(|X_j| > \varepsilon) \qquad \sum_{j=1}^{\infty} \int_{|z| \le \varepsilon} (z - \alpha_j)^2 \, dF_j(z)$$

converge. Then the distribution functions

$$P\left(\sum_{i=1}^{n} (X_i - \alpha_i) \le z\right)$$

converge weakly as $n \to \infty$.

Proof. Consider the function

$$\phi_j(t)e^{-it\alpha_j} = \int_{-\infty}^{\infty} e^{it(z-\alpha_j)} \, dF_j(z).$$

By means of the estimate

$$|e^{iw} - 1 - iw| \le \tfrac{1}{2}|w|^2,$$

which is valid for all real numbers w, we may represent it in the form

$$\int_{|z| \le \varepsilon} e^{it(z-\alpha_j)} \, dF_j(z) + \int_{|z| > \varepsilon} e^{it(z-\alpha_j)} \, dF_j(z) = 1 + k_j(t),$$

where

$$|k_j(t)| \le \frac{t^2}{2} \int_{|z| \le \varepsilon} (z - \alpha_j)^2 \, dF_j(z) + \int_{|z| > \varepsilon} dF_j(z).$$

We deduce from the hypothesis of the present lemma (1.20) that the series $\sum |k_j(t)|$, and hence the product

$$\prod_{j=1}^{n} \phi_j(t) \exp(-it\alpha_j) \qquad (n \to \infty),$$

converges uniformly on any bounded interval of t-values.

The desired result now follows from an application of lemma (1.11).

To return to the original proof that (ii) implies (iii), we see that we have already established the validity of the hypothesis of lemma (1.20). We conclude that the distributions

$$P\left(\sum_{j=1}^{n} X_j - \sum_{j=1}^{n} \alpha_j \le z\right) \qquad (n = 1, 2, \ldots),$$

converge weakly. According to lemma (1.9) this is consistent with our hypothesis that (ii) is valid only if the series

$$\sum_{j=1}^{\infty} \alpha_j = \sum_{j=1}^{\infty} \int_{|z| \leq \varepsilon} z \, dF_j(z),$$

is convergent.

We have now established all of the conditions stated in lemma (1.18), part (iii).

Proof that (iii) *implies* (ii). From the first and third conditions of lemma (1.18), (iii), applied to lemma (1.20), we deduce that the distributions

$$P\left(S_n - \sum_{j=1}^{n} \alpha_j \leq z\right) \qquad (n = 1, 2, \ldots),$$

converge weakly. Since the series $(\alpha_1 + \alpha_2 + \cdots)$ converges we see that (ii) holds.

It follows immediately from Lebesgue's theorem on dominated convergence that proposition (i) of lemma (1.18) implies proposition (ii). To complete the proof of the whole of lemma (1.18) we need only show that (iii) implies (i). This will be done by means of Kolmogorov's generalisation of Tchebycheff's inequality, which is itself of independent interest.

Lemma 1.21 (Kolmogorov). *Let ε be a positive real number. Then in the above notation*

$$P\left(\max_{k \leq n} |S_k - E(S_k)| \geq \varepsilon\right) \leq \varepsilon^{-2} \sum_{j=1}^{n} D^2(X_j).$$

Proof. Without loss of generality we may assume that each variable X_j has mean zero. Let A_0 be the total event. For each integer k in the range $1 \leq k \leq n$ define the events

$$A_k = \left\{\max_{j \leq k} |S_j| < \varepsilon\right\}$$

$$B_k = A_{k-1} - A_k = \{|S_j| < \varepsilon \qquad j = 1, \ldots, k-1 \qquad |S_k| \geq \varepsilon\}.$$

Here, and during the course of the proof of the present lemma, $-$ will denote complementation. Then we have the representation

$$A_0 - A_n = \bigcup_{k=1}^{n} B_k.$$

We first note that

$$\varepsilon^2 P(B_k) \leq \int_{B_k} S_n^2 \, dP,$$

for without loss of generality $P(B_k) > 0$. Conditioning with respect to the event B_k we note that the variables S_k and S_{n-k} are independent. According to the definition of B_k we see that in terms of the appropriately conditioned expectation

$$E(S_n^2 \mid B_k) = E(S_n^2) = E(\{S_n - S_k\}^2) + E(S_k^2) \geq E(S_k^2) \geq \varepsilon^2.$$

Hence

$$\begin{aligned}\varepsilon^2 P(A_0 - A_n) = \varepsilon^2 \sum_{k=1}^{n} P(B_k) &\leq \sum_{k=1}^{n} \int_{B_k} S_n^2 \, dP \leq \int S_n^2 \, dP \\ &= \sum_{j=1}^{n} D^2(X_j),\end{aligned}$$

and the proof of Kolmogorov's inequality is complete.

Proof that (iii) *implies* (i). Let ε be a positive real number. Define new independent random variables Z_j, $(j = 1, 2, \ldots)$, by

$$Z_j = \begin{cases} X_j & \text{if } |X_j| \leq \varepsilon, \\ 0 & \text{if } |X_j| > \varepsilon. \end{cases}$$

Then from our assumption concerning the validity of (iii)

$$\sum_{j=1}^{\infty} P(X_j \neq Z_j) \leq \sum_{j=1}^{\infty} P(|X_j| > \varepsilon) < \infty,$$

so that by the Borel–Cantelli lemma it will suffice to establish the almost sure convergence of the series $Z_1 + Z_2 + \cdots$.

In terms of our earlier notation

$$E(Z_j) = \int_{|z| \leq \varepsilon} z \, dF_j(z) = \alpha_j.$$

Let m and n be positive integers, $1 \leq m \leq n$. Then applying Kolmogorov's inequality

$$\begin{aligned}P\left(\max_{m \leq r \leq n} \left| \sum_{j=m}^{r} (Z_j - \alpha_j) \right| > \varepsilon\right) &\leq \varepsilon^{-2} \sum_{j=m}^{n} D^2(Z_j) \\ &\leq \varepsilon^{-2} \sum_{j=m}^{n} \int_{|z| \leq \varepsilon} (z - \alpha_j)^2 \, dF_j(z) \\ &\quad + \varepsilon^{-2} \sum_{j=m}^{n} \alpha_j^2 \int_{|z| > \varepsilon} dF_j(z).\end{aligned}$$

From its definition $|\alpha_j| \le \varepsilon$. It follows from (iii) that as $m \to \infty$

$$P\left(\exists n \ge m, \left|\sum_{j=m}^{n} (Z_j - \alpha_j)\right| > \varepsilon\right) \to 0.$$

Denote the probability which appears in this expression by $T(\varepsilon, m)$. For each positive integer k there is a further integer m_k so that

$$T(2^{-k}, m_k) \le 2^{-k}.$$

Since

$$\sum_{k=1}^{\infty} T(2^{-k}, m_k) \le \sum_{k=1}^{\infty} 2^{-k} < \infty,$$

it follows from the Borel–Cantelli lemma that almost surely the series $(Z_1 - \alpha_1) + \ldots$ satisfies the Cauchy-convergence criterion. Thus with probability one it converges. Since the series $\alpha_1 + \alpha_2 + \ldots$ is by hypothesis convergent, so is the series $Z_1 + Z_2 + \ldots$.

This completes the proof that (iii) implies (i), and of the whole of lemma (1.18).

Lemma 1.22. *Let $X_1, X_2, \ldots$ be a sequence of independent random variables for which the weak limit*

$$F(z) = \lim_{n\to\infty} P(X_1 + \cdots + X_n \le z)$$

exists. Let d_n denote the maximum jump of the variable X_n, $(n = 1, 2, \ldots)$. Then

(i) (Lévy) *$F(z)$ will be continuous if and only if the series*

$$\sum_{n=1}^{\infty} (1 - d_n)$$

diverges.

(ii) (Jessen and Wintner). *If each of the variables X_i is discrete then the limit law is of pure type. In other words it is discrete, absolutely continuous, or singular.*

Proof. We first establish (i).

Let the series in (i) diverge. Let ε be a positive number. Then for any pair of integers (m, n), with $m \le n$, Lévy's concentration function satisfies the inequality

$$Q_{S_n}(l) \le Q_{S_m}(l) \qquad (l > 0).$$

If Y denotes the random variable $X_1 + X_2 + \ldots$, then for any $l > 0$

$$Q_Y(l) \leq \lim_{n \to \infty} Q_{S_n}(l + \varepsilon) \leq Q_{S_m}(l + \varepsilon).$$

We choose ε so small that for $i = 1, 2, \ldots, m$,

$$Q_i(\varepsilon) = Q_{X_i}(\varepsilon) = d_i.$$

Applying lemma (1.14) with $l = \varepsilon$, $L = 2\varepsilon$, we see that

$$Q_Y(2\varepsilon) \leq c_1 \left(\sum_{i=1}^{m} (1 - d_i) \right)^{-1/2}.$$

Letting $\varepsilon \to 0+$, and taking m arbitrarily large shows that

$$Q_Y(\varepsilon) \to 0 \qquad (\varepsilon \to 0+)$$

so that the distribution $F(z) = P(Y \leq z)$, is continuous.

Assume now that the series in (i) converges. Let X_i have a jump of d_i at the value α_i, $(i = 1, 2, \ldots)$. Then

$$P(X_i = \alpha_i, i = 1, 2, \ldots) = \prod_{i=1}^{\infty} d_i = \prod_{i=1}^{\infty} (1 - \{1 - d_i\}) > 0.$$

Since the series $X_1 + X_2 + \ldots$ is assumed to be almost surely convergent it follows that the series $\alpha_1 + \alpha_2 + \ldots$ is convergent, and to a sum A, say. Moreover, the (limit) law for Y will have an atom at the point $z = A$, and so will not be continuous.

To establish part (ii) of the lemma we shall apply the zero-one law. For details of this law see Rényi [7], p. 280. Let the limit law have the decomposition

$$F(z) = a_1 F_1(z) + a_2 F_2(z) + a_3 F_3(z),$$

where $a_i \geq 0$ and $a_1 + a_2 + a_3 = 1$. In this decomposition we assume F_1 to be purely discrete, F_2 to be absolutely continuous, and F_3 to be singular. Let E be the module of real numbers generated by the values of all finite sums of the X_i. According to the hypothesis of (ii) E will be countable, and so of Lebesgue measure zero.

Suppose that $a_1 > 0$, and let G denote the set of discontinuities of $F_1(z)$. For $n = 1, 2, \ldots$, consider the event B_n defined by

$X_n + X_{n+1} + \ldots$ *converges and belongs* to $G + E$, where $G + E$ denotes the set of all sums of the form $x + y$, with x in G and y in E. It is clear that the events $B_1, B_2, \ldots$ are identical, and since the X_i are independent the probability that this common event may occur must be 0 or 1. But this event contains G, which has measure $a_1 > 0$. Therefore $a_1 = 1$, so that $a_2 = 0 = a_3$, and $F(z)$ is purely discrete.

Similarly, if $F(z)$ is continuous (so that $a_1 = 0$), and if $a_3 > 0$, then there is a set D, of Lebesgue measure zero, for which

$$\int_D dF_3(z) > 0.$$

We apply the above argument with the set D in place of the set G, and deduce that since F_2 is absolutely continuous

$$\int_W dF_2 = 0 \qquad P(W) = a_3 \int_W dF_3 \geq a_3 \int_D dF_3 > 0,$$

where W is the appropriate event corresponding to B_1. Thus, once again using the zero-one law, $P(W) = 1 = a_3$, and $a_2 = 0$.

Therefore, a continuous limiting distribution $F(z)$ will either be purely singular, or purely absolutely continuous.

Remark. No reasonable criterion seems to be known which will distinguish between these last two possibilities.

This completes the proof of lemma (1.22).

An alternative criterion which is sometimes useful in deciding whether a distribution function is continuous is the following one.

Lemma 1.23. *A distribution function $F(z)$ is continuous if and only if its characteristic function $\phi(t)$ satisfies the condition*

$$\liminf_{T \to \infty} \frac{1}{2T} \int_{-T}^{T} |\phi(t)|^2 \, dt = 0.$$

Remark. The assertion of this lemma is usually stated, in an equivalent form, with "lim" in place of "lim inf", when it is ascribed to N. Wiener. The present form has certain technical advantages, as may be seen in our application of it in Chapter 5.

Proof. We apply Lebesgue's theorem on dominated convergence to the expression on the right-hand side of the identity

$$\frac{1}{2T} \int_{-T}^{T} |\phi(t)|^2 \, dt = \int_{-\infty}^{\infty} \int_{-\infty}^{\infty} \frac{\sin T(u - v)}{T(u - v)} \, dF(u) dF(v).$$

Let T_k, $k = 1, 2, \ldots$ be an increasing and unbounded sequence of real numbers. We see that if we set $T = T_k$, then as $k \to \infty$ the left-hand side approaches the two dimensional measure of the set $u = v$, the measure being the product measure induced by $F(z)$. Let $d_1, d_2, \ldots$ be the jumps of $F(z)$, given in non-increasing order of magnitude. Then it follows that

$$\lim_{k \to \infty} \frac{1}{2T_k} \int_{-T_k}^{T} |\phi(t)|^2 \, dt = \sum_{n=1}^{\infty} d_n^2.$$

The assertion of lemma (1.23) is now justified.

Infinitely Divisible Laws

A central rôle in the theory of probability is played by those laws which have the following property: whatever the choice of the positive integer n, the law can be expressed as the convolution of n copies of some other law. Laws which have this property are said to be *infinitely divisible.*

In other terms, if $\phi(t)$ denotes the characteristic function of an infinitely divisible law, then for each positive integer n there will be a characteristic function $\phi_n(t)$ so that the equation

$$\phi(t) = (\phi_n(t))^n$$

holds identically for all real values of t.

A random variable X will be said to be infinitely divisible if its corresponding distribution law $P(X \leq z)$ is infinitely divisible.

We shall require a number of facts concerning such laws. For a general reference see Gnedenko and Kolmogorov [1]. For the proofs of the lemmas in this and the next two sections we give appropriate references to their monograph.

Lemma 1.24. *The characteristic function of an infinitely divisible law does not vanish on the real line.*

Proof. See Gnedenko and Kolmogorov [1] Chapter 3, Theorem 1, p. 72.

Lemma 1.25 (Lévy–Khinchine). *In order that the function $\phi(t)$ be the characteristic function of an infinitely divisible distribution, it is both necessary and sufficient that its logarithm be representable in the form*

$$\log \phi(t) = i\gamma t + \int_{-\infty}^{\infty} \left(e^{itu} - 1 - \frac{itu}{1 + u^2}\right) \frac{1 + u^2}{u^2} \, dG(u),$$

where γ is a real constant, $G(u)$ is a non-decreasing function of bounded variation, and the integrand at $u = 0$ is defined by the equation

$$\left(e^{itu} - 1 - \frac{itu}{1 + u^2}\right) \frac{1 + u^2}{u^2}\Bigg]_{u=0} = -\frac{t^2}{2}.$$

We choose the logarithm continuously from the principal value when $t = 0$, (see lemma 1.24). If we confine ourselves to right (or left) continuous functions $G(u)$ this representation is unique.

Proof. See Gnedenko and Kolmogorov [1], Chapter 3, Theorem 1, p. 76.

It is often more convenient to use the following, alternative representation, due to Lévy.

Define the functions

$$M(u) = \int_{-\infty}^{u} \frac{1 + z^2}{z^2}\, dG(z) \quad \text{for } u < 0,$$

$$N(u) = -\int_{u}^{\infty} \frac{1 + z^2}{z^2}\, dG(z) \quad \text{for } u > 0,$$

$$\sigma^2 = G(0+) - G(0-)$$

Then the functions $M(u)$ and $N(u)$ are non-decreasing in the intervals $(-\infty, 0)$ and $(0, \infty)$ respectively. They are continuous at a value of u if and only if $G(u)$ is continuous at the same value. They satisfy $M(-\infty) = 0 = N(\infty)$, and

$$\int_{-\varepsilon}^{0-} u^2\, dM(u) + \int_{0+}^{\varepsilon} u^2\, dN(u) < \infty$$

holds for every $\varepsilon > 0$.

Conversely, any functions $M(u)$ and $N(u)$ which possess these last properties, and any non-negative constant σ, determine by the above relations an infinitely divisible law, as in lemma (1.25). In terms of $M(u)$ and $N(u)$ we can write

$$\log \phi(t) = i\gamma t - \frac{\sigma^2 t^2}{2} + \int_{-\infty}^{0-} \left(e^{iut} - 1 - \frac{iut}{1 + u^2}\right) dM(u)$$

$$+ \int_{0+}^{\infty} \left(e^{iut} - 1 - \frac{iut}{1 + u^2}\right) dN(u).$$

This is called *Lévy's formula.*

There is a further convenient *modified-Lévy representation*

$$\log \phi(t) = i\eta t - \frac{\sigma^2 t^2}{2} + \int_{-\infty}^{-\tau} (e^{iut} - 1) dM(u) + \int_{\tau}^{\infty} (e^{iut} - 1) dN(u)$$

$$+ \int_{-\tau}^{0-} (e^{iut} - 1 - iut) dM(u) + \int_{0+}^{\tau} (e^{iut} - 1 - iut) dN(u),$$

where τ is an arbitrary constant, chosen so that $\pm\tau$ are continuity points of $M(u)$ and $N(u)$. In fact

$$\eta = \gamma + \int_{|u| \le \tau} u \, dG(u) - \int_{|u| \ge \tau} \frac{1}{u} \, dG(u).$$

Those laws for which $\sigma > 0$ are said to have a Gaussian component.

For infinitely divisible laws which have a finite mean and variance there is a simpler representation, which was known before that of lemma (1.25).

Lemma 1.25′ (Kolmogorov). *In order that a distribution function with a finite mean and variance be an infinitely divisible law it is both necessary and sufficient that its characteristic function $\phi(t)$ should be representable in the form*

$$\log \phi(t) = i\gamma t + \int_{-\infty}^{\infty} (e^{itu} - 1 - itu) \frac{1}{u^2} \, dK(u),$$

where γ is a constant, and $K(u)$ is a non-decreasing function of bounded variation which satisfies $K(-\infty) = 0$.

We choose the logarithm continuously from the principal branch at $t = 0$. If we confine ourselves to left (or right) continuous functions $K(u)$ this representation is unique.

Proof. We refer to Gnedenko and Kolmogorov [1] Chapter 3, p. 85. We remark that a little calculation shows that if the distribution function is $F(z)$, then

$$\gamma = \int_{-\infty}^{\infty} z \, dF(z) \qquad K(+\infty) = \int_{-\infty}^{\infty} (z - \gamma)^2 \, dF(z).$$

Thus γ and $K(+\infty)$ are respectively the mean and variance of the law.

Two infinitely divisible laws deserve particular mention. They are the normal law, and the Poisson law.

The normal law $N(a, \sigma)$, with mean a and variance σ^2, has the distribution function

$$\Phi(z) = \frac{1}{\sqrt{(2\pi)}} \int_{-\infty}^{z} e^{-(z-a)^2/2\sigma^2}\, dz.$$

The characteristic function of this law has the form

$$\exp\left(iat - \frac{\sigma^2 t^2}{2}\right).$$

In Kolmogorov's representation we set

$$K(u) = \begin{cases} \sigma^2 & \text{if } u \geq 0, \\ 0 & \text{if } u < 0, \end{cases}$$

and in Lévy's representation we set

$$\gamma = a \qquad \sigma = \sigma \qquad M(u) \equiv 0 \equiv N(u).$$

The degenerate case $\sigma = 0$ corresponds to the improper law with jump at the point γ.

A random variable X is said to be distributed according to the Poisson law if it can only assume values of the form $a + kh$, a, h fixed, $k = 0, 1, 2, \ldots,$ and if

$$P(X = a + kh) = \frac{\lambda^k e^{-\lambda}}{k!}$$

for some positive number λ, and all non-negative integers k. Direct computation shows that the characteristic function of this law has the form

$$\exp(iat + \lambda(e^{ith} - 1)).$$

In Kolmogorov's representation one sets $\gamma = a + h\lambda$ and

$$K(u) = \begin{cases} \lambda h^2 & \text{if } u \geq h, \\ 0 & \text{if } u < h. \end{cases}$$

In Lévy's representation, if $h > 0$ then one sets

$$\gamma = a + h\lambda(1 + h^2)^{-1} \qquad M(u) \equiv 0,$$

$$N(u) = \begin{cases} 0 & \text{if } u \geq h, \\ -\lambda & \text{if } u < h. \end{cases}$$

There is a similar representation if $h < 0$.

Convergence of Infinitely Divisible Laws

Lemma 1.26. *Any law which is the weak limit of infinitely divisible laws, is itself infinitely divisible.*

Proof. See Gnedenko and Kolmogorov [1], Chapter 3, p. 73.

Lemma 1.27 (Gnedenko). *In order that the sequence of infinitely divisible distributions $F_n(z)$, $n = 1, 2, \ldots$, converge to the limit law $F(z)$, it is both necessary and sufficient, in the notation of the Lévy–Khinchine representation, that as $n \to \infty$*

$$G_n(z) \Rightarrow G(z)$$

$$\gamma_n \to \gamma.$$

Proof. See Gnedenko and Kolmogorov [1], Chapter 3, Theorem 1, p. 87. We note here that the convergence of the G_n is that of weak convergence of measures. In particular, it includes the requirement that $G_n(+\infty) \to G(+\infty)$, as $n \to \infty$.

Lemma 1.28. *Let $F_n(z)$ be a sequence of infinitely divisible distribution functions with uniformly bounded second moments. Then in order that they converge to a limiting distribution $F(z)$, it is necessary and sufficient that it be infinitely divisible, and that, in the notation of Kolmogorov,*

$$K_n(u) \Rightarrow K(u),$$

$$\gamma_n \to \gamma,$$

as $n \to \infty$.

Moreover, when these conditions are satisfied, the mean and variance of $F_n(z)$ converge to the (finite) mean and variance, respectively, of the limit law $F(z)$.

Remarks. If we do not assume that the $F_n(z)$ are infinitely divisible then the final remark need not be valid, as the following example shows.

Let X_k, $(k = 1, 2, \ldots)$, be independent random variables, where X_k assumes the values ± 1, each with probability $(1 - k^2)/2$, and $\pm k$, each with probability $1/(2k^2)$. Then as $n \to \infty$ the variable

$$Z_n = \frac{1}{\sqrt{n}} \sum_{k=1}^{n} X_k$$

converges weakly to the normal law with mean zero and variance one, but the variance of Z_n has the limiting value 2.

We further note that the existence of a second moment for a distribution function is equivalent to the finite existence of both the mean and variance.

This ends the remarks.

Proof. For the proof of a similar proposition we refer to Gnedenko and Kolmogorov [1], Chapter 3, Theorem 3, p. 91.

Limit Theorems for Sums of Independent Infinitesimal Random Variables

In the previous section the only properties of infinitely divisible laws that were used were the existence of certain unique representations, and the closure of the class of such laws under weak convergence. Any class of probability laws with similar properties would be equally interesting. Infinitely divisible laws are of importance on account of their intimate connection with the limiting behaviour of sums of independent infinitesimal random variables. Such sums will concern us in later chapters.

Consider the array of random variables

$$\begin{array}{l} \xi_{11}, \xi_{12}, \ldots, \xi_{1k_1}, \\ \xi_{21}, \xi_{22}, \ldots, \xi_{2k_2}, \\ \vdots \\ \xi_{n1}, \xi_{n2}, \ldots, \xi_{nk_n}, \\ \vdots \end{array}$$

where for each particular value of n the k_n variables which occur in the nth row are independent. These variables are said to be *infinitesimal* if for each $\varepsilon > 0$,

$$\sup_{1 \le k \le k_n} P\{|\xi_{nk}| > \varepsilon\} \to 0 \qquad (n \to \infty).$$

It was proved by Khinchine that the class of limit laws for the sums

$$\xi_{n1} + \cdots + \xi_{nk_n} - A_n \qquad (n = 1, 2, \ldots),$$

of such variables, where the A_n are constants, coincides with the class of all infinitely divisible laws. (See, for example, Gnedenko and Kolmogorov [1], Chapter 4, Theorem 2, pp. 115–116).

It is convenient to define the distribution functions

$$F_{nk}(z) = P(\xi_{nk} \le z) \qquad (1 \le k \le k_n, n = 1, 2, \ldots).$$

Consider an array of the above type where we do not assume that the variables are infinitesimal, but that they satisfy the two conditions:

(α) *For each* $\varepsilon > 0$,

$$\sup_{1 \le k \le k_n} P(|\xi_{nk} - E\xi_{nk}| > \varepsilon) \to 0 \qquad (n \to \infty).$$

(β) *Each* ξ_{nk} *has a finite mean and variance, and the sums*

$$D^2\left(\sum_{k=1}^{k_n} \xi_{nk}\right) = \sum_{k=1}^{k_n} D^2(\xi_{nk})$$

are uniformly bounded.

Lemma 1.29 (Gnedenko). *In order that for some suitably chosen constants* A_n *the distribution laws of the sums*

$$\xi_{n1} + \cdots + \xi_{nk_n} - A_n$$

of independent random variables satisfying the conditions (α) *and* (β) *converge to a limit, and that the variances of these sums converge to the variance of the limit law, it is both necessary and sufficient that there exist a non-decreasing function* $K(u)$ *such that*

$$\sum_{k=1}^{k_n} \int_{-\infty}^{u} z^2 \, dF_{nk}(z + E\xi_{nk}) \Rightarrow K(u),$$

as $n \to \infty$.

When this condition is satisfied one may choose

$$A_n = \sum_{k=1}^{k_n} E\xi_{nk} - \gamma,$$

where γ is any constant. The logarithm of the characteristic function of the limit law, which is infinitely divisible, is then given by Kolmogorov's representation, with the constant γ and the function $K(u)$ defined above.

Proof. See Gnedenko and Kolmogorov [1], Chapter 4, Theorem 2, pp. 100–101.

A particular case of this theorem merits special attention.

Lemma 1.30 (Lindeberg, Feller). *Let*

$$\xi_1, \xi_2, \ldots$$

be a sequence of independent random variables, with corresponding distributions $F_k(z)$, $(k = 1, 2, \ldots)$.

In order that the distribution function of the normalised sum

$$\sum_{k=1}^{n} \left(\frac{\xi_k - E\xi_k}{B_n} \right) \qquad B_n^2 = \sum_{k=1}^{n} D^2 \xi_k,$$

converge to the normal law $N(0, 1)$ as $n \to \infty$, and that the summands be infinitesimal, it is both necessary and sufficient that for every $\varepsilon > 0$ the condition

$$\frac{1}{B_n^2} \sum_{k=1}^{n} \int_{|z| > \varepsilon B_n} z^2 \, dF_k(z + E\xi_k) \to 0,$$

be satisfied.

Proof. For a discussion of this and related theorems see Gnedenko and Kolmogorov [1], Chapter 4, Theorems 3 and 4, pp. 101–103.

If we do not make any assumption concerning the variances of the individual laws then the following, more general proposition holds.

Lemma 1.31 (Gnedenko). *In order that for some suitably chosen constants A_n the distribution laws of the sums*

$$\xi_{n1} + \cdots + \xi_{nk_n} - A_n$$

of independent, infinitesimal random variables converge to a limit, it is necessary and sufficient that there exist non-decreasing functions

$$M(u) \quad \text{with} \quad M(-\infty) = 0 \quad \text{and} \quad N(u) \quad \text{with} \quad N(+\infty) = 0,$$

defined over the intervals $(-\infty, 0)$ *and* $(0, +\infty)$ *respectively, and a constant* $\sigma \geq 0$, *such that*

(i) *At every continuity point of* $M(u)$, *and* $N(u)$,

$$\lim_{n\to\infty} \sum_{k=1}^{k_n} F_{nk}(u) = M(u) \qquad (u < 0),$$

$$\lim_{n\to\infty} \sum_{k=1}^{k_n} \{F_{nk}(u) - 1\} = N(u) \qquad (u > 0).$$

(ii) $$\lim_{\varepsilon\to 0} \liminf_{n\to\infty} \sum_{k=1}^{k_n} \left\{ \int_{|z|<\varepsilon} z^2\, dF_{nk}(z) - \left(\int_{|z|<\varepsilon} z\, dF_{nk}(z) \right)^2 \right\}$$
$$= \lim_{\varepsilon\to 0} \limsup_{n\to\infty} \sum_{k=1}^{k_n} \left\{ \int_{|z|<\varepsilon} z^2\, dF_{nk}(z) - \left(\int_{|z|<\varepsilon} z\, dF_{nk}(z) \right)^2 \right\} = \sigma^2.$$

When these conditions are satisfied the constants A_n *may be chosen according to the formula*

$$A_n = \sum_{k=1}^{k_n} \int_{|z|<\tau} z\, dF_{nk}(z) - \eta,$$

where $\pm\tau$ *are continuity points of both* $M(u)$ *and* $N(u)$, *and* η *is a constant. The logarithm of the characteristic function of the limit law is then given by the modified-Lévy representation, or if* γ *is suitably chosen, by the Lévy–Khinchine formula.*

Proof. For a proof of this Theorem see Gnedenko and Kolmogorov [1], Chapter 4, theorem 1, pp. 116–120.

Analytic Characteristic Functions

A characteristic function $\phi(t)$ will be said to have an analytic continuation into a domain D, if there is a function $\phi(z)$, defined and analytic on the connected open set D, and which coincides with $\phi(t)$ whenever z (in D) has the real value t. We shall only consider situations where D intersects the real axis in at least one proper open interval.

If, for some $r > 0$, $\phi(t)$ is analytically continuable into the disc $|z| < r$, then it is also continuable into the semi-infinite strip $|\text{Im}(z)| < r$. If $F(u)$ is the distribution function which corresponds to $\phi(t)$, then the representation

$$\phi(z) = \int_{-\infty}^{\infty} e^{izu}\, dF(u)$$

holds throughout the strip. Moreover, if $\phi(t)$ is the characteristic function of an infinitely divisible distribution, then the Lévy–Khinchine representation holds in the whole strip $|\text{Im}(z)| < r$. In particular $\phi(z)$ will have no zeros in this strip. For proofs of these results, all duc to Raikov, wc rcfcr to Linnik [4], Chapter 3, pp. 49–60.

There are serious restrictions upon the nature of those characteristic functions which can be analytically continued over the whole plane. It was proved by Marcinkiewicz [1] (see also Linnik [4], Chapter 3, Theorem 3.3.1, pp. 56–58) that, in the standard notation, any integral characteristic function of order ρ greater than two cannot have an index ρ_1 of convergence for the set of zeros which satisfies $\rho_1 < \rho(<\infty)$. It is a corollary of this result that if $\phi(t)$ has the form $\exp(P(t))$, where $P(t)$ is a polynomial, then P has degree at most two.

We shall need a special case of this result which we can establish simply by first principles, that is to say, without the use of the theory of integral functions. We shall instead make use of the following lemma, which can be derived from the maximum-modulus principle.

Lemma 1.41 (Borel–Carathéodory). *Let $f(z)$ be analytic in some domain which contains the disc $|z| \le R$, $(R > 0)$. Let $M(r)$ and $A(r)$ denote the maximum of $|f(z)|$ and $\text{Re}\{f(z)\}$, respectively, on the disc $|z| = r$. Then for $0 < r < R$*

$$M(r) \le \frac{2r}{R-r} A(R) + \frac{R+r}{R-r} |f(0)|.$$

Proof. For a proof of this result see Titchmarsh [2], Chapter V, 5.5, pp. 174–175.

Lemma 1.42. *Consider a law $F(u)$ which is confined to a finite interval $|u| \le U$; that is to say $F(u) = 1$ if $u > U$, and $= 0$ if $u < -U$.*

Then this law can be infinitely divisible if and only if it is improper.

Remark. The application of this lemma will allow us to conclude that certain limit laws, which arise in arithmetic problems, are not infinitely divisible.

It is clear that the characteristic function of the improper law is continuable to an integral function.

Proof. Let $F(u)$ be confined to the interval $[-U, U]$. Then its characteristic function $\phi(t)$ has the representation

$$\phi(t) = \int_{-U}^{U} e^{itu}\, dF(u).$$

Clearly $\phi(t)$ can be analytically continued to an integral function, $\phi(z)$ say. For example, one may differentiate the integral directly with respect to (complex-valued) t, applying Lebesgue's theorem on dominated convergence.

If $F(u)$ is infinitely divisible, then from the remarks made preceeding lemma (1.41), $\phi(z)$ will not vanish over the whole complex z-plane.

Consider the branch of $\log \phi(z)$ which is zero when $z = 0$. This is also an integral function. Owing to the assumption concerning $F(u)$ we have the estimate

$$\operatorname{Re}\{\log \phi(z)\} = \log|\phi(z)| \leq 2U|z|.$$

We apply lemma (1.41) with $R = 2|z|$, $r = |z|$, noting that $\log \phi(0) = \log 1 = 0$, and deduce that if $z \neq 0$, then

$$|\log \phi(z)| \leq \frac{2|z|}{2|z| - |z|} \cdot 2U2|z| + \frac{2|z| + |z|}{2|z| - |z|}|\log \phi(0)| \leq 8U|z|.$$

It follows from this estimate that $\log \phi(z)$ has the form $Az + B$ for certain constants A and B. (See, for example, Titchmarsh [2] Chapter 2, 2.52, pp. 85–86). Thus, for real values of t,

$$\phi(t) = \exp(At + B).$$

Setting $t = 0$ shows that $\exp(B) = 1$. If, furthermore, $\operatorname{Re}\{A\} > 0$, then as $t \to \infty$, $|\phi(t)|$ becomes unbounded. This contradicts the fact that $\phi(t)$ is a characteristic function. A similar contradiction can be effected if $\operatorname{Re}\{A\} < 0$. It follows that $A = i\alpha$ for some real number α.

We have now shown that $\phi(t) = \exp(i\alpha t)$, and $F(u)$ must be an improper law, as was asserted.

This completes the proof of lemma (1.42).

The Method of Moments

Lemma 1.43 (Wintner, Fréchet–Shohat). *Let $F_n(z)$, $(n = 1, 2, \ldots)$ be a sequence of distribution functions. For each non-negative integer k let*

$$\alpha_k = \lim_{n \to \infty} \int_{-\infty}^{\infty} z^k \, dF_n(z).$$

exist.

Then there is a subsequence $F_{n_j}(z)$, $(n_1 < n_2 < \ldots)$, which converges weakly to a limiting distribution $F(z)$ for which

$$\alpha_k = \int_{-\infty}^{\infty} z^k \, dF(z) \qquad (k = 0, 1, \ldots).$$

Moreover, if the set of moments α_k determine $F(z)$ uniquely, then as $n \to \infty$ the distributions $F_n(z)$ converge weakly to $F(z)$.

Proof. Assume only that α_2 exists, and is finite. Then there is a positive constant c_1 so that for each positive real number N,

$$\int_{|z|>N} dF_n(z) \leq N^{-2} \int_{-\infty}^{\infty} z^2 \, dF_n(z) \leq c_1 N^{-2} \qquad (n = 1, 2, \ldots).$$

The first assertion of the present lemma now follows from an application of lemma (1.8).

As for the second assertion of lemma (1.43), if the α_k determine $F(z)$ uniquely, then every (convergent) subsequence $\{F_{m_j}(z)\}$ will have the same limit $F(z)$.

This completes the proof of lemma (1.43).

Remark. It is clear that in the hypotheses of lemma (1.43), it is only necessary that there exist enough α_k to determine $F(z)$ uniquely. In general, the difficult part in the application of this lemma concerns the possibility that the α_k determine $F(z)$ uniquely. The following result is of some assistance in this respect.

Lemma 1.44. *In the notation of lemma* (1.43) *let the series*

$$\phi(t) = \sum_{k=0}^{\infty} \alpha_k \frac{(it)^k}{k!}$$

converge absolutely in a disc of complex t-values, $|t| < \tau$, $\tau > 0$.

Then the α_k determine the distribution function $F(u)$ uniquely. Moreover, the characteristic function $\phi(t)$ of this distribution has the above representation in the disc $|t| < \tau$, and can be analytically continued into the strip $|\mathrm{Im}(t)| < \tau$.

Remark. The final assertions of this lemma show that its hypothesis is very powerful.

Proof. It is easy to see that in the disc $|t| < \tau$,

$$\phi(t) = \int_{-\infty}^{\infty} e^{itu} \, dF(u) = \int_{-\infty}^{\infty} \sum_{k=0}^{\infty} \frac{(itu)^k}{k!} \, dF(u) = \sum_{k=0}^{\infty} \alpha_k \frac{(it)^k}{k!}.$$

Therefore the characteristic function $\phi(t)$ is analytically continuable into the disc $|t| < \tau$, and, by the remarks preceeding lemma (1.41), into the strip $|\mathrm{Im}(t)| < \tau$.

Let $G(u)$ be a further distribution function with the property that

$$\alpha_k = \int_{-\infty}^{\infty} u^k \, dG(u) \qquad (k = 0, 1, 2, \ldots).$$

Then the function

$$\psi(t) = \int_{-\infty}^{\infty} e^{itu} \, dG(u)$$

can also be analytically continued into the strip $|\mathrm{Im}(t)| < \tau$. It follows from the fact that analytic continuation is uniquely possible, if at all, that the functions $\phi(t)$ and $\psi(t)$ coincide in the strip $|\mathrm{Im}(t)| < \tau$, and therefore on the whole real line. Since their characteristic functions coincide, so do the distributions $F(u)$ and $G(u)$.

This completes the proof of lemma (1.44).

Mellin–Stieltjes Transforms

As an alternative to the standard characteristic function V. M. Zolotarev [1] suggested the use of a Mellin–Stieltjes transformation. This transform has particular advantages when the multiplication of random variables is being considered. It also has applications to the theory of numbers.

Consider the pair of functions $w_0(t)$ and $w_1(t)$, which are defined for all real numbers t by

$$w_j(t) = \int_{-\infty}^{\infty} |z|^{it} \operatorname{sign}^j z \, dF(z),$$

where it is understood that $|0|^{it} \operatorname{sign}^j 0 = 0$, $(j = 0, 1)$. It is customary to present these functions in the array

$$\mathbf{W}(t) = \begin{pmatrix} w_0(t) & 0 \\ 0 & w_1(t) \end{pmatrix}.$$

We shall call this pair the *Mellin transform*, or more shortly the *M*-transform, of the distribution function $F(z)$.

It is possible to derive analogues of the theorems concerning the standard characteristic functions; for example, that $\mathbf{W}(t)$ determines the distribution function $F(z)$ uniquely. One method to effect such results is to reduce the problem to one involving Fourier–Stieltjes transforms, in the following manner.

Define two numbers

$$c^+ = (1 - F(0)) \qquad c^- = F(0-).$$

Let $I(z)$ denote the improper law with jump at the origin. Then,

$$I(z) = \begin{cases} 1 & \text{if } z \geq 0. \\ 0 & \text{if } z < 0. \end{cases}$$

In terms of the distribution function $F(z)$ we define two new distribution functions $F^+(z)$ and $F^-(z)$ by

$$F^+(z) = \begin{cases} \dfrac{1}{c^+}(F(e^z) - F(0)) & \text{if } c^+ \neq 0, \\ I(z) & \text{if } c^+ = 0, \end{cases}$$

and

$$F^-(z) = \begin{cases} \dfrac{1}{c^-}(F(0-) - F(-e^z-)) & \text{if } c^- \neq 0, \\ I(z) & \text{if } c^- = 0. \end{cases}$$

Let their corresponding standard characteristic functions be $f^+(t)$ and $f^-(t)$. It is not difficult to see that for each value of j, $(j = 0, 1)$, the relation

$$w_j(t) = c^+ f^+(t) + (-1)^j c^- f^-(t)$$

is satisfied. It is clear from these relations that each w_j is uniformly continuous on the whole real line. Note, however, that $w_0(0) = c^+ + c^-$ and $w_1(0) = c^+ - c^-$ so that $w_0(0)$ will only have the value 1 if $F(z)$ is continuous at the point $z = 0$; whilst $w_1(0) = 0$ may hold.

We shall say that the distribution function $F(z)$ is symmetric if, whenever $\pm z$ are continuity points of $F(z)$, the relation

$$1 - F(z) = F(-z)$$

is satisfied. An equivalent condition is that $w_1(t)$ be identically zero.

We prove one theorem involving Mellin transforms, an analogue of lemma (1.11). For the purposes of this lemma we shall say that a sequence of 2×2 matrices $\mathbf{A}_n$, $(n = 1, 2, \ldots)$, with complex entries, converges to a limiting matrix $\mathbf{A}$ if and only if the respective entries of the $\mathbf{A}_n$ converge to the corresponding entries of $\mathbf{A}$.

Lemma 1.45

(i) *Let the sequence of distribution functions* $F_n(z)$, $(n = 1, 2, \ldots)$, *converge weakly to the distribution function* $F(z)$. *Let also* $F_n(0) \to F(0)$ *and* $F_n(0-) \to F(0-)$ *as* $n \to \infty$. *Then the corresponding Mellin transforms satisfy*

$$\mathbf{W}_n(t) \to \mathbf{W}(t) \qquad (n \to \infty),$$

uniformly on any compact set of t-values.

(ii) *Conversely, let* $F_n(z)$, $(n = 1, 2, \ldots)$, *be a sequence of distribution functions whose Mellin transforms* $\mathbf{W}_n(t)$ *converge to a matrix* $\mathbf{A}(t)$ *as* $n \to \infty$, *where the entries of* $\mathbf{A}(t)$ *are continuous at* $t = 0$. *Then* $\mathbf{A}(t)$ *is the Mellin transform of a distribution function* $F(z)$. *Moreover, as* $n \to \infty$, $F_n(0) \to F(0)$, $F_n(0-) \to F(0-)$, *and the distributions* $F_n(z)$ *converge weakly to* $F(z)$.

Proof. Consider first part (i) *of this lemma.*

Assume, using an obvious notation, that both c^+ and c^- are non-zero. Then according to the hypothesis of (i)

$$F_n^+(z) \Rightarrow F^+(z) \qquad F_n^-(z) \Rightarrow F^-(z) \qquad (n \to \infty).$$

It follows from an application of lemma (1.11) that $f_n^\pm(t) \to f^\pm(t)$ and therefore $\mathbf{W}_n(t) \to \mathbf{W}(t)$, uniformly over any bounded interval of t-values.

If now $c^+ \neq 0$, but $c^- = 0$, say, then we may still conclude that the $F_n^+(z)$ converge weakly to $F^+(z)$. However, if $z < 0$ then

$$F_n(z) \leq F_n(0-) = c_n^- \to 0 \qquad (n \to \infty)$$

Therefore, as $n \to \infty$ $\mathbf{W}_n(t)$ approaches the matrix

$$\begin{pmatrix} c^+ f^+(t) & 0 \\ 0 & c^- f^-(t) \end{pmatrix}.$$

Inspection shows that this is indeed the matrix $\mathbf{W}(t)$.

The remaining degenerate cases may be similarly dealt with.

Consider now part (ii) *of lemma* (1.45)

Let

$$\mathbf{W}_n(t) = \begin{pmatrix} w_{n,0}(t) & 0 \\ 0 & w_{n,1}(t) \end{pmatrix}$$

be a representation of the Mellin transforms of the distribution function $F_n(z)$. Then according to the hypothesis of (ii) the numbers c_n^+, c_n^- $(n = 1, 2, \ldots)$, which are defined by

$$c_n^+ = \tfrac{1}{2}(w_{n,0}(0) + w_{n,1}(0))$$

$$c_n^- = \tfrac{1}{2}(w_{n,0}(0) - w_{n,1}(0))$$

converge to finite limits c^+ and c^- respectively, say. Assume for the moment that both c^+ and c^- are non-zero. Then for all sufficiently large values of n, and all real numbers t,

$$f_n^+(t) = \frac{1}{2c_n^+}\{w_{n,0}(t) + w_{n,1}(t)\}$$

$$f_n^-(t) = \frac{1}{2c_n^-}\{w_{n,0}(t) - w_{n,1}(t)\}.$$

As $n \to \infty$, f_n^+, and similarly f_n^-, approaches a function which is continuous at the point $t = 0$. Let these limits be $f^+(t)$ and $f^-(t)$ respectively, with corresponding distribution functions $F^+(z)$ and $F^-(z)$. The existence of these distributions is guaranteed by lemma (1.11), which also allows us to assert that, as $n \to \infty$,

$$F_n^+(z) \Rightarrow F(z),$$

$$F_n^-(z) \Rightarrow F(z).$$

Define the distribution function $F(z)$ by

$$F(z) = \begin{cases} 1 - c^+(1 - F^+(\log z)) & \text{if } z > 0, \\ c^-(1 - F^-(\log - z)) & \text{if } z < 0. \end{cases}$$

This choice of notation is consistent with that made in the remarks immediately preceeding the statement of lemma (1.45). It is easy to see that as $n \to \infty$,

$$F_n(0-) \to F(0-)$$

and, by making use of the right-continuity of distribution functions,

$$F_n(0) \to F(0).$$

Therefore if $z > 0$, and e^z is a continuity point of F,

$$F_n(e^z) = c_n^+ F_n^+(z) + F_n(0) \to c^+ F^+(z) + F(0) = F(e^z) \qquad (n \to \infty).$$

There is a similar limiting relation involving $F_n(-e^z)$, and by means of these two relations we deduce that as $n \to \infty$ the $F_n(z)$ converge weakly to $F(z)$.

The degenerate cases, when at least one of c^+, c^- vanishes, can be established by a modification of this argument. As an example, let $c^+ \neq 0$, $c^- = 0$ hold. In this case we still have

$$f_n^+(t) \to f^+(t) \qquad F_n^+(z) \Rightarrow F^+(z) \qquad (n \to \infty),$$

say. We now define the distribution function

$$F(z) = \begin{cases} 1 - c^+(1 - F(\log z)) & \text{if } z > 0, \\ 0 & \text{if } z < 0. \end{cases}$$

Our notation is once again consistent, with $F^-(z)$ now being the distribution function $I(z)$. The only modification necessary to the previous argument concerns the limiting behaviour at points $-e^z$, $z > 0$. In fact

$$F_n(-e^z) \leq F_n(0-) \to c^- = 0 = F(-e^z),$$

so that we may conclude, as before, that the $F_n(z)$ weakly converge to $F(z)$ as $n \to \infty$.

This concludes the proof of lemma (1.45).

Distribution Functions (mod 1)

We shall also consider distribution functions (mod 1).

A *distribution function* (mod 1) will be any distribution function $F(z)$, defined on the whole real line, and such that

$$F(z) = \begin{cases} 1 & \text{if } z \geq 1, \\ 0 & \text{if } z < 0. \end{cases}$$

We say that the sequence of distribution functions (mod 1), $(n = 1, 2, \ldots)$, converges weakly to the limiting distribution $F(z)$ (mod 1) as $n \to \infty$, if for every pair of points (α, β), $0 \leq \alpha \leq \beta \leq 1$, which are continuity points of the law $F(z)$, we have

$$F_n(\beta) - F_n(\alpha) \to F(\beta) - F(\alpha).$$

The use of the pair of points (α, β) is to ensure consistency with the notion of weak convergence of measures.

If the limit law is continuous at all points then we may adopt the choice $\alpha = 0$.

We note that (historically) a sequence of non-negative real numbers $a_1, a_2, \ldots,$ is said to be *uniformly distributed* (mod 1) if

$$n^{-1} \sum_{\substack{m=1 \\ \alpha < a_m \leq \beta}}^{n} 1 \to (\beta - \alpha) \qquad (n \to \infty).$$

This is a special case of the above definition, with the choices

$$F_n(z) = n^{-1} \sum_{\substack{m=1 \\ a_m \leq z}}^{n} 1 \qquad F(z) = z \qquad 0 \leq z < 1,$$

This limit law is otherwise known as the *uniform law* (mod 1).

The characteristic function of the distribution function $F(z)$ (mod 1) is the set of Fourier series coefficients

$$c_k = \int_{0-}^{1+} e^{2\pi ikz}\, dF(z) \qquad (k = 0, \pm 1, \pm 2, \ldots).$$

Since the function $F(z)$ is of bounded variation over any finite interval the periodic function $g(z)$, with period 1, which is defined in the range $0 \leq z < 1$ by

$$g(z) = \begin{cases} \frac{1}{2}\{F(z+) + F(z-)\} & \text{if } 0 < z < 1, \\ \frac{1}{2}\{F(1-) + F(0+)\} & \text{if } z = 0, \end{cases}$$

has the Fourier representation

$$g(z) = \text{constant} + \sum_{\substack{k=-\infty \\ k \neq 0}}^{\infty} \frac{(\bar{c}_k - 1)}{2\pi ik} e^{2\pi ikz},$$

where the constant has the value

$$\int_0^1 F(z) dz,$$

and the series converges boundedly over any compact set of real z-values. In particular, $g(z) = F(z)$ whenever the latter function is continuous, $0 < z < 1$. In the range $0 < z < 1$ the c_k therefore determine $F(z)$ to within translation by a constant.

It is of interest to note that if $0 < z < 1$ then

$$-\frac{1}{2\pi i}\sum_{k\neq 0}\frac{e^{2\pi ikz}}{k} = z - \frac{1}{2},$$

so that when $F(z)$ is continuous

$$F(z) = z - \tfrac{1}{2} + \text{constant} + \sum_{\substack{k=-\infty\\ k\neq 0}}^{\infty}\frac{\bar{c}_k}{2\pi ik}\,e^{2\pi ikz}.$$

We shall need a further analogue of lemma (1.11).

Lemma 1.46. *In order that the distribution functions* (mod 1) $F_n(z)$, $(n = 1, 2, \ldots)$, *converge weakly to a limiting distribution* (mod 1), *it is both necessary and sufficient that for every integer k the limit*

$$\beta_k = \lim_{n\to\infty}\int_{0-}^{1+} e^{2\pi ikz}\,dF_n(z)$$

exists. When this is indeed the case the limit law $F(z)$ will be determined by

$$\beta_k = \int_{0-}^{1+} e^{2\pi ikz}\,dF(z) \qquad (k = 0, \pm 1, \pm 2, \ldots),$$

together with the requirement that $F(1-) = 1$. Moreover, this particular limit law will then be continuous if and only if

$$\liminf_{N\to\infty}(2N+1)^{-1}\sum_{k=-N}^{N}|\beta_k|^2 = 0,$$

and absolutely continuous with a derivative which belongs to the class $L^2(0, 1)$ if and only if the series

$$\sum_{k=-\infty}^{\infty}|\beta_k|^2$$

converges.

Proof. If the $F_n(z)$ converge weakly to a law $F(z)$, then the application of Lebesgue's dominated convergence theorem shows that for each integer k

$$\int_{0-}^{1+} e^{2\pi ikz}\,dF_n(z) \to \int_{0-}^{1+} e^{2\pi ikz}\,dF(z) \qquad (n\to\infty).$$

Conversely, assume that the limits β_k all exist. It is clear that any subsequence of distribution functions $F_{n_1}(z), F_{n_2}(z), \ldots$, with $n_1 < n_2 < \ldots$, is compact in the sense of lemma (1.8). Considered (mod 1), it will therefore contain a subsequence which converges weakly to a distribution $G(z)$ (mod 1), say. Applying Lebesgue's dominated convergence theorem once again, we see that

$$\beta_k = \int_{0-}^{1+} e^{2\pi ikz}\, dG(z) \qquad (k = 0, \pm 1, \pm 2, \ldots).$$

If we impose the condition that $G(1-) = 1$, then $G(z)$ is uniquely determined by the β_k, and it follows that the F_{n_j}(mod 1) converge weakly to G(mod 1) for any increasing sequence $\{n_j\}$. In particular, $F_n(z)$(mod 1) converges weakly to $G(z)$(mod 1) as $n \to \infty$.

The assertion concerning the continuity of the limit law may be justified in a manner exactly similar to that for lemma (1.23), making use of the relation

$$\sum_{k=-N}^{N} e^{2\pi ikw} = \begin{cases} \dfrac{\sin(2N+1)\pi w}{\sin \pi w} & w \neq 0, \pm 1, \pm 2, \ldots, \\ 2N+1, & \text{otherwise.} \end{cases}$$

Owing to the restriction $G(1-) = 1$ we have

$$\int_{0-}^{1+} e^{2\pi ikz}\, dG(z) = \int_{0-}^{1-} e^{2\pi ikz}\, dG(z),$$

so that when considering the $|\beta_k|^2$ there is no possibility of counting the joint contributions of discontinuities at $z = 0$ *and* $z = 1$. By making use of the fact that each $|\beta_k| \leq 1$, and applying the Cauchy–Schwarz inequality, one may readily show that the limit law is continuous if and only if

$$\lim_{N\to\infty} \frac{1}{2N+1} \sum_{k=-N}^{N} |\beta_k| = 0,$$

and in this form this particular result is due to N. Wiener.

The assertion involving the absolute continuity of the limit law is proved in Edwards [1], (8.5.3), pp. 129–130.

Remarks. Note that the $F_n(z)$(mod 1) converge weakly to the uniform law (mod 1) if and only if, as $n \to \infty$, the limiting relation

$$\int_{0-}^{1+} e^{2\pi ikz}\, dF_n(z) \to 0$$

holds for every non-zero integer k.

In the special case of a sequence, mentioned earlier, it follows that $a_1, a_2, \ldots,$ will be uniformly distributed (mod 1) if and only if for every non-zero integer k,

$$n^{-1} \sum_{m=1}^{n} e^{2\pi i k a_m} \to 0 \qquad (n \to \infty).$$

This is the celebrated criterion of H. Weyl [1].

It is possible to define random variables which take values in the additive group of real numbers (mod 1), that is to say, according to the group law. If a variable X is the sum of n independent copies of a variable Y, then the Fourier coefficients c_k and d_k of their corresponding distribution functions are connected by the relation $c_k = d_k^n$. Thus if $|d_k| < 1$, then c_k might well be small for large values of n. In this way X will be approximately uniformly distributed. This is a circumstance which has no proper analogue in the standard theory of probability.

Quantitative Fourier Inversion

It is often useful to have an estimate of the distance between two distribution functions in terms of some measure of closeness of their respective characteristic functions. Such an estimate is usually bought at the expense of assumptions concerning the smoothness of at least one of the distributions.

Lemma 1.47. *Let $F(z)$ and $G(z)$ be two distribution functions with corresponding characteristic functions $f(t)$ and $g(t)$. Then there is a positive absolute constant c_1, so that for any $T > 0$ the inequality*

$$\sup_z |F(z) - G(z)| \leq c_1 S_G\left(\frac{1}{T}\right) + c_1 \int_{-T}^{T} \left| \frac{f(t) - g(t)}{t} \right| dt$$

is satisfied, where

$$S_G(h) = \sup_z \frac{1}{2h} \int_0^h (G(z + u) - G(z - u))du \qquad (h > 0).$$

Proof. Consider the function $P(x) = P_T(x)$ which is defined by

$$P_T(x) = T \cdot \frac{1 - \cos(Tx - 3)}{\pi(Tx - 3)^2} = \frac{T}{2\pi}\left(\frac{\sin(Tx - 3)/2}{(Tx - 3)/2}\right)^2.$$

It is straightforward to check that this function, which is a form of Fejér kernel, has the properties

(i) $0 \le P(x) \le \dfrac{T}{2\pi}$ for all real x,

(ii) $\displaystyle\int_{-\infty}^{\infty} P(x)dx = 1$,

(iii) $\displaystyle\int_{0}^{8/T} P(x)dx = w > \tfrac{1}{2}$.

Thus the distribution function

$$H(u) = \int_{-\infty}^{u} P(x)dx$$

is a good approximation to the improper law with a jump at the origin. Direct computation shows that $H(u)$ has the characteristic function

$$\phi(t) = \phi_T(t) = \int_{-\infty}^{\infty} e^{itx}P(x)dx = \begin{cases} \left(1 - \dfrac{|t|}{T}\right)e^{3it/T} & \text{if } |t| \le T, \\ 0 & \text{if } |t| > T. \end{cases}$$

We shall make essential use of the existence for $H(u)$ of a probability density, and of the fact that $\phi(t)$ vanishes for all sufficiently large values of t.

Let $F^*(z)$ denote the convolution of the distributions $F(z)$ and $1 - H(-u - 0)$. Thus

$$F^*(z) = \int_{-\infty}^{\infty} F(u)P(u - z)du,$$

and F^* has the characteristic function $f(t)\bar{\phi}(t)$.

We similarly define $G^*(z)$.

Applying the inversion formula (lemma (1.10)) we see that for any pair of real numbers z and y,

$$F^*(z) - G^*(z) - (F^*(y) - G^*(y))$$
$$= \frac{i}{2\pi}\int_{-T}^{T} \frac{f(t) - g(t)}{t} \cdot \bar{\phi}(t)e^{-itz}\,dt - \frac{i}{2\pi}\int_{-T}^{T} \frac{f(t) - g(t)}{t} \cdot \bar{\phi}(t)e^{-ity}\,dt.$$

If the function $|f(t) - g(t)|/|t|$ does not belong to the Lebesgue class $L(0, T)$, then the result of the lemma is trivially valid. If it does then we let $y \to -\infty$, and apply the Riemann–Lebesgue theorem (see, for example, Titchmarsh [3] 1.8, pp. 11–12) to deduce that

$$F^*(z) - G^*(z) = \frac{i}{2\pi}\int_{-T}^{T} \frac{f(t) - g(t)}{t} \cdot \bar{\phi}(t)e^{-itz}\,dt.$$

In particular,

$$|F^*(z) - G^*(z)| \leq \frac{1}{\pi} \int_0^T |f(t) - g(t)| \frac{dt}{t}.$$

We must now unravel this statement by removing the function $H(u)$. For this step we implicitly assume $G(z)$ to be smooth in the sense that $S_G(h)$, more conveniently $S(h)$, is small when h is small.

Let $h = 8/T$. Then we have

$$\begin{aligned} F(z) &\leq \frac{1}{w} \int_z^{z+h} F(u)P(u - z)du \\ &= G(z) + \frac{1}{w} \int_z^{z+h} \{G(u) - G(z)\}P(u - z)du \\ &\quad + \frac{1}{w} \int_z^{z+h} \{F(u) - G(u)\}P(u - z)du \\ &= G(z) + w^{-1}I_1 + w^{-1}I_2, \end{aligned}$$

say. Clearly

$$I_1 \leq \frac{T}{2\pi} \int_0^h \{G(z + u) - G(z)\}du \leq \frac{8}{\pi} S(h).$$

Next, let

$$\Delta = \sup_z |F(z) - G(z)|.$$

Then

$$\begin{aligned} |I_2| &= \left| \int_{-\infty}^{\infty} - \int_{u<z} - \int_{u>z+h} \{F(u) - G(u)\}P(u - z)du \right| \\ &\leq |F^*(z) - G^*(z)| + \Delta\left(\int_{u<z} + \int_{u>z+h} P(u - z)du\right) \\ &\leq \frac{1}{\pi} \int_0^T |f(t) - g(t)| \frac{dt}{t} + \Delta(1 - w). \end{aligned}$$

Altogether, therefore,

$$F(z) - G(z) \leq \frac{8}{\pi w} S(h) + \frac{1}{\pi w} \int_0^T |f(t) - g(t)| \frac{dt}{t} + \Delta\left(\frac{1}{w} - 1\right).$$

We can obtain a similar inequality, but going in the other direction, by beginning with the inequality

$$F(z) \geq \frac{1}{w} \int_{z-h}^{z} F(u)P(z-u)du.$$

In this way we see that

$$\Delta \leq \frac{8}{\pi w} S(h) + \frac{1}{\pi w} \int_0^T |f(t) - g(t)| \frac{dt}{t} + \Delta\left(\frac{1}{w} - 1\right),$$

and therefore

$$\Delta\left(2 - \frac{1}{w}\right) \leq \frac{8}{\pi w} S(h) + \frac{1}{\pi w} \int_0^T |f(t) - g(t)| \frac{dt}{t}.$$

Since $2w > 1$ we have a result of the type stated in the present lemma, but with $S(8/T)$ in place of $S(1/T)$. Our proof will therefore be complete if we prove that for any $l > 0$, the inequality

$$S(2l) \leq 2S(l)$$

is satisfied. This requires only a little manipulation. In fact, for any z,

$$\frac{1}{2l}\int_0^{2l} \{G(z+u) - G(z-u)\}du = \frac{1}{2l}\left(\left(\int_0^l + \int_l^{2l}\right)\{G(z+u) - G(z-u)\}du\right)$$

$$\leq \frac{1}{2} S(l) + \frac{1}{2l}\int_0^l \{G(z+y+l) - G(z-l-y)\}dy$$

$$\leq \frac{1}{2} S(l) + \frac{1}{2l}\int_0^l \{G(z+y+l) - G(z+l-y)\}$$

$$+ \{G(z+l-y) - G(z-l-y)\}dy$$

$$\leq S(l) + \frac{1}{2l}\int_0^l \{G(z+l-y) - G(z-l-y)\}dy$$

$$\leq S(l) + \frac{1}{2l}\int_0^l \{G(z+l-y) - G(z-l+y)\}$$

$$+ \{G(z-l+y) - G(z-l-y)\}dy$$

$$\leq S(l) + \frac{1}{2l}\int_0^l \{G(z+u) - G(z-u)\}du$$

$$+ \frac{1}{2l}\int_0^l \{G(z-l+y) - G(z-l-y)\}dy$$

$$\leq 2S(l).$$

We conclude that $S(8/T) \leq 8S(1/T)$, and the proof of lemma (1.47) is complete.

Remark. The function

$$Q(l) = \sup_z \{G(z + l) - G(z)\}$$

is a generalisation to distribution functions of Lévy's notion of a *concentration function.* In terms of this function

$$S(l) \leq Q(l).$$

However, $S(l)$ involves an average, and so is more readily dealt with than $Q(l)$. In fact, there is a convenient representation,

$$\frac{1}{2l}\int_0^l (G(z + u) - G(z - u))du = \frac{1}{2\pi}\int_{-\infty}^{\infty}\left(\frac{\sin t}{t}\right)^2 e^{-2itz/l}\cdot g\left(\frac{2t}{l}\right)dt,$$

where the integrand is clearly integrable over the whole real line.

This representation may be obtained by means of the well-known device of convoluting with a normal law.

Indeed, the normal law

$$M(\varepsilon, z) = \frac{1}{\varepsilon\sqrt{2\pi}}\int_{-\infty}^{z} e^{-w^2/(2\varepsilon^2)}\, dw$$

has the characteristic function $\exp(-\varepsilon^2 t^2/2)$. Define the convolution $H(z) = G(z) * M(\varepsilon, z)$. Its characteristic function $g(t)\exp(-\varepsilon^2 t^2/2)$ belongs to the class $L(-\infty, \infty)$ for every fixed $\varepsilon > 0$. We may therefore apply the inversion formula lemma (1.10) and let $N \to \infty$ to obtain

$$H(z + u) - H(z - u) = \frac{1}{2\pi i}\int_{-\infty}^{\infty} g(t)e^{-\varepsilon^2 t^2/2}\cdot e^{itz}\left(\frac{e^{itu} - e^{-itu}}{t}\right)dt.$$

Integrating with respect to u over the range $0 \leq u \leq l$ gives

$$\frac{1}{2l}\int_0^l (H(z + u) - H(z - u))du$$

$$= \frac{1}{2\pi}\int_{-\infty}^{\infty} g\left(\frac{2\tau}{l}\right)\cdot\exp\left(-\frac{2\varepsilon^2\tau^2}{l^2}\right)\cdot\exp\left(-\frac{2i\tau z}{l}\right)\cdot\left(\frac{\sin\tau}{\tau}\right)^2 d\tau,$$

this last integral being obtained by means of the substitution $t = 2\tau/l$. We let $\varepsilon \to 0+$ and apply Lebesgue's theorem on dominated convergence. It is not difficult to show that when z is a continuity point of $G(z)$

$$\lim_{\varepsilon \to 0+} G(z) * M(\varepsilon, z) = G(z),$$

and the desired relation follows readily.

If the distribution function $G(z)$ is everywhere differentiable on the real line, and has a uniformly bounded derivative there, say $|G'(z)| \leq A$, then one has $Q(l) \leq Al$ for every $l \geq 0$. With this assumption, but with particular care concerning the constants involved, lemma (1.47) was proved by Esseen [2]. The presentation given here is due to Faĭnleĭb [2], (see also Postnikov [1], Chapter one, pp. 94–101).

Lemma 1.48 (Berry, Esseen). *Let $X_1, \ldots, X_n$ be independent random variables, with mean zero, variance D_m and absolute third moment L_m, $(m = 1, \ldots, n)$, respectively. Let*

$$\sigma^2 = D_1 + D_2 + \cdots + D_n \qquad \sigma > 0.$$

Then uniformly for all real values of z

$$P\left(\sigma^{-1} \sum_{m=1}^{n} X_m \leq z\right) = \frac{1}{\sqrt{(2\pi)}} \int_{-\infty}^{z} e^{-w^2/2}\, dw + O\left(\sigma^{-3} \sum_{m=1}^{n} L_m\right),$$

where the implied constant is absolute.

Proof. This lemma was proved by Esseen [2], as an example in the use of his form of lemma (1.47). It was also proved independently by Berry [1]. The essential ingredient in both of their proofs, as in the above proof of lemma (1.47), is the convolution of the distribution functions under consideration with a suitably smooth approximation to an improper law.

Analogues of these results may also be established for distribution functions (mod 1).

Lemma 1.49. *Let $F(z)$ and $G(z)$ be distribution functions* (mod 1) *with corresponding characteristic functions c_k, $(k = 0, \pm 1, \pm 2, \ldots)$, and d_k, $(k = 0, \pm 1, \pm 2, \ldots)$. Then there is a positive absolute constant B so that the inequality*

$$|F(z) - G(z)| \leq BQ_G(m^{-1}) + B \sum_{k=1}^{m} k^{-1} |c_k - d_k|$$

holds uniformly for all real numbers z, and positive integers m. Here, for any $l \geq 0$,

$$Q_G(l) = \sup_z \{G(z + l) - G(z)\}.$$

Proof. A proof of this result may be found in Faĭnleĭb [2], (see also Elliott [10]).

Remark. The application of lemma (1.49) is useful only if $Q_G(m^{-1})$ becomes small as $m \to \infty$. In other words, the distribution $G(z)$ must be essentially continuous.

The following special case is of independent interest. Let $a_1, a_2, \ldots,$ be a sequence of rational integers. For each integer k define the sum

$$S_k = S_k(n) = n^{-1} \sum_{m=1}^{n} e^{2\pi i k a_m} \qquad (n = 1, 2, \ldots).$$

We set

$$F_n(z) = n^{-1} \sum_{\substack{m=1 \\ a_m \leq z \pmod 1}}^{n} 1$$

and take $G(z)$ to be the uniform law (mod 1). Then for every positive integer n,

$$\sup_{0 \leq z \leq 1} |F_n(z) - z| \leq Bn^{-1} + B \sum_{k=1}^{n} k^{-1} |S_k(n)|.$$

This represents a quantitative formulation of Weyl's criterion for uniform distribution. It was obtained by Erdös and Turán [1], earlier than the result of lemma (1.49).

A pretty result of LeVeque [2], asserts that

$$\sup_{0 \leq z \leq 1} |F_n(z) - z| \leq \left(\frac{6}{\pi^2} \sum_{k=-\infty}^{\infty}{}' k^{-2} |S_k|^2 \right)^{1/3}$$

where $'$ indicates that $k = 0$ is omitted for the range of summation. A general form of this inequality can also be proved, (see, for example, Elliott [10]). In applications inequalities of this type usually give a result which is inferior to that which is obtainable from lemma (1.49). However, the following analogue of LeVeque's result may be applied to distribution functions (mod 1) which are not everywhere continuous.

Lemma 1.50. *Let* $F(z)$ *and* $G(z)$ *be distribution functions* (mod 1), *with corresponding characteristic functions* c_k *and* d_k. *Let* m *be a positive integer, and let* α *and* β *be real numbers which satisfy* $m^{-1} \le \alpha \le \beta \le 1 - m^{-1}$. *Then the inequality*

$$|F(\beta) - F(\alpha) - \{G(\beta) - G(\alpha)\}| \le 2\left\{G\left(\beta + \frac{1}{2m}\right) - G\left(\beta - \frac{1}{2m}\right)\right\}$$

$$+ 2\left\{G\left(\alpha + \frac{1}{2m}\right) - G\left(\alpha - \frac{1}{2m}\right)\right\} + 4m\left(\frac{1}{2\pi^2} \sum_{k=-\infty}^{\infty}{}' k^{-2}|c_k - d_k|^2\right)^{1/2}.$$

holds uniformly for all positive integers m consistent with the above conditions.

Proof. See Elliott [10], Theorem 3, pp. 517–520.

Remark. There is some loss of precision in this theorem when compared to the above inequality of LeVeque. For example, if G is continuously differentiable in some compact neighbourhood of the points α, β, $(0 < \alpha < \beta < 1)$, then the best that lemma (1.50) will yield is, in general,

$$|F(\beta) - F(\alpha) - \{G(\beta) - G(\alpha)\}| \le B\left(\sum_{k=-\infty}^{\infty}{}' k^{-2}|c_k - d_k|^2\right)^{1/4}.$$

Here we have replaced the exponent $\frac{1}{3}$ by $\frac{1}{4}$.

Concluding Remarks

Cauchy solved his functional equation, for a continuous function, in 1821. His original account, which includes the consideration of certain related functional equations, may be found in pages 98–113 of his collected works [1] deuxième Série, Vol. 3, Chapter 5. It should be mentioned that Cauchy was the first to give the currently accepted definition of continuity.

Subsequently, a number of authors showed that it is possible to obtain Cauchy's conclusion, (lemma (1.2)), by making somewhat weaker restrictions upon $f(x)$. During the course of a paper related to such considerations, Steinhaus [1], Théorème VIII, p. 99, proved lemma (1.1). Sierpinski [1], showed that measurability suffices, without the axiom of choice.

In 1905 G. Hamel [1], proved that, if a sufficiently strong axiom of choice is allowed, then there exist solutions to Cauchy's functional equation which are not of the form $f(x) \doteq Ax$. Such solutions are, of course, nowhere continuous.

For further details concerning these results we refer to Aczél [1], Chapter 2, pp. 31–37. This reference also contains a very large bibliography concerning Cauchy's functional equation.

In a discrete form, and with a different notation, slowly oscillating functions were introduced by R. Schmidt (ref. [1], Definition 9, p. 127), who also established appropriate analogues of lemmas (1.3) and (1.4).

In the form given in the present chapter, slowly oscillating functions were studied by Karamata. Both Schmidt and Karamata were concerned with applications to the Hardy–Littlewood tauberian theorem, (see lemma (2.18) of the following chapter). According to Karamata's definition, a slowly oscillating function, $L(x)$, is assumed to be defined, *real and positive*, for all real $x > 0$. He established an analogue of lemma (1.3) in his paper [1]. Moreover, he showed that there is a continuous function $\delta(t)$, which is $o(1)$ as $t \to \infty$, so that

$$L(x)\exp\left(-\int_1^x \delta(t)dt\right) \to 1 \qquad (x \to \infty).$$

This representation then gives an alternative characterisation of slowly oscillating functions.

The extension of these results to the case of measurable positive real functions $L(x)$ was carried out by Korevaar, van Aardenne-Ehrenfest and de Bruijn [1], and de Bruijn [1], respectively.

For a survey, with proofs, of results concerning slowly oscillating functions, up to 1974, see the monograph by Chi-hsing Yong [1].

One cannot but be struck by the analogy between lemma (1.6) of Halász [5], (p. 148), and the (α, β) theorem of Mann. We shall take advantage of this similarity in theorem (8.2) of Chapter 8. Let $a_1 < a_2 < \ldots$ be a sequence of positive integers. For each positive integer n let $A(n)$ denote the number of integers, belonging to the sequence, which do not exceed n. Define the Schnirelmann density α of the sequence by

$$\alpha = \inf_{n \geq 1} n^{-1}A(n).$$

Let β be the density of a second sequence of positive integers, $b_1 < b_2 < \ldots$. We define the Schnirelmann sum $A + B$ to be the sequence of those integers which are of the form a_i, b_j, or $a_i + b_j$. Then Mann's theorem states that *either* $A + B$ contains every positive integer, *or* it has a Schnirelmann density of at least $(\alpha + \beta)$. For a proof of this result we refer to Halberstam and Roth [1] §4. Lemma (1.6) corresponds to the case of the Schnirelmann addition of k identical sequences. This special case of Mann's theorem had, in fact, already been established by Khinchine [1].

The equivalence of propositions (i) and (iii) of lemma (1.18) is known as "Kolmogorov's Three Series Theorem".

We reconsider Sperner's lemma, (lemma (1.15)), in the remarks at the end of Chapter 5, where we say more concerning both its application to the theory of probability, and to the theory of numbers. The present short proof of Sperner's result is due to Lubell [1]. The particular result lemma (1.14) is due to Kolmogorov [3], and Rogozin [1]. See, also, Linnik and Ibragimov [1], Chapter 15, §2, pp. 268–273.

Lévy's theorem concerning the possible continuity of infinite convolutions appears in Lévy [1]. Jessen and Wintner's theorem on the purity of type appears as Theorem 35 on page 86 of their paper [1].

There is a considerable literature concerning both analytic, and non-infinitely-divisible laws. For proofs of most of these results we refer to Linnik [4], and Lukacs [1]. This area of the theory of probability is only partially developed.

Lemma (1.43) occurs as Satz II of Wintner [1]. A little while later it was established, independently, by Fréchet and Shohat [1], as they acknowledge in the last footnote on page 543 of their paper.

A slightly more detailed treatment of the application of Mellin transforms to the theory of probability, as embodied in lemma (1.45), can be found in Postnikov [1], Chapter IV, §4.9, pp. 367–371. The treatment given in the present chapter is based upon his, together with certain changes necessary because of a different convention concerning the continuity of distribution functions.

Ultimately, results of the form of lemmas (1.47), (1.48), and (1.49) are based upon the original method of Berry [1], and Esseen [2]; namely, the convolution of the distribution functions under study with a suitably smooth approximation to the improper law. An appropriate use of a Fejér kernel will often suffice.

For further results concerning the particular notion of uniform distribution, and its generalisations to locally compact groups, we refer to Kuipers and Niederreiter [1].

Chapter 2

Arithmetical Results, Dirichlet Series

We begin this chapter with a discussion of the sieve method of Selberg [1]. We shall confine ourselves to a result which is sufficient for our future needs. Its proof is not long.

Lemma 2.1. *Let $f(n)$ be a real-valued non-negative arithmetic function. Let a_n, $n = 1, \ldots, N$, be a sequence of rational integers. Let r be a positive real number, and let $p_1 < p_2 < \cdots < p_s \leq r$ be rational primes. Set $Q = p_1 \cdots p_s$. If $d \mid Q$ then let*

(i) $$\sum_{\substack{n=1 \\ a_n \equiv 0 \pmod{d}}}^{N} f(n) = \eta(d)X + R(N, d),$$

where X, R are real numbers, $X \geq 0$, and $\eta(d_1 d_2) = \eta(d_1)\eta(d_2)$ whenever d_1 and d_2 are coprime divisors of Q.

Assume that for each prime p, $0 \leq \eta(p) < 1$.

Let $I(N, Q)$ denote the sum

$$\sum_{\substack{n=1 \\ (a_n, Q)=1}}^{N} f(n).$$

Then the estimate

$$I(N, Q) = \{1 + 2\theta_1 H\}X \prod_{p \mid Q} (1 + \eta(p)) + 2\theta_2 \sum_{\substack{d \mid Q \\ d \leq z^3}} 3^{\omega(d)} |R(N, d)|$$

holds uniformly for $r \geq 2$, $\max(\log r, S) \leq \frac{1}{8} \log z$, where $|\theta_1| \leq 1$, $|\theta_2| \leq 1$, and

$$H = \exp\left(-\frac{\log z}{\log r}\left\{\log\left(\frac{\log z}{S}\right) - \log\log\left(\frac{\log z}{S}\right) - \frac{2S}{\log z}\right\}\right)$$

$$S = \sum_{p \mid Q} \frac{\eta(p)}{1 - \eta(p)} \log p.$$

When these conditions are satisfied there is a positive absolute constant c so that $2H \le c < 1$.

Remarks. If only an upper bound for $I(N, Q)$ is desired then one may replace the condition $d \le z^3$ by $d \le z^2$, and change the definition of S to

$$\sum_{p|Q} \eta(p)\log p.$$

This then allows the possibility that $\eta(p) = 1$.

For the duration of the present lemma, and its proof, $\omega(d)$ denotes the number of distinct prime divisors of the integer d.

In the course of the proof of lemma (2.1) it is shown that a permissible value for the constant c is $\exp(-0.006)$.

The present lemma (2.1) represents a general form of the so-called *Fundamental Lemma* of Kubilius, ref. [5], Chapter one, lemma (1.4). For further results in this direction see Barban [5], Halberstam and Richert [1].

A detailed account of the sieve methods of Brun and Selberg may be found in Halberstam and Richert [2].

In each of these accounts it is assumed that on average $p\eta(p)$ is absolutely bounded. We do not make such an assumption.

This ends our remarks.

We need a few preliminary results. We shall assume, without loss of generality, that $\eta(p) > 0$ for each prime p.

Lemma 2.2. *If $w \ge e$ then*

$$\min_{\rho \ge 0} \rho(e^\rho - w) \le -w\left(\log w - \log\log w + \frac{\log\log w}{\log w} - 1\right).$$

Proof. When

$$\rho = \log w - \log\log w$$

the function which is to be minimized attains the value given as an upper bound in the statement of the lemma.

Remark. A more elaborate argument shows that the upper bound which we give for the minimum differs from its actual value by an amount which is $O(w(\log w)^{-1})$.

Lemma 2.3. *For divisors d of Q define the function*

$$h(d) = \frac{1}{\eta(d)} \prod_{p|d} (1 - \eta(p)).$$

Set

$$Z = \sum_{d|Q} \frac{1}{h(d)}$$

and

$$\Delta = \log\left(\frac{\log z}{S}\right) - \log\log\left(\frac{\log z}{S}\right) - 1.$$

Then, uniformly for $r \geq 2$, $\max.(\log r, S) \leq \frac{1}{8}\log z$ *the inequalities*

$$\sum_{\substack{d|Q \\ d>z}} \frac{1}{h(d)} \leq Z\exp\left(-\frac{\log z}{\log r}\cdot\Delta\right)$$

$$2\exp\left(-\frac{\log z}{\log r}\cdot\Delta\right) < 1.$$

are satisfied.

Proof. By making use of the fact that $h(d)$ is multiplicative on the divisors of Q, we see that

$$Z = \prod_{p|Q}(1-\eta(p))^{-1},$$

and that for each real number $\lambda \geq 0$

$$\sum_{d|Q} \frac{d^\lambda}{h(d)} = Z\prod_{p|Q}(1+\eta(p)(p^\lambda - 1)).$$

Since $1 + t \leq e^t$ and $e^t - 1 \leq te^t$ when $t \geq 0$, this last product does not exceed

$$\exp\left(\sum_{p|Q}\eta(p)\{p^\lambda - 1\}\right) \leq \exp\left(\lambda\sum_{p|Q}\eta(p)\log p\cdot p^\lambda\right) \leq \exp(\lambda S r^\lambda).$$

We set $\lambda = \rho(\log r)^{-1}$, so that $r^\lambda = e^\rho$, and note that for any $\rho \geq 0$ the value of the sum which we wish to estimate does not exceed

$$\sum_{d|Q}\frac{1}{h(d)}\left(\frac{d}{z}\right)^\lambda \leq Z\exp\left(\frac{\rho S}{\log r}\left\{e^\rho - \frac{\log z}{S}\right\}\right)$$

The first assertion of lemma (2.3) is now obtained by an appeal to lemma (2.2).

As for the second assertion, differentiation shows that the function

$$K(t) = \log t - \log\log t - 1$$

is increasing if $t > e$. Therefore, in our present circumstances,

$$\frac{\log z}{\log r}\Delta \geq 8K(8) \geq e^2 K(e^2) = e^2 \log\frac{e}{2}.$$

Since $e > 2,6$ we have

$$\left(\frac{e}{2}\right)^3 > (1.3)^3 > 2,\ e^2\log\frac{e}{2} > \frac{1}{3}e^2\log 2 > \log 2,$$

and

$$2\exp\left(-\frac{\log z}{\log r}\Delta\right) < 2\exp(-\log 2) = 1.$$

This completes the proof of lemma (2.3).

Remarks. It is clear from the proof that the result of lemma (2.3) remains valid if we replace S by any number S' which satisfies $S \leq S' \leq \frac{1}{8}\log z$.

It is not difficult to show that the function

$$K(t) - \frac{2}{t}$$

is non-decreasing for $t \geq e$, so that uniformly for $t \geq 8$,

$$8(K(t) - 2t^{-1}) \geq 8K(8) - 2 > 0.70 > \log 2.$$

This involves a better lower bound for $8K(8)$ than that given above, but is established by recourse to mathematical tables. We shall make use of it in the following result.

Lemma 2.4. *In the notation of lemma* (2.1) *and lemma* (2.3)

$$\sum_{i=0}^{s-1}\eta(p_{i+1})\prod_{j=i+1}^{s}(1-\eta(p_j))^{-1}\exp\left(-\frac{\log z}{\log p_i}\Delta\right) \leq \exp\left(-\frac{\log z}{\log r}\left(\Delta - \frac{2S}{\log z}\right)\right).$$

Moreover, there is an absolute constant c so that

$$2\exp\left(-\frac{\log z}{\log r}\left(\Delta - \frac{2S}{\log z}\right)\right) \leq c < 1.$$

Proof. If $0 \leq t < 1$, then $(1 - t)^{-1} \leq \exp(t\{1 - t\}^{-1})$, so that

$$\prod_{j=i+1}^{s} (1 - \eta(p_j))^{-1} \leq \exp\left(\sum_{j=i+1}^{s} \frac{\eta(p_j)}{1 - \eta(p_j)}\right)$$

$$\leq \exp\left(\sum_{j=i+1}^{s} \frac{\eta(p_j)}{1 - \eta(p_j)} \cdot \frac{\log p_j}{\log p_{i+1}}\right) \leq \exp\left(\frac{S}{\log p_{i+1}}\right)$$

$$= \exp\left(\frac{\log z}{\log p_{i+1}} \cdot \frac{S}{\log z}\right).$$

Therefore, the sum which we wish to estimate does not exceed

$$\sum_{p|Q} \eta(p)\exp\left(-\frac{\log z}{\log p}\left\{\Delta - \frac{S}{\log z}\right\}\right)$$

$$= \sum_{p|Q} \eta(p)\exp\left(-\frac{S}{\log p}\right) \cdot \exp\left(-\frac{\log z}{\log p}\left\{\Delta - \frac{2S}{\log z}\right\}\right).$$

According to the second of the two remarks made following lemma (2.3), under the conditions of the present lemma $\Delta - 2S(\log z)^{-1}$ is positive. Hence the exponential factor which occurs on the extreme right-hand side of the equation immediately above does not exceed

$$\exp\left(-\frac{\log z}{\log r}\left\{\Delta - \frac{2S}{\log z}\right\}\right)$$

Next

$$\exp\left(\frac{S}{\log p}\right) > \frac{S}{\log p},$$

so that

$$\sum_{p|Q} \eta(p)\exp\left(-\frac{S}{\log p}\right) \leq \frac{1}{S} \sum_{p|Q} \eta(p)\log p \leq \frac{1}{S} \cdot S = 1.$$

This justifies the first assertion of the present lemma.

The validity of the second assertion follows from the same remark, which shows that one may indeed choose $c = \exp(-0.006)$.

We are now ready to establish lemma (2.1).

Upper Bound

We follow the (by now classical) approach of Selberg (ref. [1]) to obtain an upper bound.

Let λ_d be real numbers, defined for integers d in the range $1 \leq d \leq z$, $d | Q$, and restricted only in that $\lambda_1 = 1$. Then, since f is non-negative,

$$\sum_{\substack{n=1 \\ (a_n, Q)=1}}^{N} f(n) \leq \sum_{n=1}^{N} f(n) \Bigg(\sum_{d | (a_n, Q)} \lambda_d \Bigg)^2 .$$

Expanding the square, and inverting the order of summation, the sum on the right-hand side of this inequality becomes

$$\sum_{d_1 \leq z}' \sum_{d_2 \leq z}' \lambda_{d_1} \lambda_{d_2} \sum_{\substack{n=1 \\ a_n \equiv 0 (\mathrm{mod}\, d_j),\, j=1,2}}^{N} f(n)$$

where $'$ indicates that the d_j, $(j = 1, 2)$, divide Q.

By making use of hypothesis (i) of lemma (2.1) we may express this as

$$X\Gamma + E$$

where Γ denotes the quadratic form

$$\sum_{d_1 \leq z}' \sum_{d_2 \leq z}' \lambda_{d_1} \lambda_{d_2} \eta([d_1, d_2])$$

and where

$$E = \sum_{d_1 \leq z}' \sum_{d_2 \leq z}' \lambda_{d_1} \lambda_{d_2} R(N, [d_1, d_2]).$$

The λ_d are now chosen so as to minimize Γ, consistent with the requirement that $\lambda_1 = 1$. This may be effected by reducing Γ to a sum of squares, which we shall do. We note that $h(d)$ is multiplicative on the divisors d of Q, and satisfies

$$\sum_{m | d} h(m) = \frac{1}{\eta(d)}.$$

Hence

$$\eta([d_1, d_2]) = \frac{\eta(d_1)\eta(d_2)}{\eta((d_1, d_2))} = \eta(d_1)\eta(d_2) \sum_{\substack{m | d_1 \\ m | d_2}} h(m)$$

and

$$\Gamma = \sum_{d_1 \le z}' \sum_{d_2 \le z}' \lambda_{d_1}\eta(d_1)\lambda_{d_2}\eta(d_2)\sum_{\substack{m|d_1 \\ m|d_2}} h(m)$$

$$= \sum_{m \le z}' h(m)\left\{ \sum_{\substack{d \le z \\ d \equiv 0 (\mathrm{mod}\ m)}}' \eta(d)\lambda_d \right\}^2$$

If we set

$$y_m = \sum_{\substack{d \le z,\, d|Q \\ d \equiv 0 (\mathrm{mod}\ m)}} \eta(d)\lambda_d,\ m \mid Q,$$

then Γ assumes the form

$$\Gamma = \sum_{m \le z}' h(m) y_m^2.$$

It is easy to check that if $\lambda_1 = 1$ then

$$\sum_{m \le z}' \mu(m) y_m = 1.$$

By means of this relation we can write

$$\Gamma = \sum_{m \le z}' h(m)\left(y_m - \frac{\mu(m)}{h(m)} L^{-1} \right)^2 + L^{-1}$$

where

$$L = \sum_{m \le z}' \frac{1}{h(m)}.$$

It is now clear that to minimize Γ we should choose the y_m (if possible) so that

$$y_m = \frac{\mu(m)}{h(m)} L^{-1}.$$

In fact, a Möbius inversion (see Hardy and Wright [1], Chapter XVI, §§16.4, 16.5, pp. 236–237) shows that the definition of the y_m is equivalent to

$$\eta(d)\lambda_d = \sum_{\substack{m \le z \\ m \equiv 0 (\mathrm{mod}\ d)}}' \mu\left(\frac{m}{d}\right) y_m = \sum_{\substack{w \le z/d \\ (w,d) = 1}}' \mu(w) y_{wd}.$$

In order to obtain the above values of y_m we therefore set

$$\lambda_d = \frac{\mu(d)}{Lh(d)\eta(d)} \sum_{\substack{w \le z/d \\ (w,d)=1}}' \frac{1}{h(w)}.$$

This definition is meaningful only when d is a (squarefree) divisor of Q. With this choice of the λ_d, Γ assumes its minimum value L^{-1}.

Although the expression for λ_d is somewhat awkward, an upper bound for it, which will be sufficient for our purposes, can be found by the following simple argument.

For each divisor d of Q

$$L = \sum_{k|d} \sum_{\substack{m \le z \\ (m,d)=k}}' \frac{1}{h(m)} = \sum_{k|d} \frac{1}{h(k)} \sum_{\substack{w \le z/k \\ (w,d)=1}}' \frac{1}{h(w)}$$

$$\ge \sum_{k|d} \frac{1}{h(k)} \cdot \sum_{\substack{w \le z/d \\ (w,d)=1}}' \frac{1}{h(w)}.$$

Since $h(k)$ is multiplicative on Q we have

$$\sum_{k|d} \frac{1}{h(k)} = \frac{1}{h(d)\eta(d)}$$

and

$$|\lambda_d| \le 1, (d|Q).$$

We have now proved that

$$I(N, Q) \le XL^{-1} + \sum_{d_1 \le z}' \sum_{d_2 \le z}' |R(N, [d_1, d_2])|.$$

From lemma (2.3), using the fact that $(1-y)^{-1} \le 1 + 2y$ when $0 \le y \le \frac{1}{2}$,

$$L = \left\{1 + \theta_3 \exp\left(-\frac{\log z}{\log r}\Delta\right)\right\} \sum_{d|Q} \frac{1}{h(d)} \qquad |\theta_3| \le 1,$$

so that

$$I(N, Q) \le \left\{1 + 2\exp\left(-\frac{\log z}{\log r}\Delta\right)\right\} \prod_{i=1}^{s} (1 - \eta(p_i)) + E_1,$$

where the error term E_1 is given by the double sum, over d_1 and d_2, which appears a few lines earlier in the inequality involving $I(N, Q)$.

Lower Bound

To obtain a lower bound, and so complete the proof of lemma (2.1), we begin with the representation

$$\text{(ii)} \qquad I(N, Q) = \sum_{n=1}^{N} f(n) - \sum_{i=0}^{s-1} \sum_{\substack{n=1 \\ a_n \equiv 0 (\mathrm{mod}\ p_{i+1}) \\ p_j \nmid a_n,\ 1 \le j \le i}}^{N} f(n),$$

where it is to be understood that the condition $p_j \nmid a_n$, $1 \le j \le 0$, which is meaningless, is to be ignored. If $1 \le i \le s - 1$ then we have

$$\sum_{\substack{n=1 \\ a_n \equiv 0 (\mathrm{mod}\ p_{i+1}) \\ p_j \nmid a_n,\ 1 \le j \le i}}^{N} f(n) = \sum_{\substack{n=1 \\ p_j \nmid a_n,\ 1 \le j \le i}}^{N} g(n)$$

where

$$g(n) = \begin{cases} f(n) & \text{if } p_{i+1} \mid a_n, \\ 0 & \text{otherwise.} \end{cases}$$

If $d \mid (p_1 \ldots p_i)$ then

$$\sum_{\substack{n=1 \\ a_n \equiv 0 (\mathrm{mod}\ d)}}^{N} g(n) = \sum_{\substack{n=1 \\ a_n \equiv 0 (\mathrm{mod}\ p_{i+1} d)}}^{N} f(n)$$
$$= \eta(p_{i+1} d) X + R(N, p_{i+1} d).$$

This estimate makes sense, even if $i = 0$, provided that we adopt the usual convention that empty products have the value 1. In such a case we would have $d = 1$. Applying the upper bound inequality with $f(n)$ replaced by $g(n)$, X replaced by $\eta(p_{i+1})X$, and r by p_i, we deduce that if $i \ge 1$ then

$$\sum_{\substack{n=1 \\ a_n \equiv 0 (\mathrm{mod}\ p_{i+1}) \\ p_j \nmid a_n,\ 1 \le j \le i}}^{N} f(n) \le \eta(p_{i+1}) X \left\{ 1 + 2 \exp\left(-\frac{\log z}{\log p_i} \Delta \right) \right\} \prod_{j=1}^{i} (1 - \eta(p_j))$$
$$+ \sum_{\substack{d_1 \le z \\ d_t \mid (p_1 \ldots p_i),\ (t = 1, 2)}} \sum_{d_2 \le z} |R(N, p_{i+1}[d_1, d_2])|$$

Note that we have here made an application of the first of the two remarks which follow lemma (2.3).

It is straightforward to check the identity

$$1 - \sum_{i=0}^{s-1} \eta(p_{i+1}) \prod_{j=1}^{i} (1 - \eta(p_j)) = \prod_{k=1}^{s} (1 - \eta(p_k)).$$

Applying the above inequalities in the equation (ii), and making use of this last identity, we obtain for $I(N, Q)$ the lower bound

$$\{1 + E_2\} X \prod_{k=1}^{s} (1 - \eta(p_k)) - E_3$$

where

$$|E_2| \le 2 \sum_{i=0}^{s-1} \eta(p_{i+1}) \prod_{j=i+1}^{s} \{1 - \eta(p_j)\}^{-1} \exp\left(-\frac{\log z}{\log p_i} \Delta\right)$$

and

$$\text{(iii)} \qquad |E_3| \le \sum_{i=0}^{s-1} \sum_{\substack{d_1 \le z \\ d_t | (p_1 \dots p_i)}} \sum_{d_2 \le z} |R(N, p_{i+1}[d_1, d_2])| + |R(N, 1)|.$$

Considering together the upper and lower bounds for $I(N, Q)$, and making use of lemma (2.4), leads to the estimate

$$I(N, Q) = \{1 + 2\theta_1 H\} X \prod_{k=1}^{s} (1 - \eta(p_k)) + \theta_3 \max\{|E_1|, |E_3|\}$$

for some $|\theta_1| \le 1$ and $|\theta_3| \le 1$.

It only remains for us to give upper bounds for the error terms $|E_1|$ and $|E_3|$.

Consider the triple sum which occurs in the inequality (iii). Let us collect together those terms for which $p_{i+1}[d_1, d_2]$ has a given value d. Then the sum may be expressed in the form

$$\sum_{d|Q} |R(N, d)| \sum_{i=0}^{s-1} \sum_{\substack{d_1 \le z \\ p_{i+1}[d_1, d_2] = d}} \sum_{d_2 \le z} 1$$

where, in the innermost double-sum, each d_t, $(t = 1, 2)$, is a divisor of the product of primes $(p_1 \dots p_i)$. The innermost triple-sum is zero unless $d \le p_{i+1} z^2 \le z^3$. Otherwise it does not exceed the number of solutions to the equation $[u_1, u_2] = d$ where we count, in turn, those u_t which have as their largest prime divisor p_{i+1}, $(i = 0, 1, \ldots, s-1)$. We can estimate this number to be at most

$$\sum_{k|d} \sum_{\substack{(u_1, u_2)=k \\ [u_1, u_2]=d}} 1 = \sum_{k|d} 2^{\omega(dk^{-1})} = 3^{\omega(d)}.$$

It is now clear that

$$|E_3| \leq \sum_{\substack{d|Q \\ d \leq z^3}} 3^{\omega(d)} |R(N, d)|.$$

Since $|E_1|$ satisfies a similar inequality, lemma (2.1) is proved.

Distribution of Prime Numbers

Lemma 2.5. *The following estimates hold for* $x \geq 2$,

$$\sum_{p \leq x} \frac{\log p}{p} = \log x + O(1),$$

$$\sum_{p \leq x} \frac{1}{p} = \log \log x + B + O\left(\frac{1}{\log x}\right),$$

$$\prod_{p \leq x} \left(1 - \frac{1}{p}\right) = \frac{e^{-\gamma}}{\log x} \left\{1 + O\left(\frac{1}{\log x}\right)\right\}.$$

Remarks. Elementary proofs of these estimates may be found in Hardy and Wright [1], Chapter XXII, pp. 340–353. Here γ denotes Euler's constant, and the constant B has the representation

$$B = \gamma + \sum_p \left\{\log\left(1 - \frac{1}{p}\right) + \frac{1}{p}\right\}.$$

The values of these constants will not concern us.

The third estimate of the present lemma is due to Mertens [1]. As a corollary we note the pretty, asymptotic formula

$$\prod_{p \leq x} \left(1 + \frac{1}{p}\right) \sim \frac{6e^{\gamma} \log x}{\pi^2} \qquad (x \to \infty)$$

which contains many of the objects upon which mathematicians have traditionally broken their teeth.

It is convenient to include the following (simple) form of the Prime Number Theorem.

Lemma 2.6. *Let m be a positive number. For* $x \geq 2$, $\pi(x)$, *the number of primes not exceeding* x, *has the estimate*

$$\pi(x) = \int_2^x \frac{dy}{\log y} + O\left(\frac{x}{(\log x)^m}\right).$$

Proof. An elementary proof of this result may be found in Wirsing [3] or Bombieri [2]. A much stronger result may be found in Chapter 3, pp. 55–70 of Prachar [1]. In fact we shall apply this lemma only a few times, mostly in examples, and will need the cases $m = 2$ or $m = 3$.

Several times in this monograph results will be established from which a form of the prime-number theorem may be deduced.

For integers $D > 0$, l, and real numbers x, let $\pi(x, D, l)$ denote the number of primes not exceeding x which satisfy the congruence relation $p \equiv l(\text{mod } D)$.

Lemma 2.7 (Brun–Titchmarsh). *Let* α *be a real number,* $0 \leq \alpha < 1$. *Then there is a further real number* c_1, *which may depend upon* α, *so that the inequality*

$$\pi(x, D, l) \leq \frac{c_1 x}{\varphi(D)\log x}$$

holds uniformly for all real $x \geq 2$, *and integers* l, *and* D *with* $1 \leq D \leq x^\alpha$.

Lemma 2.8 (Erdös). *The number of solutions to the equation*

$$p + 1 = Dq$$

where p *and* q *are prime numbers which do not exceed* x, $x \geq 2$, *and* D *is a particular integer in the range* $1 \leq D \leq x^\alpha$, *is not more than*

$$\frac{c_2 x}{\varphi(D)\log^2 x}.$$

We sketch a proof of lemma (2.7) which was proved by Titchmarsh [1]. We apply lemma (2.1) with $f(n) \equiv 1$. For the sequence $\{a_n\}$ we take those integers n in the interval $1 \leq n \leq x$ which satisfy $n \equiv l(\text{mod } D)$. For the primes p_i we take all rational primes p in the interval $2 \leq p \leq z$ which do not divide D; z will be chosen presently. Then, by the Chinese Remainder Theorem, the estimate

$$\left| \sum_{a_n \equiv 0(\text{mod } d)} 1 - \frac{x}{dD} \right| \leq 2$$

will hold uniformly for all divisors d of Q. We set $X = xD^{-1}$, and $\eta(p) = p^{-1}$. Then $|R(N, d)| \le 2$ is satisfied. Clearly $0 < \eta(p) < 1$ and for any $y \ge 2$ the second of the three results in lemma (2.5) allows us to conclude that

$$\sum_{p \le y} \frac{\eta(p)}{1 - \eta(p)} \log p = \sum_{p \le y} \frac{\log p}{p - 1} = \log y + O(1).$$

If $z < x^{1/2}$ then

$$\pi(x, D, l) \le z + I(N, Q),$$

and from lemma (2.1) we obtain the estimate, valid when $r = z^{1/9}$, z large, c_3 absolute:

$$I(N, Q) \le \frac{c_3 x}{D} \prod_{\substack{p \le r \\ p \nmid D}} \left(1 - \frac{1}{p}\right) + 4 \sum_{\substack{d \le z^3 \\ d | Q}} 3^{\omega(d)}.$$

According to lemma (2.5) the first of these majorising terms is

$$O\left(\frac{x}{\varphi(D)\log z}\right).$$

As for the second majorising term, it does not exceed

$$\sum_{\substack{d \le z^3 \\ d | Q}} \frac{z^3}{d} 3^{\omega(d)} \le z^3 \prod_{p \le z} \left(1 + \frac{3}{p}\right) = O((z \log z)^3).$$

It is now straightforward to check that the choice $z = x^{(1-\alpha)/4}$ leads to the result stated in lemma (2.7).

A proof of lemma (2.8) may be constructed upon similar lines. In this case one considers, in place of the a_n, the sequence of integers $n(n + 1)/D$, where $1 \le n \le x$, and $n \equiv -1 (\mathrm{mod}\ D)$ is satisfied. Lemma (2.8) appears in Erdös [3].

A slightly different working of the proofs of lemmas (2.7) and (2.8) may be found in Prachar [1], Kap. II, pp. 35–45, 51–52.

For use in Chapter 22 and in some examples we state the following lemma.

Lemma 2.9 (Siegel–Walfisz). *Let c be a fixed real number. Then the estimate*

$$\pi(x, D, l) = \frac{1}{\varphi(D)} \int_2^x \frac{dt}{\log t} + O(xe^{-c_1\sqrt{\log x}})$$

holds uniformly for all integers l and D such that $(l, D) = 1$ and $1 \le D \le (\log x)^c$.

Proof. See, for example, Prachar [1], Kap. IV, Satz 8.3, p. 144. Lemma (2.9) contains the prime-number theorem as a special case, but its proof is somewhat more complicated than a direct proof of lemma (2.6).

Lemma 2.10 (Rényi, Barban, Vinogradov, Bombieri). *Let A be a positive real number. Then there is a further real number B such that the inequality*

$$\sum_{D \le x^{1/2}(\log x)^{-B}} \max_{(l,D)=1} \max_{y \le x} \left| \pi(y, D, l) - \frac{Li(y)}{\varphi(D)} \right| = O(x(\log x)^{-A})$$

holds uniformly for all $x \ge 2$.

Remarks. An important precursor to this result, but involving weighting the primes, was proved by Rényi [2].

In 1961 Barban [1] wrote a paper asserting that there exists a (small) constant δ_0 so that

$$\sum_{D \le x^{\delta_0}} \mu^2(D) \max_{(l, D)=1} \left| \pi(x, D, l) - \frac{Li(x)}{\varphi(D)} \right| = O(x(\log x)^{-A})$$

and that he will show that any $\delta_0 < 1/6$ is permissible. (See page 1 and page 14 of his paper, both results are labelled *Theorem* 1.) This paper was subject to considerable criticism, perhaps because the result was somewhat unexpected. Apart from a few simple slips in calculation, the only contentious statement is that

$$\sum_{\chi (\mathrm{mod}\ D)} \sum_{|\gamma| \le D^2} x^{\beta - 1} = O((\log xD)^8)$$

where χ runs through all the Dirichlet characters (mod D), and, for each χ, $\beta + i\gamma$ runs through the zeros of the associated L-series $L(s, \chi)$, $0 \le \beta \le 1$. It is also implicit in his statement that this result is to be uniform for D not exceeding $x^{(1/6)-\varepsilon}$, $\varepsilon > 0$. For proof an unspecific reference to Prachar [1] is given. Whilst all other references to Prachar [1] which are given in this paper can readily be substantiated, it is not clear where to look for this one. However, in the following six lines of his paper (page 13) the author proves that this relation certainly holds if $D \le x^{(3/23)-\varepsilon}$.

If we change the assertion $\delta_0 < 1/6$ to $\delta_0 < 3/23$ this paper of Barban is, in all essentials, correct.

Although it is not usual, the Russian review of this paper of Barban is anonymous, (see Ref. Zhurnal. (1962) 10A80). An inaccurate translation of this same review serves in *Math. Reviews* (M.R. **30**(1965), # 1990). Barban's paper was apparently not reviewed in *Zentralblatt*.

Barban [8] himself indicated how to slightly modify his argument so as to restore the range $D \le x^{(1/6)-\varepsilon}$. In a later paper [9] he removed the restriction that the moduli D be squarefree, and improved the condition on D to $D \le x^{(3/8)-\varepsilon}$.

Meanwhile, by supplying an additional result, a form of lemma (2.10) was established by A. I. Vinogradov, but with the uniformity over D restricted to $D \le x^{1/2-\varepsilon}$ for any fixed $\varepsilon > 0$. See A. I. Vinogradov [1], [2], and Barban [9], note added in proof, p. 94.

Independently, Bombieri [3], and with a somewhat different proof, involving a more appropriate and exact formulation of the large sieve, together with an effective use of it, established the precise form which we have stated above. See also the concluding remarks of Chapter 4.

For an exposition along Bombieri's lines, with certain improvements, we refer to Montgomery [3], Chapter 15, pp. 133–140. A proof which relies more on first principles, and which is short, may be found in Gallagher [1], [2].

In 1961, Leningrad University published U. V. Linnik's monograph on "The Dispersion method in Binary Additive Problems." On p. 164 (Chapter VIII) of the American translation [5] the following interesting remark may be found:

"It should be remarked that it was apparently not noticed that

$$\sum_{p \le n} \tau(p-1) > cn$$

can be derived without difficulty from theorems of the "large sieve" (Ju. V. Linnik [9]) type, developed by Rényi [50–52]."

Here $\tau(m)$ denotes the number of divisors of the integer m, and the references are those of Linnik himself.

Such an application may be found in Barban's paper [1], Theorem 2, p. 17.

The next lemma will be useful in certain examples. Let $g(x)$ be a polynomial with integral coefficients which is irreducible over the rational number field. For each positive integer r let $\rho(r)$ denote the number of residue class $k(\text{mod } r)$ for which $g(k) \equiv 0(\text{mod } r)$.

Lemma 2.11. *As $x \to \infty$*

$$\sum_{p \le x} \rho(p) = \frac{x}{\log x} + O\left(\frac{x}{(\log x)^2}\right)$$

and

$$\sum_{p \le x} \frac{\rho(p)}{p} = \log \log x + C + O\left(\frac{1}{\log x}\right).$$

Proof. The first of these results is a consequence of the Prime Ideal Theorem, (Landau [1]). For a sketch of the proof, with references, see Heilbronn, Chapter VIII p. 229 of ref. [1].

The second result may be obtained from the first using integration by parts. The value of the constant C will not concern us.

Dirichlet Series

For the purposes of this monograph a *Dirichlet series* will have the form

$$f(s) = \sum_{n=1}^{\infty} a_n n^{-s}$$

where the a_n may have complex values, and $s = \sigma + i\tau$ is a complex variable, with $\sigma = Re(s)$. We have replaced the traditional symbol t for $Im(s)$ by τ, since the use of t as a dummy variable has already been usurped by the characteristic functions on the real line. I well remember a lecture on the theory of complex variables, given some ten years ago in Cambridge by Miss M. Cartwright, in which she poured wonderful scorn upon the traditional hybrid notation $\sigma + it$. Pace! Miss Cartwright, for the duration of this volume at least, the sensible will prevail.

As was indicated in Chapter one, the symbol σ^2 is traditionally associated in the theory of probability with the Gaussian component of an infinitely divisible law. Only in Chapter 18 might these two usages have occurred together, and to avoid this we shall there use D^2 to represent the Gaussian component.

The derivation of various basic properties of Dirichlet series may be found in Titchmarsh [2], Chapter IX, pp. 289–317. We note here a few of these properties.

If a Dirichlet series converges at a point $s = s_0 = \sigma_0 + i\tau_0$, then it converges in the half-plane $\sigma > \sigma_0$, and absolutely in the half-plane $\sigma > \sigma_0 + 1$.

If a function $f(s)$ is representable by a Dirichlet series in some half-plane $\sigma > \sigma_0$ then the coefficients a_n are uniquely determined. We shall in fact need a quantitative form of this result.

Lemma 2.12 (a Perron's theorem). *Let*

$$f(s) = \sum_{n=1}^{\infty} a_n n^{-s} \qquad (\sigma > 1),$$

where $|a_n| \le c_1 \psi(n)$ *for some non-decreasing function* $\psi(n)$, *and as* $\sigma \to 1+$

$$\sum_{n=1}^{\infty} |a_n| n^{-\sigma} = O\left(\frac{1}{(\sigma - 1)^{\alpha}}\right)$$

for some fixed number α.

Then if $c > 0, \sigma + c > 1$, x is positive but not an integer, and N is the nearest integer to x,

$$\sum_{n<x} a_n n^{-s} = \frac{1}{2\pi i} \int_{c-iT}^{c+iT} f(s+w) \frac{x^w}{w} \, dw + O\left(\frac{x^c}{T(\sigma + c - 1)^\alpha}\right)$$
$$+ O\left(\frac{\psi(2x)x^{1-\sigma} \log x}{T}\right) + O\left(\frac{\psi(N)x^{1-\sigma}}{T|x-N|}\right).$$

If x is an integer we add the term $a_x/(2x^s)$ to the sum which appears on the left-hand side of this equation, and delete the third of the three error terms.

Proof. A proof of this form of Perron's theorem may be found in Titchmarsh [4], Chapter 3, Lemma 3.12, pp. 53–55; (see also Titchmarsh [2] 9.42, pp. 300–301, and Prachar [1], Anhang §3, Satz 3.1, pp. 376–379). It is based upon the contour integral

$$\frac{1}{2\pi i} \int_{c-i\infty}^{c+i\infty} \frac{x^s}{s} \, ds = \begin{cases} 1 & \text{if } 0 < x < 1, \\ \frac{1}{2} & \text{if } x = 1, \\ 0 & \text{if } x > 1, \end{cases}$$

which holds whenever $c > 0$.

Euler Product

A Dirichlet series will be said to have an *Euler product* if there is a representation

$$f(s) = \prod_p (1 + k(p)p^{-s} + k(p^2)p^{-2s} + \dots)$$

valid in some half-plane $\sigma > \sigma_0$. Here the product runs over all rational primes p. The following result concerning Euler products is very useful.

Lemma 2.13. *Let $g(n)$ be a multiplicative arithmetic function. Then the relation*

$$\sum_{n=1}^{\infty} g(n)n^{-s} = \prod_p (1 + g(p)p^{-s} + g(p^2)p^{-2s} + \dots)$$

holds in the sense that if for a given s at least one side is absolutely convergent then so is the other, and both sides have the same value.

Proof. Let the product be absolutely convergent, so that

$$S = \sum_p \sum_{m=1}^{\infty} |g(p^m)| p^{-m\sigma} < \infty.$$

Then, since each integer n has a unique representation as the product of primes,

$$\sum_{n=1}^{\infty} |g(n)| n^{-\sigma} \leq \prod_{p} (1 + |g(p)| p^{-\sigma} + \ldots) \leq \exp(S) < \infty.$$

This establishes the absolute convergence of the series. Moreover, it follows from the above remark that, for any $N > 0$,

$$\left| \sum_{n=1}^{\infty} g(n) n^{-s} - \prod_{p \leq N} (1 + g(p) p^{-s} + \ldots) \right| \leq 2 \sum_{n > N} |g(n)| n^{-\sigma}.$$

It is clear from this that by letting $N \to \infty$ we obtain the equality of the series and the product.

The proof of lemma (2.12) is now readily completed.

It is not difficult to show that if a function $f(s)$ is represented by a (convergent) Dirichlet series in some proper half-plane $\sigma > \sigma_0$ then there will be a further half-plane $\sigma > \sigma_1$ in which $f(s)$ does not vanish. The study of such half-planes is important in certain parts of analytic number theory, and the existence of an Euler product often helps to reduce the size of σ_1. Suppose, for example, that the series $\sum g(n) n^{-s}$ converges absolutely for $\sigma > \sigma_2$, and that the function $g(n)$ is multiplicative. Then from lemma (2.13) we see that for a large enough value of N we may assert, in the above notation, that the inequalities

$$\left| \prod_{p > N} (1 + g(p) p^{-s} + \ldots) \right| \geq \prod_{p > N} (1 - |g(p)| p^{-\sigma} - \ldots) \geq \exp(-S) > 0$$

hold over the half-plane $\sigma > \sigma_2$. Thus to discover if $f(s)$ may vanish for $\sigma > \sigma_2$ we need only examine the finitely many functions

$$1 + g(p) p^{-s} + \ldots, \qquad 2 \leq p \leq N,$$

which are, hopefully, simple.

We remark that equality may hold in the equation of lemma (2.13) even if neither side converges absolutely, but it is sometimes difficult to prove this.

Riemann Zeta Function

An important Dirichlet series is the case when all the a_n have the value 1:

$$\zeta(s) = \sum_{n=1}^{\infty} n^{-s}.$$

The sum function is called the *Riemann zeta function*. It can be analytically continued over the whole complex s-plane, where it is regular save for a simple pole, with residue 1, at the point $s = 1$. It satisfies the functional equation

$$\zeta(s) = 2^s \pi^{s-1} \sin \tfrac{1}{2} s\pi \cdot \Gamma(1 - s)\zeta(1 - s),$$

although this will not concern us here. It has an Euler product representation

$$\zeta(s) = \prod_p (1 + p^{-s} + p^{-2s} + \ldots) = \prod_p (1 - p^s)^{-1}$$

which, by lemma (2.13), is certainly valid if $\sigma > 1$. It follows from the remark just made that $\zeta(s) \neq 0$ in the half-plane $\sigma > 1$. The Riemann Hypothesis amounts to the assertion that $\zeta(s)$ has no zeros in the half-plane $\sigma > 1/2$.

Let $\Lambda(n)$ denote von Mangoldt's function:

$$\Lambda(n) = \begin{cases} \log p & \text{if } n = p^m, m = 1, 2, \ldots, \\ 0 & \text{otherwise.} \end{cases}$$

Then when $\sigma > 1$ we have

$$-\frac{\zeta'(s)}{\zeta(s)} = \sum_{n=1}^{\infty} \frac{\Lambda(n)}{n^s}.$$

This formula may be obtained by logarithmically differentiating the Euler product representation of $\zeta(s)$. Concerning this function we shall need the following estimates:

Lemma 2.14. *In the half-plane $\sigma > 1$ the estimates*

$$\left| \frac{\zeta'}{\zeta} (\sigma + i\tau) \right| \leq \frac{c_1}{|s - 1|} + c_2 \log(2 + |\tau|)$$

$$|\zeta(\sigma + i\tau)| + \left| \frac{1}{\zeta(\sigma + i\tau)} \right| \leq c_2 \log(2 + |\tau|) \qquad |\tau| \geq 1,$$

hold.

Proof. In Titchmarsh [4], Theorem 3.11 pp. 52–53, it is shown that for some $A > 0$ inequalities of these types hold in the larger region

$$\sigma > 1 - \frac{A}{\log(2 + |\tau|)},$$

provided that the point $s = 1$ is deleted from the domain.

It will be seen that in most of our applications we only need to know the behaviour of $\zeta(s)$ in the half-plane $\sigma > 1$ to the extent of the upper bound for $|\zeta(s)|$ given in lemma (2.14). Such an upper bound may be obtained as follows:

Integration by parts shows that for every $\sigma > 1$ and positive integer N

$$\zeta(s) - \sum_{n=1}^{N} n^{-s} = \int_{N+}^{\infty} y^{-s}\, d[y] = -N^{1-s} + s \int_{N}^{\infty} y^{-1-s}[y]dy$$

$$= \frac{N^{1-s}}{s-1} + s \int_{N}^{\infty} y^{-1-s}([y] - y)dy.$$

Hence

$$|\zeta(s)| \le \sum_{n=1}^{N} n^{-1} + \frac{1}{|s-1|} + |s| \int_{N}^{\infty} y^{-1-\sigma}\, dy$$

$$\le \log N + \frac{1}{|s-1|} + \frac{|s|}{\sigma} N^{-\sigma} + \text{constant}$$

and the desired result is obtained by choosing N suitably.

It is possible to view Dirichlet series from the prospect of the theory of probability. Let $a_1, a_2, \ldots$ be a sequence of complex numbers. For positive real numbers x define

$$A(x) = \sum_{n \le x} a_n.$$

Then we have (formally) the identity

$$f(s) = \sum_{n=1}^{\infty} a_n n^{-s} = \int_{1-}^{\infty} x^{-s}\, dA(x),$$

and we can regard $f(s)$ as a kind of characteristic function of $A(x)$. Of course, $|A(x)|$ need not be bounded as $x \to \infty$. The analogy with probability will be extended if we introduce the notion of Dirichlet multiplication, or convolution.

Let $b_1, b_2, \ldots$ be a further sequence of complex numbers. Then the Dirichlet convolution of the sequences $\{a_n\}$ and $\{b_n\}$ is the sequence $\{c_n\}$ which is defined by

$$c_k = \sum_{mn=k} a_m b_n \qquad (k = 1, 2, \ldots).$$

We now need

Lemma 2.15. *Let the series* $f(s) = \sum a_n n^{-s}$, $g(s) = \sum b_n n^{-s}$ *converge absolutely. Define the sequence* $\{c_k\}$ *in the above manner.*

Then the series $h(s) = \sum c_k k^{-s}$ *converges absolutely, and the relation*

$$h(s) = f(s)g(s)$$

is satisfied.

Proof. For any $N \geq 1$, making use of the definition of the c_k, we have

$$\left| \sum_{k \leq N} c_k k^{-s} - \left\{ \sum_{m \leq N} a_m m^{-s} \right\} \left\{ \sum_{n \leq N} b_n n^{-s} \right\} \right|$$

$$\leq \sum_{\substack{m \leq N,\, n \leq N \\ mn > N}} \sum |a_m b_n| m^{-\sigma} n^{-\sigma}$$

$$\leq \sum_{m > N^{1/2}} |a_m| m^{-\sigma} \sum_{n=1}^{\infty} |b_n| n^{-\sigma} + \sum_{m=1}^{\infty} |a_m| m^{-\sigma} \sum_{n > N^{1/2}} |b_n| n^{-\sigma}.$$

Since the series $\sum |a_m| m^{-\sigma}$, $\sum |b_n| n^{-\sigma}$ converge, we may let $N \to \infty$ to obtain the desired result.

This completes the proof of lemma (2.15).

Thus, if we define

$$B(x) = \sum_{n \leq x} b_n \qquad C(x) = \sum_{k \leq x} c_k,$$

we may view $C(x)$ as a 'convolution' of $A(x)$ and $B(x)$, and which has 'characteristic function' $f(s)g(s)$.

However, serious difficulties quickly arise. In most applications to number theory the coefficients a_n are complex. Even if they are not, any measure induced on the semi-infinite line-segment $[1, \infty)$ by the corresponding function $A(x)$ will usually be unbounded. Worse, the 'characteristic function' $f(s)$ may not be well defined for all complex values of s. Indeed it may be nowhere defined. For example, $f(s)$ will be defined for $\sigma > 1$ when

$$a_n = \begin{cases} 1 & \text{if } n \text{ is prime,} \\ 0 & \text{otherwise.} \end{cases}$$

In this case the function

$$f(s) = \sum p^{-s}$$

can be analytically continued into the half-plane $\sigma > 0$, but it has as a natural boundary the line $\sigma = 0$. (See, for example, Landau and Walfisz [1]).

If we set $a_n = n!$ then the series

$$\sum_{n=1}^{\infty} n!n^{-s}$$

does not converge for any complex value of s.

Thus we can expect to lose almost all of the advantages of the traditional theory of probability.

Nonetheless, by lowering his sights somewhat, Wiener was able to establish a valuable result concerning Dirichlet series, which has several probabilistic overtones. Although we shall not have need of it we state it here for comparison with certain related results which we shall prove in later chapters. We give it in two forms which together comprise the so-called *Wiener–Ikehara Tauberian Theorem.*

Lemma 2.16. *Let a_n, $(n = 1, 2, \ldots)$, be a sequence of non-negative real numbers. For real numbers $x > 0$ let*

$$A(x) = \sum_{n \le x} a_n.$$

Let the corresponding Dirichlet series $f(s) = \sum a_n n^{-s}$ have the following properties:

(i) *The integral in the representation*

$$f(s) = \int_{1-}^{\infty} x^{-s}\, dA(x)$$

converges for $\sigma > 1$.

(ii) *For each $\lambda > 0$*

$$\lim_{\sigma \to 1+} \left\{ f(s) - \frac{1}{s-1} \right\} \qquad s = \sigma + i\tau,$$

exists uniformly for $|\tau| \le \lambda$.

Then as $x \to \infty$ we have

$$x^{-1}A(x) \to 1.$$

Lemma 2.17. *Let the situation in Lemma* (2.16) *be in force. Let* b_n, $(n = 1, 2, \ldots)$, *be a sequence of complex numbers which satisfy* $|b_n| \leq Ca_n$ *for some constant* C, *and all positive integers* n. *Define the Dirichlet series*

$$g(s) = \sum_{n=1}^{\infty} b_n n^{-s},$$

and define

$$B(x) = \sum_{n \leq x} b_n.$$

Let the hypothesis (i) *of lemma* (2.16) *be satisfied with* $A(x)$ *replaced by* $B(x)$, *and* $f(s)$ *by* $g(s)$. *Corresponding to hypothesis* (ii) *of that lemma let*

$$\lim_{\sigma \to 1+} \left\{ g(s) - \frac{\alpha}{s-1} \right\}$$

exist uniformly on each τ-*interval* $[-\lambda, \lambda]$. *Here the value of the constant* α *may be zero.*

Then as $x \to \infty$ *we have*

$$x^{-1}B(x) \to \alpha.$$

Detailed proofs of these two results may be found in Lang, ref. [1], Chapter 8, pp. 116–123. It is interesting to compare the detailed proof of lemma (2.16), which is given in that reference, to the proof of lemma (1.47) which was given in Chapter one. The essential features of both proofs are the judicious use of non-decreasing functions, and of the Fejér kernel; and they are built upon very similar lines.

Underlying the Berry–Esseen theorem, as embodied in lemma (1.47), is the notion that the behaviour of a distribution function $F(z)$ as $|z| \to \infty$, is directly related to the behaviour of its characteristic function $\phi(t)$ as $t \to 0$. (See, for example, lemma (1.12) of Chapter one). Lemma (2.16) has the same form, save that the rôle of $F(z)$ is played by $A(z)$, and that of the characteristic function by $f(s)$, $(s \to 1)$.

An advantage of the Wiener–Ikehara tauberian theorem is that the a_n need not have many *a priori* properties. It seems fair to view lemma (2.16) as more basic than lemma (2.17). Indeed, the second of these two results may be obtained directly from the first by considering the series $\sum c_n n^{-s}$ with

$$c_n = \{a_n - Re(b_n/2C)\}/\{1 - Re(\alpha/2C)\},$$

and so on.

We shall need the following tauberian theorem.

Lemma 2.18 (Hardy and Littlewood). *Let $w(t)$ be a real-valued function, defined and of bounded variation on each finite interval $[0, t]$ with $t > 0$, and satisfying $w(0) = 0$. Let the following two properties hold:*

(i) *The integral in the representation*

$$I(y) = \int_0^\infty e^{-yt}\, dw(t)$$

converges for all real y, $y > 0$. Moreover, if $y > x$, $x \to \infty$ and $y/x \to 1$, then

$$\liminf_{x, y \to \infty}(w(y) - w(x)) \geq 0.$$

(ii) *There is a finite number l so that as $y \to 0+$*

$$I(y) \to l.$$

Then as $t \to \infty$ we have

$$w(t) \to l.$$

Remarks. The second requirement in (i) is that $w(t)$ be a so-called *slowly decreasing function* of t as $t \to \infty$.

In his volume on Divergent Series [2], Hardy considers a large number of tauberian theorems as being under the same roof. In particular he discusses Wiener's tauberian theorem, and certain theorems related to lemma (2.18).

The Wiener–Ikehara theorem may be put into a form more clearly related to that of the Hardy–Littlewood tauberian theorem. Notice that in the latter result one needs only to know the behaviour of $I(y)$ for *real* values of y. We can reconcile this with the circumstances of the Wiener–Ikehara theorem by noting that in order to apply the Hardy–Littlewood theorem we should need the function $x^{-1}A(x)$ to be slowly decreasing as $x \to \infty$. Thus lemma (2.18) does not supersede lemma (2.17).

Proof: This result appears as Theorem 105 in Hardy [2], Chapter VII, p. 164. There is a discussion of what is called *slowly decreasing functions* in §6.2, pp. 124–125 of the same reference.

Since we make use of this result a few times in the following chapters, we sketch a proof. It is built in three steps.

Assume first that $z(t)$ is a non-decreasing function of t for $t > 0$, and that as $y \to 0+$

$$y \int_0^\infty e^{-yt}\, dz(t) \to C.$$

Thus $C \geq 0$ must hold. We assert that as $t \to \infty$, $z(t)t^{-1} \to C$.

Indeed, for each non-negative integer m

$$\lim_{y \to 0} y \int_0^\infty e^{-yt} \cdot e^{-myt}\, dz(t) = \frac{C}{m+1} = C \int_0^\infty e^{t} \cdot e^{-mt}\, dt.$$

Hence the relation

$$\lim_{y \to 0} y \int_0^\infty e^{-yt} g(e^{-yt}) dz(t) = C \int_0^\infty e^{-t} g(e^{-t}) dt$$

holds for every polynomial $g(x)$.

Define the function

$$h(x) = \begin{cases} 0 & \text{if } 0 \leq x < \dfrac{1}{e}, \\ \dfrac{1}{x} & \text{if } \dfrac{1}{e} \leq x \leq 1. \end{cases}$$

Let ε be a positive number. Then there are polynomials $g_j(x), j = 1, 2$, which satisfy $g_1(x) \leq h(x) \leq g_2(x)$ uniformly for $0 \leq x \leq 1$, and for which

$$\int_0^1 (g_2(x) - g_1(x)) dx < \varepsilon.$$

In terms of these polynomials

$$\limsup_{y \to 0} y \int_0^1 e^{-yt} h(e^{-yt}) dz(t) \leq \limsup_{y \to 0} y \int_0^\infty e^{-yt} g_2(e^{-yt}) dz(t)$$

$$= C \int_0^\infty e^{-t} g_2(e^{-t}) dt \leq C \int_0^\infty e^{-t} h(e^{-t}) dt + C \int_0^\infty e^{-t} \{g_2(e^{-t}) - g_1(e^{-t})\} dt,$$

so that

$$\limsup_{y \to 0} y \int_0^{1/y} dz(t) \leq C(1 + \varepsilon).$$

A similar result may be obtained involving 'lim inf,' and in tne other direction. Since ε may be chosen arbitrarily small, and $y^{-1} \to \infty$ as $y \to 0+$, $z(t)t^{-1} \to C$ as $t \to \infty$, as was asserted. This completes the first step.

As the second step let $r(t)$ be of bounded variation in each finite interval, satisfy $r(t) \geq -Ht^{-1}$ for some positive constant H and all $t > 0$, and for any fixed $\varepsilon > 0$ be $O(\exp(\varepsilon t))$ as $t \to \infty$. Assume further that

$$\lim_{y \to 0} \int_0^\infty r(t)e^{-yt}\, dt = 0.$$

We assert that

$$\lim_{T \to \infty} \int_0^T r(t)dt = 0.$$

The function

$$\gamma(t) = Ht + \int_0^t ur(u)du$$

is non-decreasing, and we have

$$\int_0^\infty e^{-yt}\, d\gamma(t) = Hy^{-1} + \int_0^\infty te^{-yt}r(t)dt.$$

Let

$$f(y) = \int_0^\infty e^{-yt}r(t)dt.$$

According to our (temporary) hypothesis $f(y) \to 0$ as $y \to 0+$, and

$$f''(y) = \int_0^\infty t^2 e^{-yt}r(t)dt \geq -H\int_0^\infty te^{-yt}\, dt = -Hy^{-2}.$$

We next prove that $yf'(y) \to 0$ as $y \to 0+$.

If v is a positive real number, $0 < v < y$, then by Taylor's theorem

$$f(y+v) = f(y) + \frac{v}{1!}f'(y) + \frac{v^2}{2!}f''(y+\theta v)$$

for some $|\theta| \leq 1$. Choosing $v = \delta y$ with δ fixed, $0 < \delta \leq 1/2$, we see that

$$\begin{aligned}\limsup_{y \to 0+} yf'(y) &\leq \limsup_{y \to 0+} \left\{\frac{f(y(1+\delta)) - f(y)}{\delta} + \frac{\delta y^2}{2!}\frac{H}{(y(1-\delta))^2}\right\}\\ &\leq 2\delta H.\end{aligned}$$

A similar result concerning the function $-yf'(y)$ may be obtained by replacing v with $-v$. Since δ may be chosen arbitrarily close to zero we have $yf'(y) \to 0$, or what is the same thing,

$$y\int_0^\infty te^{-yt}r(t)dt \to 0 \qquad y \to 0+.$$

Hence

$$y\int_0^\infty e^{-yt}\,d\gamma(t) \to H \qquad y \to 0+.$$

The function $\gamma(t)$ is non-decreasing so we may apply the result of the first step to deduce that $\gamma(t)t^{-1} \to H$ as $t \to \infty$, and therefore

$$\lim_{t\to\infty} \frac{1}{t}\int_0^t ur(u)du = 0.$$

This is not quite what we want.

Define the functions

$$\lambda(t) = \int_0^t (u+1)r(u)du \qquad \rho(t) = \int_0^t ur(u)du.$$

Let u_1, u_2 be continuity points of $r(u)$. An integration by parts shows that

$$\begin{aligned}\lambda(u_2) - \lambda(u_1) &= \int_{u_1}^{u_2} \frac{u+1}{u}\,d\rho(u) \\ &= \left(1 + \frac{1}{u_2}\right)\rho(u_2) - \left(1 + \frac{1}{u_1}\right)\rho(u_1) + \int_{u_1}^{u_2} \frac{\rho(u)}{u^2}\,du.\end{aligned}$$

Fixing u_1 and letting u_2 become unbounded we see that $\lambda(t) = o(t)$ as $t \to \infty$. This result is first obtained by restricting the values of t to continuity points of $r(t)$. Since $r(t)$ has bounded variation it is almost surely differentiable (and so continuous). Its continuity points are therefore everywhere dense in the half-line $t > 0$. The function

$$\eta(t) = \int_1^t \left\{r(u) + \frac{H}{u}\right\}(u+1)du$$

satisfies $\eta(t) \sim Ht$ as $t \to \infty$ through these continuity points. Moreover, $\eta(t)$ is non-decreasing for $t \geq 1$. Thus $\eta(t) \sim Ht$ as $t \to \infty$ with otherwise no restriction upon the values of t. Hence $\lambda(t)t^{-1} \to 0$ as $t \to \infty$ on the same terms.

Integrating by parts we have

$$\int_0^\infty r(t)e^{-yt}\,dt = \int_0^\infty \frac{e^{-yt}}{t+1}\,d\lambda(t)$$
$$= y\int_0^\infty \frac{\lambda(t)}{t+1}\cdot e^{-yt}\,dt + \int_0^\infty \frac{\lambda(t)}{(t+1)^2}\cdot e^{-yt}\,dt.$$

As $y \to 0+$ the first of these four integrals is by hypothesis $o(1)$; the third is also $o(1)$ since $\lambda(t)t^{-1} \to 0$ as $t \to \infty$. Hence

$$\int_0^\infty \frac{\lambda(t)}{(t+1)^2}\,e^{-yt}\,dt \to 0 \qquad y \to 0+.$$

Therefore, as $y \to 0+$

$$\int_0^{1/y} \frac{\lambda(t)}{(t+1)^2}\,dt = \int_0^{1/y} \frac{\lambda(t)}{(t+1)^2}\cdot(1-e^{-yt})dt - \int_{1/y}^\infty \frac{\lambda(t)}{(t+1)^2}\,e^{-yt}\,dt + o(1).$$

For any fixed $\varepsilon > 0$ we have $|\lambda(t)| \leq \varepsilon t$ for all $t \geq N$, say. Hence

$$\limsup_{y\to 0}\left|\int_0^{1/y} \frac{\lambda(t)}{(t+1)^2}\,dt\right| \leq \limsup_{y\to 0}\int_0^N \frac{|\lambda(t)|}{(t+1)^2}\,yt\,dt$$
$$+ \limsup_{y\to 0} y\varepsilon\int_N^{1/y}\left(\frac{t}{t+1}\right)^2 dt + \limsup_{y\to 0}\int_{1/y}^\infty \frac{\varepsilon}{t}\,e^{-yt}\,dt \leq 2\varepsilon.$$

Since ε may be chosen arbitrarily small

$$\lim_{y\to 0}\int_0^{1/y} r(u)du = \lim_{y\to 0}\int_0^{1/y}\frac{d\lambda(u)}{u+1} = \lim_{y\to 0}\int_0^{1/y}\frac{\lambda(u)}{(u+1)^2}\,du = 0.$$

This completes our second step.

For our third step let

$$I(y) = \int_0^\infty e^{-yt}\,dw(t).$$

A translation of $w(t)$ by a suitable step function, if necessary, ensures that without loss of generality $w(t) = 0$ for $0 \leq t \leq 1$, and that $l = 0$. Hence we may assume that $I(y) \to 0$ as $y \to 0+$.

If $0 < v < x < \infty$ then

$$\int_v^x \frac{I(u)}{u}\,du = \int_v^x \frac{du}{u}\int_0^\infty e^{-ut}\,dw(t) = \int_0^\infty dw(t)\int_v^x \frac{e^{-ut}}{u}\,du$$
$$= -\int_0^\infty w(t)\left\{\frac{d}{dt}\int_v^x \frac{e^{-ut}}{u}\,du\right\}dt$$
$$= \int_0^\infty w(t)\int_v^x e^{-ut}\,du\,dt = \int_0^\infty \frac{e^{-vt}-e^{-xt}}{t}\,w(t)dt.$$

If $w(t)$ is of bounded variation on each finite interval these inversions may be justified by applying Lebesgue's dominated convergence theorem, noting that the hypothesis that $w(t)$ be slowly decreasing and that $I(y)$ converges for each $y > 0$ ensures that, for each fixed $\varepsilon > 0$, $w(t) = O(\exp(\varepsilon t))$ as $t \to \infty$.

In particular, for any (fixed) $p > 1$, setting $v = y/p$, $x = y$ we see that as $y \to 0$,

$$\int_0^\infty \frac{w(pt)-w(t)}{t}\,e^{-yt}\,dt = \int_0^\infty \frac{e^{-yt}}{t}\,w(pt)dt - \int_0^\infty \frac{e^{-yt}}{t}\,w(t)dt$$
$$= \int_0^\infty \frac{e^{-yt/p}}{t}\,w(t)dt - \int_0^\infty \frac{e^{-yt}}{t}\,w(t)dt = \int_{y/p}^y \frac{I(u)}{u}\,du = o(1).$$

We are now in a position to apply the result of step two with

$$r(u) = \{w(pu) - w(u)\}/u.$$

We deduce that

$$\lim_{T\to\infty}\int_0^T \frac{w(pt)-w(t)}{t}\,dt = 0$$

so that

$$\lim_{T\to\infty}\int_T^{pT} \frac{w(u)}{u}\,du = 0.$$

Assume now that $w(T_k) \geq M > 0$ for some positive number M and a sequence of values T_k unbounded with k. Choosing $p > 1$ so that the inequality $w(u) - w(T) \geq -M/2$ uniformly for $T \leq u \leq pT$ is assured by the slowly-decreasing property of $w(u)$, we see that for all sufficiently large values of k

$$\int_{T_k}^{pT_k} \frac{w(u)}{u}\,du \geq \frac{M}{2}\int_{T_k}^{pT_k} \frac{du}{u} = \frac{M}{2}\log p > 0.$$

This contradicts the previous result. Hence

$$\limsup_{T\to\infty} w(T) \leq 0.$$

Likewise $\liminf w(T) \geq 0$, and, finally, $w(T) \to 0$ as $T \to \infty$.

This completes our third step, and the proof of lemma (2.18).

There is a useful converse to lemma (2.15). We shall indicate it with an example.

For each integer $k \geq 2$, let $d_k(n)$ denote the number of ways of expressing the positive integer n as a product of k integral factors, $n = y_1 \ldots y_k$. When $k = 2$ we set $d_2(n) = d(n)$, the Dirichlet divisor function.

It is not difficult to establish the following identities, valid in the half-plane $\sigma > 1$:

$$\sum_{n=1}^{\infty} \frac{d(n)^2}{n^s} = \frac{\zeta^4(s)}{\zeta(2s)}, \quad \sum_{n=1}^{\infty} \frac{|\mu(n)|}{n^s} = \frac{\zeta(s)}{\zeta(2s)}, \quad \sum_{n=1}^{\infty} \frac{d_3(n)}{n^s} = \zeta^3(s).$$

From the identity

$$\frac{\zeta^4(s)}{\zeta(2s)} = \zeta^3(s)\cdot\frac{\zeta(s)}{\zeta(2s)},$$

and the fact that there is at most one representation by a Dirichlet series in any half-plane $\sigma > \sigma_0$, we deduce the identity

$$d^2(n) = \sum_{mr=n} d_3(m)|\mu(r)|.$$

By means of this identity it is straightforward to obtain the estimate

$$\sum_{n\leq x} d^2(n) = \sum_{r\leq x} |\mu(r)| \sum_{m\leq x/r} d_3(m)$$

$$= \frac{x}{\pi^2}(\log x)^3 + O(x(\log x)^2).$$

For details regarding the above identities we refer to Hardy and Wright [1] Chapter XVII, pp. 244–259. See, also, the same reference, Chapter XVIII, §(18.2), pp. 263–266.

Another example of such an argument occurs sufficiently often to warrant its inclusion as a lemma.

Lemma 2.19. *Let the arithmetic function $f(n)$ be the Dirichlet convolution of the arithmetic functions $g(n)$ and $h(n)$. Let the finite limit*

$$A = \lim_{x \to \infty} (xL(x))^{-1} \sum_{n \le x} h(n)$$

exist, where $L(x)$ is a slowly oscillating function of x, which satisfies $|L(x)| = 1$ for all sufficiently large values of x. Let the series $\sum g(n)n^{-1}$ converge absolutely.

Then, as $x \to \infty$,

$$(xL(x))^{-1} \sum_{n \le x} f(n) \to A \sum_{n=1}^{\infty} \frac{g(n)}{n}.$$

Proof. Let N, x be positive real numbers, exceeding two in value. Then we have

$$x^{-1} \sum_{n \le x} f(n) = x^{-1} \sum_{\substack{mr \le x \\ m \le N}} g(m)h(r) + x^{-1} \sum_{\substack{mr \le x \\ m > N}} g(m)h(r).$$

If we fix N, then, as $x \to \infty$, the first of the two sums which appear on the right-hand side may be estimated by

$$\sum_{m \le N} g(m)m^{-1} \cdot \frac{m}{x} \sum_{r \le x/m} h(r) = AL(x) \sum_{m \le N} g(m)m^{-1} + o(1),$$

since $L(x/m) = L(x) + o(1)$ for each fixed positive integer m.

Moreover, the second sum does not exceed (in absolute value)

$$\sum_{N < m \le x} |g(m)|^{-1} \left| \frac{m}{x} \sum_{r \le x/m} h(r) \right| \le c \sum_{m > N} |g(m)| m^{-1}$$

for a certain constant c, which does not depend upon N.

Thus

$$\limsup_{x \to \infty} \left| (xL(x))^{-1} \sum_{n \le x} f(n) - A \sum_{m=1}^{\infty} g(m)m^{-1} \right| \le (|A| + c) \sum_{m > N} |g(m)| m^{-1},$$

and since N may be chosen arbitrarily large, lemma (2.19) is established.

In a later chapter we shall discuss the global distribution of the quadratic class number. For use in that chapter we shall need a well-known identity of Dirichlet.

Let d be an integer which is not a square, and which satisfies one of the conditions $d \equiv 0 \pmod 4$ or $d \equiv 1 \pmod 4$. Corresponding to such integers d consider

$$\chi(n) = \left(\frac{d}{n}\right),$$

where the symbol on the right-hand side is the so-called *Kronecker symbol.* For positive integers n this symbol is defined as follows

$$\left(\frac{d}{p}\right) = 0 \quad \text{if } p \mid d,\ p \text{ prime},$$

$$\left(\frac{d}{2}\right) = \begin{cases} 1 & \text{if } d \equiv 1 \pmod 8, \\ -1 & \text{if } d \equiv 5 \pmod 8, \end{cases}$$

$$\left(\frac{d}{p}\right) \text{ is the Legendre symbol if } p > 2 \text{ and } p \nmid d,$$

$$\left(\frac{d}{n}\right) = \prod_{j=1}^{k} \left(\frac{d}{p_j}\right) \quad \text{if } n = p_1 \cdots p_k,$$

$$\left(\frac{d}{1}\right) = 1.$$

It is not difficult to see that when $|d| > 1$ $\chi(n)$ is a non-principal character mod $(|d|)$. Moreover, this character is primitive if and only if d is the product of relatively prime factors of the form

$$-4,\ 8,\ -8,\ (-1)^{(p-1)/2}p \quad (p > 2),$$

and then $\chi(n)$ will be primitive (mod $|d|$).

In terms of the (not necessarily primitive) character $\chi(n)$ we define the Dirichlet series

$$L(s, \chi) = \sum_{n=1}^{\infty} \chi(n) n^{-s}.$$

This series converges in the half-plane $\sigma > 0$.

The numbers d for which $\chi(n)$ is a primitive character (mod $|d|$) are precisely those numbers which occur as *discriminants* of quadratic fields, or as *fundamental discriminants* in the theory of binary quadratic forms. We shall only consider those d which are negative.

If N is a squarefree integer, negative and $N \equiv 2$ or 3 (mod 4), then 1 and $N^{1/2}$ form an integral basis for the field generated over the rational numbers by $N^{1/2}$. In this case the discriminant is $4N$ $(=d)$.

If N is squarefree, negative and $N \equiv 1$ (mod 4), then 1 and $(1 + N^{1/2})/2$ form an integral basis for the field generated by $N^{1/2}$. In this case the discriminant is N $(=d)$.

For such integers d let $h(d)$ denote the number of ideal classes in the field $Q(N^{1/2})$.

Define the number $w = w(d)$ by

$$w = \begin{cases} 2 & \text{if } d \leq -4, \\ 4 & \text{if } d = -4, \\ 6 & \text{if } d = -3. \end{cases}$$

Lemma 2.20. *If d is a negative discriminant of a quadratic field, then*

$$h(d) = \frac{w|d|^{1/2}}{2\pi} L(1, \chi)$$

where χ is defined in the above manner.

Proof. An outline of the proof of this result, together with a careful discussion of the connections between binary quadratic forms and quadratic extensions of the rational number field, may be found in Davenport [2] Chapter 5, pp. 37–44, and Chapter 6, pp. 45–55. A fully detailed account, in particular the consideration of the Kronecker symbol, may be found in Landau [3], [4].

Concluding Remarks

The composer Felix Mendelssohn had a sister Fanny, of whom he was extremely fond. She was herself very talented, an excellent pianist, and wrote songs which were indistinguishable from those of Felix. His biographers devote most of their considerations to this relationship. He had a second sister, Rebecka, scarcely mentioned. She, too, was unusual; she married the mathematician Gustav Peter Lejeune-Dirichlet. It seems certain that these two men would have met at one of the many Mendelssohn soirées. One wonders what they talked about.

For a life of F. Mendelssohn see Marek [1]. On page 193 we read "... he received the joyful news that Rebecka had become engaged to a professor of mathematics ..."

But justice is had in the obituary to Dirichlet which was given by E. E. Kummer, and which appears as pp. 311–344 of Dirichlet's collected works

[1]. On page 324 we discover, in one sentence, that Dirichlet married Rebecka Mendelssohn-Bartholdy, the grandchild of Moses Mendelssohn, whose house is distinguished in intellect and art.

For an elementary discussion of Dirichlet series see Hardy and Wright [1], Chapter XVII.

It is clear that if $a_n \geq 0$, $(n = 1, 2, \ldots)$, and if the series $\sum a_n n^{-c}$ converges, then the Dirichlet series

$$f(s) = \sum_{n=1}^{\infty} a_n n^{-s}$$

is well-defined, and regular, in the half-plane $\sigma > c$. As a further example of the analogy between such Dirichlet series and characteristic functions, we shall indicate a characteristic function proof that the converse is true. We shall sketch a proof that *if $f(s)$ is representable by the above series in some half-plane $\sigma > \sigma_0$, and if $f(s)$ is regular in the half-plane $\sigma > c$ $(c < \sigma_0)$, then the series which defines $f(s)$ actually converges in the same half-plane $\sigma > c$.*

By effecting the change of variable $x = e^u$ in the representation of $f(s)$ in lemma (2.16) we see that

$$\begin{aligned} f(s) &= \int_{-\infty}^{\infty} e^{-us}\, d\{A(e^u)\} \\ &= \int_{-\infty}^{\infty} e^{-uit} \cdot e^{-u\sigma}\, d\{A(e^u)\}. \end{aligned}$$

This last integral is in every respect a characteristic function, save that the integral

$$\int_{-\infty}^{\infty} e^{-u\sigma}\, d\{A(e^u)\}$$

is only known to be finite if $\sigma > \sigma_0$.

One can now follow the proof of Raikov's theorem concerning the possible analytic continuation of characteristic functions. (See Linnik [4], Theorem 3.1.2., pp. 51–53). According to this result, if the characteristic function

$$\phi(t) = \int_{-\infty}^{\infty} e^{itz}\, dF(z)$$

has an analytic continuation into the disc $|t| < r$, (t complex), then as $z \to \infty$

$$1 - F(z) + F(-z) = O(e^{-\beta z})$$

holds for every fixed $\beta < r$. In fact, for such values of β the integral

$$\int_{-\infty}^{\infty} e^{\beta|z|}\, dF(z)$$

exists, and $\phi(t)$ is continuable into the strip $|Im(t)| < r$.

An exactly similar proof may be applied to our present circumstances. Let $c < \sigma_0$ hold, and let $f(s)$ be regular in the half-plane $\sigma > c$. Let ε be a positive real number. Then, considered as a function of the (now) *complex* variable τ, the function

$$f(\sigma_0 + \varepsilon + i\tau) = \int_{-\infty}^{\infty} e^{-i\tau u} \cdot e^{-|u|(\sigma_0+\varepsilon)}\, dA(e^u)$$

may be analytically continued into the strip $|\mathrm{Im}(\tau)| < \sigma_0 - c$. By following Raikov's argument we see that the integral

$$\int_{-\infty}^{\infty} e^{|u|(\sigma_0 - c - \varepsilon) - |u|(\sigma_0+\varepsilon)}\, dA(e^u) = \int_{-\infty}^{\infty} e^{-|u|(c+2\varepsilon)}\, dA(e^u)$$

exists. Note that $A(e^u) = 0$ if $u < 0$.

Therefore, the integral

$$\int_{1-}^{\infty} x^{-\sigma}\, dA(x)$$

exists if $\sigma \geq c + 2\varepsilon$. Hence, if $\sigma > c$ the series

$$\sum_{n=1}^{\infty} a_n n^{-\sigma}$$

converges, and this is what we wished to prove.

A characteristic function $\psi(t)$ is called a *component* of the characteristic function $\phi(t)$ if there is a further characteristic function $\eta(t)$, say, so that the relation $\phi(t) = \psi(t)\eta(t)$ is satisfied for all real values of t. Two distribution functions (or laws) are defined to be components one of the other if and only if their corresponding characteristic functions are. These are standard definitions from the theory of probability, although they are usually given in the reverse order. It is difficult (and largely an unsolved problem) to determine the components of almost any given characteristic function; See, for example, Linnik [4].

To some extent the existence of an Euler product may be viewed as the decomposition of a Dirichlet series into the product of simpler components. For example

$$\zeta(s) = \prod_p (1 + p^{-s} + p^{-2s} + \ldots)$$

where each series

$$\sum_{m=1}^{\infty} p^{-ms}$$

has non-negative coefficients. As may be seen in Chapters 6 and 9 the success of Halász' analytic treatment of Dirichlet series $f(s)$ depends upon the existence of a suitable Dirichlet series representation for $f'(s)/f(s)$. In the applications which we shall consider an Euler product is always available. However, it would be feasible to employ other representations which are not Euler products, say

$$f(s) = \prod_{j=1}^{\infty} f_j(s),$$

$$\frac{f'(s)}{f(s)} = \sum_{j=1}^{\infty} \frac{f'_j(s)}{f_j(s)}$$

provided that the individual $f_j(s)$ were 'simple' enough.

The possible representation of Dedekind (and other) Dirichlet series in terms of simpler (component) Dirichlet series is in an even more hopeless state. For a survey of some results in this direction see Heilbronn [1], Theorems 6 and 7, pp. 217–226. We note here only the example

$$\zeta_K(s) = \zeta(s) \prod_{\chi \neq \chi_0 (\text{mod } k)} L(s, \chi)$$

where k is an odd (rational) prime, K is the cyclotomic field generated over the rational field by the kth root of unity $\exp(2\pi i/k)$, and the product runs over the non-principal Dirichlet characters (mod k). As is usual

$$\zeta_K(s) = \sum_{\mathfrak{A}} N(\mathfrak{A})^{-s},$$

the norm $N(\ \ )$ being taken from K down to the rational field. Here the 'components' $L(s, \chi)$ do not have non-negative coefficients but are 'simpler' in the sense that they are defined in terms of concepts involving only rational numbers, and do not involve an understanding of the field K.

Chapter 3

Finite Probability Spaces

In this chapter we study certain finite probability spaces, paying particular attention to a model of Kubilius, of which we shall make much use in later chapters.

We begin with a very simple, but nonetheless useful, model.

Let x be a real number, $x \geq 2$. Let $q_1 \leq q_2 \leq \ldots \leq q_k$ be (not necessarily distinct) prime numbers, and let Q denote their product.

Denote by $E(x, l)$ the set of positive integers not exceeding x which satisfy the relation $n \equiv l \pmod{Q}$. For values of l which are distinct (mod Q) these sets will be disjoint.

Let $\mathfrak{A}$ denote the least algebra of sets which contains the sets $E(x, l)$, where l is allowed to assume any value in the range $1 \leq l \leq Q$. On this algebra we define a simple frequency measure, typically by

$$\nu E = [x]^{-1} \sum_{n \leq x}' 1 = [x]^{-1} |E|$$

where $[x]$ denotes the greatest integer not exceeding x, and $'$ indicates that summation is confined to those integers which belong to the set E. If S denotes the set of all positive integers not exceeding x, then $\nu S = 1$ and the pair

$$(\mathfrak{A}, \nu)$$

forms a finite probability space.

This particular model is useful because the values of many arithmetic functions depend only upon the residue class (mod Q) to which they belong. To take advantage of this fact we define a second measure μ, on $\mathfrak{A}$, by

$$\mu \bigcup_{j=1}^{s} E(x, l_j) = \frac{s}{Q},$$

where $l_1, \ldots, l_s$ are distinct (mod Q). In particular, $\mu S = 1$.

The measures ν and μ approximate each other to the following extent: if

$$E = \bigcup_{j=1}^{s} E(x, l_j)$$

then

$$\nu E = [x]^{-1} \sum_{j=1}^{s} \left\{ \left[\frac{x - l_j}{Q} \right] + 1 \right\} = \mu E + \theta s [x]^{-1},$$

where $|\theta| \le 1$. The estimate

$$|\nu E - \mu E| < 2Qx^{-1}$$

therefore holds uniformly on the algebra $\mathfrak{A}$. This enables us to transfer certain problems from the space $(\mathfrak{A}, \nu)$ to the space $(\mathfrak{A}, \mu)$.

As an example in the use of this construction, consider the particular model where the primes q_i run exactly once through the odd prime numbers not exceeding $\frac{1}{2} \log x$. If x is sufficiently large, then according to lemma (2.9), we shall have

$$Q = \exp\left(\sum_{2 < p \le (\log x)/2} \log p \right) \le \exp(\tfrac{3}{4} \log x) = x^{3/4}.$$

In terms of the Legendre symbol define the functions

$$Y_q = \left(\frac{n}{q} \right) \qquad (n = 1, 2, \ldots, [x]),$$

one such function for each odd prime q not exceeding $\frac{1}{2} \log x$. These functions are random variables on the space $(\mathfrak{A}, \mu)$. With respect to the measure μ

$$Y_q = \begin{cases} 1 & \text{with probability } \dfrac{1}{2}\left(1 - \dfrac{1}{q}\right), \\ -1 & \text{with probability } \dfrac{1}{2}\left(1 - \dfrac{1}{q}\right), \\ 0 & \text{with probability } \dfrac{1}{q}. \end{cases}$$

and for varying values of q these variables are easily seen to be independent. Each of them has mean zero, whilst

$$\text{Expect}\,|Y_q|^2 = \text{Expect}\,|Y_q|^3 = 1 - \frac{1}{q}.$$

We apply the Berry–Esseen theorem, (lemma (1.48)), and obtain, uniformly for $2 \le H \le \frac{1}{2}\log x$,

$$\mu\left(\sum_{2<q\le H} Y_q \le z\sigma\right) = G(z) + O(\pi(H)^{-1/2})$$

where

$$\sigma^2 = \sum_{2<q\le H}\left(1 - \frac{1}{q}\right),$$

and

$$G(z) = \frac{1}{\sqrt{(2\pi)}}\int_{-\infty}^{z} e^{-w^2/2}\,dw.$$

It is now straightforward to adjust this estimate to prove that, uniformly for $2 < H \le \frac{1}{2}\log x$, and all real numbers z,

$$\nu_x\left(n;\ \sum_{2<p\le H}\left(\frac{n}{p}\right) \le z(\pi(H))^{1/2}\right) = G(z) + O((\pi(H))^{-1/2}).$$

In Chapter 22 we shall return to this simple model in order to study the distribution of the values of the quadratic class number.

Lemma 3.1. *Let r and x be real numbers, $2 \le r \le x$. Let N be a positive integer. Define the additive function*

$$g(n) = \sum_{\substack{p^k \| n,\ p \le r \\ k<N}} f(p^k),$$

where the $f(p^k)$ assume real values. Define independent random variables X_p, one for each prime p not exceeding r, by

$$X_p = \begin{cases} f(p^j) & \text{with probability } \left(1 - \dfrac{1}{p}\right)\dfrac{1}{p^j},\ j = 1, \dots, N-1, \\[2ex] 0 & \text{with probability } 1 - \dfrac{1}{p} + \dfrac{1}{p^N}. \end{cases}$$

Then the estimate

$$\nu_x(n; g(n) \le z) = P\left(\sum_{p\le r} X_p \le z\right) + O(e^{4Nr}x^{-1})$$

holds uniformly for all real numbers $f(p^k)$, z, x (≥ 2), and r $(2 \le r \le x)$.

Proof. We consider the above probability models $(\mathfrak{A}, \nu)$ and $(\mathfrak{A}, \mu)$, where the q_i now run through all the prime numbers not exceeding r, each taken with multiplicity N.

The variable X_p is defined on those integers n, not exceeding x, and belonging to the residue class $l \pmod{Q}$, by

$$X_p(n) = \begin{cases} f(p^j) & \text{if } p^j \| l,\ 0 \le j \le N-1, \\ 0 & \text{if } p^N \mid l. \end{cases}$$

For differing values of p these variables are easily seen to be independent with respect to the measure μ. This is a direct consequence of the fact that an integer can be divisible by each of a number of distinct primes if and only if it is divisible by all of them. Moreover, for each positive integer n,

$$\sum_{p \le r} X_p(n) = \sum_{p^k \| n,\ p \le r,\ k < N} f(p^k).$$

According to the approximation of ν by μ, the result of lemma (3.1) will certainly be valid if we can prove that in our present circumstances $Q \le e^{4Nr}$. We shall do this directly, by means of an argument of Erdös (see Hardy and Wright [1], Chapter XXII, Theorem 415.).

Let m be an even positive integer. The integral binomial coefficient

$$\binom{m}{\frac{m}{2}} = \frac{m!}{\left(\frac{m}{2}!\right)^2}$$

is divisible by every prime p in the range $\frac{1}{2}m < p \le m$, since such a prime will divide the numerator $m!$, but not the denominator. Moreover, this binomial coefficient certainly does not exceed 2^m in size. Therefore

$$\sum_{(1/2)m < p \le m} \log p \le \log \binom{m}{\frac{m}{2}} \le m \log 2 < m.$$

If m is a power of two then we may replace m by $m/2$, $m/2^2, \ldots,$ and so on, and add, to obtain

$$\sum_{p \le m} \log p < 2m$$

Since $r \geq 2$ there is a positive integer k so that $2^k \leq r < 2^{k+1}$. Therefore

$$Q = \exp\left(N \sum_{p \leq r} \log p\right) \leq \exp\left(N \sum_{p \leq 2^{k+1}} \log p\right) < \exp(N2^{k+2})$$
$$\leq \exp(4Nr).$$

This completes the proof of lemma (3.1).

Remarks. The result of this lemma is clearly valid if the condition $g(n) \leq z$ is replaced by the requirement that $g(n)$ belong to some given set.

If one makes use of the prime number theorem in the above proof, the coefficient 4 in e^{4r} may essentially be replaced by any fixed number greater than one.

This simple lemma will not suffice for the study of some of the more subtle problems connected with the value-distribution theory of arithmetic functions. For use in certain of these we introduce a model of Kubilius, which is established by means of an application of the Selberg sieve method.

The Model of Kubilius

Once again let r and x be real numbers, subject only to the restriction that $2 \leq r \leq x$. Let D denote the product of the primes not exceeding r, and let S denote the set of positive integers which do not exceed x.

For each prime p which divides D, let $E(p)$ be the set of those integers, not exceeding x, which are divisible by p. Let $\bar{E}(p) = S - E(p)$ be its complement in S.

Corresponding to each integer k which divides D we define the set

$$E_k = \bigcap_{p|k} E(p) \bigcap_{p|(D/k)} \bar{E}(p)$$

For differing values of k these sets are disjoint. By taking unions of finitely many of them we form $\mathfrak{B}$, the least σ-algebra which contains the $E(p)$.

In terms of the usual frequency function we introduce a measure v: *If*

$$A = \bigcup_{j=1}^{m} E_{k_j}$$

then

$$vA = \sum_{j=1}^{m} [x]^{-1} |E_{k_j}|.$$

The pair $(\mathfrak{B}, v)$ then becomes a finite probability space.

We shall now construct a second measure, μ, on $\mathfrak{B}$, with respect to which additive functions will behave like sums of independent random variables.

The essential ingredient of this construction is the estimate

$$|E_k| = \{1 + O(L)\} \cdot \frac{x}{k} \prod_{p|(D/k)} \left(1 - \frac{1}{p}\right), \tag{1}$$

which is valid, whenever k does not exceed $x^{1/2}$, with

$$L = \exp\left(-\frac{1}{8}\frac{\log x}{\log r} \log\left(\frac{\log x}{\log r}\right)\right) + x^{-1/15},$$

where the positive constant which is implicit in the error term $O(L)$ is absolute. We shall obtain this estimate by an application of lemma (2.1).

An integer m ($\leq x$) can belong to E_k if and only if it has the form $m = kn$, where n is prime to D/k. We apply lemma (2.1) with

$$f(n) \equiv 1 \qquad N = \left[\frac{x}{k}\right]; \quad a_n = n \qquad (n = 1, \ldots, N),$$

and for the primes $p_1, \ldots, p_s$ we take the prime divisors of D/k. Note that none of these primes exceeds r.

For any positive integer d

$$\sum_{\substack{n=1 \\ a_n \equiv 0 (\text{mod } d)}}^{N} 1 = \left[\frac{x}{kd}\right] = \frac{x}{kd} + \theta,$$

where $|\theta| \leq 1$. We set

$$X = \frac{x}{k} \qquad \eta(d) = \frac{1}{d},$$

so that

$$|R(N, d)| \leq 1.$$

The application of lemma (2.1) now follows closely that used in the proof of the Brun–Titchmarsh lemma, lemma (2.7). In the notation of lemma (2.1) we obtain

$$|E_k| = I(N, Q) = \{1 + O(H)\} \cdot \frac{x}{k} \prod_{p|(D/k)} \left(1 - \frac{1}{p}\right) + O\left(\sum_{d \leq z^3} \mu^2(d) 3^{\omega(d)}\right)$$

with

$$H = \exp\left(-\frac{\log z}{\log r}\left\{\log\left(\frac{\log z}{S}\right) - \log\log\left(\frac{\log z}{S}\right) - \frac{2S}{\log z}\right\}\right)$$

We set $z = x^{1/7}$. Then if $2 \leq r \leq x^{\delta}$, for a sufficiently small (but fixed) positive value of δ, we obtain the desired estimate (1).

Since the constant implicit in the error term $O(L)$ may be deemed to be as (absolutely) large as we wish, the same estimate is trivially valid when $x^{\delta} < r \leq x$.

Regarding D and k as fixed in (1), we see that as $x \to \infty$ the integers n which are of the form kw, where $(w, D/k) = 1$, have the asymptotic density

$$\frac{1}{k}\prod_{p|(D/k)}\left(1 - \frac{1}{p}\right).$$

Define the measure μ by

$$\mu E_k = \frac{1}{k}\prod_{p|(D/k)}\left(1 - \frac{1}{p}\right).$$

This definition makes sense for *every* divisor k of D.

We have proved that if $k \leq x^{1/2}$ then

$$\nu E_k = \{1 + O(L)\}\mu E_k. \tag{2}$$

It is clear that

$$S = \bigcup E_k \qquad \sum \mu E_k = 1,$$

where both union and sum run over all the positive integers k permitted by the restriction that k divides D. Since $\mu S = 1$ the pair

$$(\mathfrak{B}, \mu)$$

is also a finite probability space.

Before proving that the measures ν and μ closely approximate one another, we make two remarks.

The first is that, in the notation of lemma (2.3) of Chapter two,

$$Z = \prod_{p|D}\left(1 - \frac{1}{p}\right)^{-1}$$

and

$$Z^{-1}\sum_{\substack{d|D\\ d>z}} \frac{1}{h(d)} = \sum_{\substack{d|D\\ d>z}} \frac{1}{d} \prod_{p|(D/d)} \left(1 - \frac{1}{p}\right) = \sum_{d>z} \mu E_d.$$

Applying that same lemma we see that

$$\sum_{k>x^{1/2}} \mu E_k = O(L).$$

The second is that, since $\nu S = \mu S = 1$,

$$\begin{aligned}\sum_{k>x^{1/2}} \nu E_k &= 1 - \sum_{k\le x^{1/2}} \nu E_k \\ &= 1 - \{1 + O(L\} \sum_{k\le x^{1/2}} \mu E_k = \sum_{k>x^{1/2}} \mu E_k + O(L) = O(L).\end{aligned}$$

Here we have made essential use of the estimate (2).

This ends the remarks.

Finally, let

$$A = \bigcup_{j=1}^{m} E_{k_j}.$$

Then, making use of our two remarks, and the estimate (2),

$$\begin{aligned}\nu A &= \sum_{k_j\le x^{1/2}} \nu E_{k_j} + \sum_{k_j>x^{1/2}} \nu E_{k_j} \\ &= \{1 + O(L)\} \sum_{k_j\le x^{1/2}} \mu E_{k_j} + O(L) \\ &= \sum_{j=1}^{m} \mu E_{k_j} + O(L) \\ &= \mu A + O(L)\end{aligned}$$

so that the estimate

$$\nu A = \mu A + O(L)$$

holds uniformly for all sets A in the algebra $\mathfrak{B}$.

The appropriate generalisation of lemma (3.1) is now:

Lemma 3.2 (J. Kubilius; M. B. Barban and A. I. Vinogradov). *Let r and x be real numbers, $2 \le r \le x$. Define the strongly additive function*

$$g(n) = \sum_{p|n,\, p\le r} f(p),$$

where the $f(p)$ assume real values. Define the independent random variables X_p, one for each prime not exceeding r, by

$$X_p = \begin{cases} f(p) & \text{with probability } \dfrac{1}{p}, \\ 0 & \text{with probability } 1 - \dfrac{1}{p}. \end{cases}$$

Then the estimate

$$\nu_x(n; g(n) \le z) = P\left(\sum_{p \le r} X_p \le z\right) + O\left(\exp\left(-\frac{1}{8}\frac{\log x}{\log r}\log\left(\frac{\log x}{\log r}\right)\right)\right) + O(x^{-1/15})$$

holds uniformly for all real numbers $f(p)$, z, x (≥ 2), and r ($2 \le r \le x$).

Proof. Consider the probability models $(\mathfrak{B}, \nu)$ and $(\mathfrak{B}, \mu)$.

The variable X_p is defined on the integers n not exceeding x, by

$$X_p(n) = \begin{cases} f(p) & \text{if } p \mid n, \\ 0 & \text{if } p \nmid n. \end{cases}$$

For differing values of p these variables are independent with respect to the measure μ.

Lemma 3.2 now follows from the approximation of ν by μ.

Our exposition of Kubilius' model differs somewhat from that which appears in Chapter 2, pp. 25–29 of his monograph [5]. It is worthwhile to dwell upon these differences.

Let us follow Kubilius' account and define

$$\gamma_p = \left[\frac{\log r}{\log p}\right].$$

In terms of the canonical factorisation of an integer n which is given by

$$n = \prod p^{\alpha_p(n)}$$

one further defines

$$\beta_p(n) = \min(\alpha_p(n), \gamma_p).$$

Sets $E(p^\alpha)$ are now defined to consist of those integers n, not exceeding x, for which $\beta_p(n) = \alpha$.

For integers k which are of the form

$$k = \prod_{p \le r} p^{\alpha_p} \qquad 0 \le \alpha_p \le \gamma_p, \tag{3}$$

one defines

$$E_k = \bigcap_{p^\alpha \| k} E(p^\alpha)$$

Finally, an algebra $\mathfrak{C}$ is formed by taking unions of finitely many of the E_k.

The motivation for this procedure is the construction of an analogue to lemma (3.2) which involves the *additive* function

$$\sum_{p^\alpha \| n,\ p^\alpha \le r} f(p^\alpha)$$

If we define a measure ν by

$$\nu E_k = [x]^{-1} |E_k|$$

then the pair $(\mathfrak{C}, \nu)$ becomes a finite probability space.

Define

$$\Delta(p^\alpha) = \begin{cases} \dfrac{1}{p^\alpha}\left(1 - \dfrac{1}{p}\right) & \text{if } 0 \le \alpha < \gamma_p, \\[2ex] \dfrac{1}{p^\alpha} & \text{if } \alpha = \gamma_p. \end{cases}$$

With this definition

$$\sum_{\alpha=0}^{\gamma_p} \Delta(p^\alpha) = 1.$$

Applying lemma (2.1) we see that for permissible integers k which do not exceed $x^{1/2}$

$$|E_k| = \{1 + O(L)\} \prod_{p^\alpha \| k} \Delta(p^\alpha),$$

with

$$L = \exp\left(-\frac{\log x}{8 \log r} \log\left(\frac{\log x}{\log r}\right)\right) + x^{-1/15}.$$

Accordingly, we define a measure μ on $\mathfrak{C}$ by

$$\mu E_k = \prod_{p^\alpha \| k} \Delta(p^\alpha).$$

This definition makes sense for each permissible value of k. Moreover

$$\mu E_k = \prod_{p \le r} \left\{ \sum_{\alpha=0}^{\gamma_p} \Delta(p^\alpha) \right\} = 1,$$

so that the pair $(\mathfrak{C}, \mu)$ is also a probability space.

We have shown that

$$\nu E_k = \{1 + O(L)\} \mu E_k \tag{4}$$

holds uniformly for $k \le x^{1/2}$. In order to obtain an estimate for

$$\sup |\nu A - \mu A|$$

taken over all the members A of the algebra $\mathfrak{C}$, we need an upper bound for the sum

$$\sum_{k > x^{1/2}} \nu E_k.$$

This Kubilius obtains by noting that it has the same value as the frequency

$$\nu_x(n; n_r > x^{1/2}),$$

where

$$n_r = \prod_{p \le r} p^{\beta_p(n)}.$$

He estimates this frequency by a combination of elementary number theory, and an application of the method of moments from the theory of probability. In this way he obtains the uniform estimate

$$\nu A - \mu A = O\left(\exp\left(-c \frac{\log x}{\log r}\right)\right),$$

with a positive absolute constant c. Actually, he also obtains this weaker error term in place of L in the estimate for $|E_k|$ which is given earlier.

By setting $f(p) = f(p^2) = \dots$, for each prime p, one now obtains lemma (3.2), but with the weaker error term

$$O\left(\exp\left(-c\frac{\log x}{\log r}\right)\right).$$

A way to sharpen this error term to one similar to L was sketched by M. B. Barban and A. I. Vinogradov [1]. It depended upon the possession of a good estimate for the number of integers, not exceeding x, which are made up entirely of primes $p \le r$. Such estimates are usually difficult to come by.

We can treat this present model of Kubilius exactly as we did the above simpler model by considering, instead, the sum

$$\sum_{k > x^{1/2}} \mu E_k.$$

Since $(\mathfrak{C}, \nu)$ and $(\mathfrak{C}, \mu)$ are both probability spaces, estimate (4) shows that this sum differs from

$$\sum_{k > x^{1/2}} \nu E_k$$

by an amount $O(L)$.

Define independent random variables Z_p, one for each prime p in the range $2 \le p \le r$, by

$$Z_p = \log p^\alpha \text{ with probability } \Delta(p^\alpha) \qquad 0 \le \alpha \le \gamma_p.$$

If k is an integer of the form (3), then the probability that

$$\sum_{p \le r} Z_p = \sum_{p \le r} \alpha_p \log p = \log k$$

is

$$\prod_{p^\alpha \| k} \Delta(p^\alpha) = \mu E_k.$$

Hence, the sum which we wish to estimate is

$$\sum_{k > x^{1/2}} \mu E_k = P\left(\sum_{p \le r} Z_p > \tfrac{1}{2} \log x\right).$$

We have now reached a problem in the theory of probability proper; one involving large deviations.

An application of the theory of probability to the estimation of arithmetic sums which appear in connection with the Selberg sieve may be found in Philipp [1] pp. 74–79. However, the result which is obtained there is not sufficiently sharp for our present purposes, and leads, once again, to an estimate of the form

$$\sum_{k > x^{1/2}} \mu E_k = O\left(\exp\left(-\frac{c \log x}{\log r}\right)\right).$$

This is apparently due to the lack of a suitable result in the theory of probability. Traditionally one compares the sum of the Z_p with its variance. See, for example, Loève [1], §18.1, p. 254. In our present circumstance, however, the following simple result, which is perhaps new, is preferable.

Lemma 3.3. *Let $X_j, j = 1, \ldots, n$ be independent random variables. Let $|X_j| \le c$ hold almost surely for each variable. Define*

$$S_n = X_1 + \cdots + X_n \qquad m = \sum_{j=1}^{n} \text{Expect}(|X_j|).$$

Then the inequality

$$P(S_n \ge u) \le \exp\left(-\frac{u}{c}\left\{\log\frac{u}{m} - \log\log\frac{u}{m} - 1\right\}\right)$$

holds uniformly for $u \ge me$.

Proof. For each non-negative real number λ,

$$\begin{aligned}\text{Expect}(e^{\lambda X_j}) &= 1 + \text{Expect}(e^{\lambda X_j} - 1)\\ &\le 1 + \text{Expect}(|e^{\lambda X_j} - 1|) \le 1 + \text{Expect}(\lambda |X_j| e^{\lambda |X_j|})\\ &\le 1 + \lambda e^{\lambda c}\, \text{Expect}(|X_j|),\end{aligned}$$

where, in the penultimate step, we have made use of the inequality $e^t - 1 \le |t| e^{|t|}$. Since $1 + t \le e^t$ when $t \ge 0$, we deduce that

$$\text{Expect}(e^{\lambda S_n}) \le \exp\left(\lambda e^{\lambda c} \sum_{j=1}^{n} \text{Expect}(|X_j|)\right) = \exp(\lambda m e^{\lambda c}).$$

Hence

$$P(S_n \ge u) \le \text{Expect}(e^{\lambda(S_n - u)}) \le \exp(\lambda m e^{\lambda c} - \lambda u).$$

We set $\lambda = \rho c^{-1}$ and apply lemma (2.2) to complete the proof of lemma (3.3).

In our present application we may take $c = \log r$. According to lemma (2.5):

$$m = \sum_{p^\alpha \le r} \Delta(p^\alpha)\log p^\alpha = \sum_{p \le r} \frac{\log p}{p} + O\left(\sum_{p,\,\alpha \ge 2}\sum \frac{\log p^\alpha}{p^\alpha}\right) = \log r + O(1).$$

It follows at once that

$$\sum_{k > x^{1/2}} \mu E_k = O(L)$$

and, indeed, a slightly better result can be obtained if desired.

Returning to the model of Kubilius; we have proved that uniformly for all members A of the algebra $\mathfrak{C}$,

$$\nu A = \mu A + O(L).$$

We can now construct an analogue to lemma (3.2).

Lemma 3.4. *Let r and x be real numbers, $2 \le r \le x$. Define the additive function*

$$g(n) = \sum_{\substack{p^\alpha \| n \\ p^\alpha < r}} f(p^\alpha) + \sum_{\substack{p^\alpha | n \\ p^\alpha = r}} f(p^\alpha)$$

where the $f(p^\alpha)$ assume real values. Define independent random variables Y_p, one for each prime p not exceeding r, by

$$Y_p = f(p^\alpha) \text{ with probability } \Delta(p^\alpha) \qquad (\alpha = 0, 1, \ldots, \gamma_p).$$

Then the estimate

$$\nu_x(n; g(n) \le z) = P\left(\sum_{p \le r} Y_p \le z\right) + O\left(\exp\left(-\frac{1}{8}\frac{\log x}{\log r}\log\left(\frac{\log x}{\log r}\right)\right)\right)$$
$$+ O(x^{-1/15}).$$

holds uniformly for all real numbers $f(p^\alpha)$, z, x (≥ 2), and r $(2 \le r \le x)$.

Proof. Define functions

$$Y_p(m) = f(p^\alpha) \quad \text{if} \quad \beta_p(m) = \alpha \qquad 0 \le \alpha \le \gamma_p.$$

These are clearly random variables in both the models $(\mathfrak{C}, \nu)$ and $(\mathfrak{C}, \mu)$, and are independent with respect to the measure μ.

The peculiar shape of the function $g(n)$ is due to the roundoff error which is introduced in order that v and μ give to the whole space S, of the algebra $\mathfrak{C}$, the measure 1. If we change the definition of the γ_p to

$$\gamma_p = \left[\frac{\log x}{\log p}\right]$$

then we may replace $g(n)$ by the function

$$\sum_{\substack{p^\alpha \| n \\ p \le r}} f(p^\alpha)$$

and the result of lemma (3.4) is still valid provided that the additional condition $r \ge (\log x)^3$ is in force. The proof of this modified result follows that of lemma (3.4) save that one estimates

$$P\left(\sum_{p \le r} Z_p > \tfrac{1}{2} \log x\right)$$

by following the *proof* of lemma (3.3) in this particular case, paying better attention to detail, rather than applying that lemma directly.

A General Model

It is clear that one can construct a Kubilius model whenever it is possible to apply Selberg's (or any other) sieve method to a sequence of integers.

In the notation of lemma (2.1), of Chapter two, let $a_1, \ldots, a_N$ be positive integers, possibly with repetition. Let $p_1 < p_2 < \cdots < p_s \le r$ be prime numbers, and set $Q = p_1 \ldots p_s$.

For each prime p that divides Q we define

$$E_p = \{a_n; n = 1, \ldots, N, a_n \equiv 0 (\text{mod } p)\}.$$

Let

$$S = \{a_n; n = 1, \ldots, N\}$$

and

$$\bar{E}_p = S - E(p)$$

the complement of $E(p)$ with respect to S. For each divisor k of Q let

$$E_k = \bigcap_{p|k} E(p) \bigcap_{p|Q/k} \bar{E}(p).$$

We form an algebra $\mathfrak{D}$ by taking unions of finitely many of the E_k, and define a measure ν on it by

$$\nu E_k = \sum_{n=1}^{N}{}' f(n) \Big/ \sum_{n=1}^{N} f(n)$$

where $'$ indicates that summation is restricted to those integers a_n which belong to the set E_k, and where we assume that

$$\sum_{n=1}^{N} f(n) > 0.$$

The pair $(\mathfrak{D}, \nu)$ is then a finite probability space.

Suppose now that for each divisor d of Q

$$\sum_{a_n \equiv 0 (\mathrm{mod}\, d)} f(n) = \eta(d)X + R(N, d),$$

where $X \geq 0$, and $\eta(d)$ is a multiplicative function on Q. We shall assume that $0 \leq \eta(p) < 1$ whenever $\eta(p)$ is defined.

Let k be a divisor of Q, and let z be a real number which satisfies

$$\log z \geq 8 \max(\log r, S)$$

where

$$S = \sum_{p \mid Q} \frac{\eta(p)}{1 - \eta(p)} \log p.$$

Then according to lemma (2.1)

$$|E_k| = \{1 + 2\theta_1 H\} X \eta(k) \prod_{p \mid Q/k} (1 - \eta(p)) + 2\theta_2 \sum_{\substack{d \mid Q/k \\ d \leq z^3}} 3^{\omega(d)} |R(N, [k, d])|$$

where

$$H = \exp\left(-\frac{\log z}{\log r}\left\{\log\left(\frac{\log z}{S}\right) - \log\log\left(\frac{\log z}{S}\right) - \frac{2S}{\log z}\right\}\right)$$

and $|\theta_1| \leq 1$, $|\theta_2| \leq 1$.

We define a second measure, μ, on $\mathfrak{D}$ by

$$\mu E_k = \eta(k) \prod_{p|Q/k} (1 - \eta(p)).$$

Then, summing over all divisors k of Q,

$$\sum \mu E_k = \prod_{p|Q} (\eta(p) + 1 - \eta(p)) = 1,$$

so that the pair $(\mathfrak{D}, \mu)$ is also a finite probability space.

We estimate the sum

$$\sum_{k>z} \mu E_k$$

by means of lemma (3.3). Define the independent random variables Z_p by

$$Z_p = \begin{cases} \log p & \text{with probability } \eta(p), \\ 0 & \text{with probability } 1 - \eta(p). \end{cases}$$

Then the above sum is

$$P\left(\sum_{p|Q} Z_p > \log z\right)$$

which by lemma (3.3) is certainly not more than

$$\exp\left(-\frac{\log z}{8 \log r} \log\left(\frac{\log z}{S}\right)\right).$$

One now proceeds in the manner described in the earlier models. We confine ourselves to one remark. Assume for the moment that $|R(N, 1)| \leq X/2$. Then

$$\begin{aligned} \left|\sum_{k \leq z} \nu E_k - \sum_{k \leq z} \mu E_k\right| &\leq (4H + 2X^{-1}|R(N, 1)|) \sum_{k \leq z} \mu E_k \\ &\quad + 4X^{-1} \sum_{\substack{k \leq z \\ k|Q}} \sum_{\substack{d \leq z^3 \\ d|Q/k}} 3^{\omega(d)} |R(N, [k, d])| \\ &\leq 4 \exp\left(-\frac{\log z}{8 \log r} \log\left(\frac{\log z}{S}\right)\right) \\ &\quad + 6X^{-1} \sum_{\substack{m \leq z^4 \\ m|Q}} |R(N, m)| \sum_{k \leq z} \sum_{\substack{d|Q/k \\ [k, d] = m}} 3^{\omega(d)}. \end{aligned}$$

Note that if d divides Q/k then $(k, d) = 1$, so that the innermost sum which appears in the last line in this chain of inequalities does not exceed

$$\sum_{kd=m} 3^{\omega(d)} = 4^{\omega(m)}.$$

It is straightforward to establish the estimate

$$|\nu A - \mu A| \le 10 \exp\left(- \frac{\log z}{8 \log r} \log\left(\frac{\log z}{S}\right)\right) + 12X^{-1} \sum_{\substack{m \le z^4 \\ m|Q}} 4^{\omega(m)} |R(N, m)|$$

which holds uniformly for all sets A in the algebra $\mathfrak{D}$.

The temporary hypothesis that $|R(N, 1)| \le X/2$ may now be removed, since if it fails

$$|\nu A - \mu A| \le 2 \le 4X^{-1} |R(N, 1)|.$$

This completes the construction of the general model involving the algebra $\mathfrak{D}$ and its mutually approximating measures ν and μ.

Let us consider the particular case when the function f (of lemma (2.1)) is identically one. For convenience of exposition, when F is an arbitrary set let $\nu(a_n \in F)$ denote the frequency

$$\nu\{a_n; n = 1, \ldots, N, a_n \in F\} = N^{-1} \sum_{a_n \in F} 1.$$

An analogue of lemma (3.2) is now

Lemma 3.5. *Let the notation of lemma* (2.1) *be in force, with f identically one. Let $r \ge 2$ and* $8 \max(\log r, S) \le \log z$, *where*

$$S = \sum_{p|Q} \frac{\eta(p) \log p}{1 - \eta(p)}.$$

Define the strongly additive function

$$g(n) = \sum_{p|n,\, p|Q} l(p),$$

where the $l(p)$ are real numbers. Define independent random variables W_p, one for each prime divisor p of Q, by

$$W_p = \begin{cases} l(p) & \text{with probability } \eta(p), \\ 0 & \text{with probability } 1 - \eta(p). \end{cases}$$

Then

$$\nu(g(a_n) \in F) \quad \text{and} \quad P\left(\sum_{p|Q} W_p \in F\right)$$

do not differ by more than

$$10 \exp\left(-\frac{\log z}{8 \log r} \log\left(\frac{\log z}{S}\right)\right) + 12X^{-1} \sum_{\substack{m \leq z^4 \\ m|Q}} 4^{\omega(m)} |R(N, m)|,$$

and this uniformly in all sets F.

We give a few examples of the application of lemma (3.5).

Let $h(w)$ be a polynomial in w with rational integer coefficients; for example $h(w) = w(w - 2)$. Let w_0 be an integer with the property that $h(w)$ is positive for $w > w_0 \geq 0$.

In the notation of lemma (2.1) let a_n run through the (not necessarily distinct) integers $h(k)$ which are generated when k runs through the integers in the interval $w_0 < k \leq x$. In the same notation, we have

$$|N - x| \leq 1 + w_0.$$

Let $\rho(p)$ denote the number of distinct residue class solutions to the congruence $h(k) \equiv 0 (\text{mod } p)$. Since the polynomial is not identically zero there is a number p_0 so that $\rho(p) < p$ for all primes $p > p_0$.

Let r be a real number, $p_0 < r \leq x$. Let Q denote the product of those primes p in the range $p_0 < p \leq r$.

Let d be a divisor of Q (Q assumed ≥ 2), and let $k_j, j = 1, \ldots, \rho(p)$, be a representative set of solutions to the congruence $h(k) \equiv 0 (\text{mod } d)$. Without loss of generality each k_j satisfies $1 \leq k_j \leq d$. Clearly, in our present situation

$$\sum_{a_n \equiv 0 (\text{mod } d)} 1 = \sum_j \sum_{\substack{w_0 < k \leq x \\ k \equiv k_j (\text{mod } d)}} 1 = \sum_j \left(\left[\frac{x - k_j}{d}\right] - \left[\frac{w_0 - k_j}{d}\right]\right)$$

provided that $x \geq d$, and provided that when $w_0 < k_j$ we omit that part of the final summand which involves w_0. We set $X = x$, $\eta(d) = \rho(d)d^{-1}$, so that

$$|R(N, d)| \leq (2 + w_0)\rho(d) \qquad (d \leq x).$$

Let s be the degree of $h(w)$. If $h(w)$ does not vanish identically mod p then we must have $\rho(p) \leq s$ (see, for example, Hardy and Wright [1] Theorem

107, p. 84). For divisors d of Q, therefore, $\rho(d) \leq s^{\omega(d)}$ where $\omega(d)$ denotes the number of (distinct) prime divisors of d. Hence

$$X^{-1} \sum_{\substack{m|Q \\ m \leq z^4}} 4^{\omega(m)} |R(N, m)| \leq (2 + w_0) X^{-1} \sum_{\substack{m|Q \\ m \leq z^4}} (4s)^{\omega(m)}$$

$$\leq (2 + w_0) X^{-1} z^4 \sum_{m|Q} (4s)^{\omega(m)} m^{-1} \leq (2 + w_0) X^{-1} z^4 \prod_{p \leq r} \left(1 + \frac{4s}{p}\right)$$

and if we set $z = x^{1/6}$ and apply lemma (2.5) the last expression is $O(x^{-1/4})$.

We have now proved the following result, which, apart from the quality of the error term, was first established implicitly by Uzdavinis [1], using Brun's sieve method.

Lemma 3.6. *In the notation of lemma* (3.5),

$$\nu(g(h(k)) \in F) = P\left(\sum_{p_0 < p \leq r} W_p \in F\right) + O\left(\exp\left(-\frac{1}{50} \frac{\log x}{\log r} \log\left(\frac{\log x}{\log r}\right)\right)\right) + O(x^{-1/4}).$$

Here the independent random variables W_p *satisfy*

$$W_p = \begin{cases} l(p) & \text{with probability } \dfrac{\rho(p)}{p}, \\ 0 & \text{with probability } 1 - \dfrac{\rho(p)}{p}. \end{cases}$$

In the study of the arithmetic properties of polynomials it is not usually necessary to have detailed knowledge of the individual values of the $\rho(p)$. This is fortunate, since they are often difficult to determine. If $h(w) = w^2 + 1$ then we may take $w_0 = 0 = p_0$. A simple application of Euler's criterion for quadratic residuacity (see, for example, Hardy and Wright [1], theorem 83, p. 69) shows that in this case

$$\omega(2) = 1$$

$$\omega(p) = \begin{cases} 2 & \text{if } p \equiv 1 (\text{mod } 4), \\ 0 & \text{if } p \equiv 3 (\text{mod } 4). \end{cases}$$

As another application of the general model let a_n run through the integers $h(q)$ where q runs through the *prime* numbers in the interval $w_0 < q \leq x$. Here one replaces $\rho(p)$ by $\eta(p)$, the number of reduced residue class solutions to the congruence $h(k) \equiv 0 (\text{mod } p)$. Thus $\eta(p) \leq \rho(p)$, and there will certainly be equality when $p > p_0$, if $h(0) \neq 0$.

In this example we take the same Q as before, and set $X = Li(x)$, $\omega(d) = \eta(d)\varphi(d)^{-1}$. Let

$$M(x, d) = \max_{(l, d)=1} \left| \pi(x, d, l) - \frac{Li(x)}{\varphi(d)} \right|$$

If $d \mid Q$, and k_j, $j = 1, \ldots, \eta(p)$, runs over a representative set of reduced solutions (mod d), then

$$\begin{aligned} |R(N, d)| &\le \sum_j \left(\left| \pi(x, d, k_j) - \frac{Li(x)}{\varphi(d)} \right| + w_0 \right) + \rho(d) \\ &\le \rho(d)\{M(x, d) + w_0 + 1\}. \end{aligned}$$

We estimate the sum

$$S = \sum_{\substack{m \mid Q \\ m \le z^4}} 4^{\omega(m)} |R(N, m)|$$

by means of lemma (2.10). Applying the Cauchy–Schwarz inequality

$$\begin{aligned} S^2 &= \left(\sum_{\substack{m \mid Q \\ m \le z^4}} \{4^{\omega(m)}\rho(m)\varphi(m)^{-1/2}\}\{\rho(m)^{-1}\varphi(m)^{1/2}|R(N, m)|\right)^2 \\ &\le \sum_{m \mid Q} 8^{\omega(m)}\rho(m)^2\varphi(m)^{-1} \sum_{m \le z^4} \rho(m)^{-2}\varphi(m)|R(N, m)|^2. \end{aligned}$$

Let $z = x^{1/9}$, so that $r \le x$. The first of the two sums which appear in this last line is then

$$\prod_{p_0 < p \le r} \left(1 + \frac{8\rho(p)^2}{p - 1}\right) = O((\log r)^{8s^2}),$$

where, as before, s is the degree of the polynomial $h(w)$. From the Brun–Titchmarsh lemma, since $m \le x^{4/9}$,

$$\varphi(m)|R(N, m)| = O\left(\rho(m)\frac{x}{\log x}\right)$$

so that the second sum in the same line is, now using lemma (2.10),

$$O\left(\frac{x}{\log x} \sum_{d \le x^{4/9}} M(x, d)\right) = O(x^2(\log x)^{-2A-10s^2})$$

for any fixed $A > 0$.

Hence

$$X^{-1}S = O((Li(x))^{-1}\{(\log x)^{8s^2} \cdot x^2(\log x)^{-2A-10s^2}\}^{1/2})$$
$$= O((\log x)^{-A}).$$

We have now proved the following result:

Lemma 3.7. *Let A be a fixed positive number. Then in the notation of lemma* (3.5), *where q denotes a prime,*

$$v(g(h(q)) \in F) = P\left(\sum_{p_0 < p \leq r} W_p \in F\right) + O\left(\exp\left(-\frac{1}{75}\frac{\log x}{\log r}\log\left(\frac{\log x}{\log r}\right)\right)\right) + O((\log x)^{-A}),$$

where the independent random variables W_p satisfy

$$W_p = \begin{cases} l(p) & \text{with probability } \dfrac{\eta(p)}{p-1}, \\ 0 & \text{with probability } 1 - \dfrac{\eta(p)}{p-1}. \end{cases}$$

Apart from the quality of the error term, this result was implicitly proved by Barban [2]. For this purpose he constructed an early form of lemma (2.10). (See the notes which immediately follow the statement of lemma (2.10), Chapter two).

An attractive particular case of this last lemma is obtained with the choice $h(w) = w + 1$, $(w_0 = p_0 = 1)$. Then

$$v(g(q+1) \in F) = v_x(q; g(q+1) \in F) = \frac{1}{\pi(x)} \sum_{q \leq x}' 1,$$

where $'$ denotes that summation ranges over prime numbers q; whilst

$$W_p = \begin{cases} l(p) & \text{with probability } \dfrac{1}{p-1}, \\ 0 & \text{with probability } 1 - \dfrac{1}{p-1}. \end{cases}$$

The following result may readily be obtained by the method used to establish lemma (3.7):

Lemma 3.8. *Let D be a positive integer, $D \geq 2$. Define the strongly additive function*

$$g(n) = \sum_{\substack{p|n,\, p\nmid D \\ 2<p\leq r}} l(p).$$

Let U_p be independent random variables, one defined for each odd prime p not exceeding r and not dividing D, by

$$U_p = \begin{cases} l(p) & \text{with probability } \dfrac{1}{p-1}, \\ 0 & \text{with probability } 1 - \dfrac{1}{p-1}. \end{cases}$$

Then,

$$\nu_D(q; q \text{ prime}, g(D-q) \in F) = \nu(g(D-q) \in F) = P\left(\sum_{\substack{2<p\leq r \\ p\nmid D}} U_p \in F\right)$$

$$+ O\left(\exp\left(-\frac{1}{75}\frac{\log D}{\log r}\log\left(\frac{\log D}{\log r}\right)\right)\right) + O((\log D)^{-A})$$

uniformly for all $2 \leq r \leq D$, for all sets F, and where the implied constants depend at most upon A.

The final result of this chapter is useful in the study of sums of translated additive arithmetic functions, such as $\omega(m+1) - \omega(m)$, where $\omega(m)$ denotes the number of prime divisors of the integer m.

Lemma 3.9. *Let $0 \leq a_1 < a_2 < \cdots < a_s$ be integers. Let p_0 be a prime which exceeds both $2s$ and the greatest prime divisor of the differences $a_j - a_i$, $1 \leq i < j \leq s$. Define s strongly additive functions by*

$$g_j(n) = \sum_{\substack{p|n \\ p_0<p\leq r}} l_j(p) \qquad (j = 1, \ldots, s),$$

where the $l_j(p)$ are real numbers, and $r \geq 2$.

Define independent random variables V_p, one for each prime p in the range $p_0 < p \leq r$, by

$$V_p = \begin{cases} l_j(p) & \text{with probability } \dfrac{1}{p}, (j = 1, \ldots, s), \\ 0 & \text{with probability } 1 - \dfrac{s}{p}. \end{cases}$$

Then, for fixed integers a_j, the estimate

$$\nu_x(n; g_1(n + a_1) + \cdots + g_s(n + a_s) \in F)$$

$$= P\left(\sum_{p_0 < p \le r} V_p \in F\right) + O\left(\exp\left(-\frac{1}{8}\frac{\log x}{\log r}\log\left(\frac{\log x}{\log r}\right)\right)\right) + O(x^{-1/15})$$

holds uniformly for $2 \le r \le x$, *and all sets* F.

Apart from the quality of the error term this result was proved by Kubilius (see, for example, Kubilius [5]).

We sketch the construction of a finite probability space on which to study sums of translated additive functions.

Let $(\delta_1, \ldots, \delta_s)$ be an ordered s-tuple where each δ_j has the value zero or one, and where at most one of the δ_j is non-zero. For each prime p in the range $p_0 < p \le r$, let $E(p; \delta_1, \ldots, \delta_s)$ denote the set of positive integers n, not exceeding x, with the property that p divides $(n + a_j)$ if $\delta_j = 1$, and p does not divide $(n + a_j)$ if $\delta_j = 0, (j = 1, \ldots, s)$.

Let $\mathfrak{F}$ be the smallest algebra of sets which contain the $E(p; \delta_1, \ldots, \delta_s)$, as the primes p, and the integers δ_j, range over their permissible values. For each member, A, of this algebra define a measure

$$\nu A = [x]^{-1}|A| = [x]^{-1} \sum_{n \le x,\, n \in A} 1.$$

The pair $(\mathfrak{F}, \nu)$ is then a finite probability space.

Let S be the set of all positive integers n in the interval $1 \le n \le x$. Then for each prime p

$$S = \bigcup E(p; \delta_1, \ldots, \delta_s)$$

where the union is taken over all permissible s-tuples $(\delta_1, \ldots, \delta_s)$. Note that any two sets in this union are disjoint. Therefore, each member of the algebra $\mathfrak{F}$ has a representation as a union of finitely many sets of the type

$$A(k_1, \ldots, k_s) = \bigcap_{p_0 < p \le r} E(p; \delta_1(p), \ldots, \delta_s(p)),$$

where

$$k_j = \prod_{p_0 < p \le r} p^{\delta_j(p)} \qquad (j = 1, \ldots, s).$$

In this representation the k_j are pairwise coprime.

We apply lemma (2.1) to estimate the number of integers which are contained in a typical set $A(k_1, \ldots, k_s)$ where the product $k_1 \ldots k_s$ does not exceed $x^{1/2}$.

Let Q denote the product of the primes p, $p_0 < p \leq r$.

An integer n belongs to $A(k_1, \ldots, k_s)$ if and only if, for each values of j, $1 \leq j \leq s$, it satisfies the pair of conditions

$$n + a_j \equiv 0 (\bmod k_j) \qquad (n + a_j, Qk_j^{-1}) = 1.$$

This pair of conditions is equivalent to the pair

$$n + a_j \equiv 0 (\bmod k_j) \qquad (n + a_j, Qk^{-1}) = 1.$$

To see this suppose, for example, that a prime q divides k/k_1 and $n + a_1$. Since $k/k_1 = k_2 \ldots k_s$ at least one of the integers $k_2, \ldots, k_s$ must be divisible by q. Suppose, further, that q divides k_2. Then since n belongs to $A(k_1, \ldots, k_s)$ we have $k_2 | (n + a_2)$ so that $q | (n + a_2)$. Therefore q divides

$$a_2 - a_1 (= n + a_2 - (n + a_1)),$$

which is impossible, since $q > p_0$. By the Chinese Remainder Theorem there is an a_0, $0 \leq a_0 < k$, so that the above pairs of conditions are altogether equivalent to

$$n \equiv a_0 (\bmod k) \qquad (n + a_j, Qk^{-1}) = 1 \qquad (j = 1, \ldots, s).$$

We apply lemma (2.1) with the a_n of that lemma replaced by those integers n which do not exceed x, and which satisfy $n \equiv a_0 (\bmod k)$. We set $X = xk^{-1}$, and define $\eta(d)$ by

$$\eta(p) = \begin{cases} s & \text{if } p \text{ divides } Qk^{-1} \\ 0 & \text{if } p \text{ divides } k. \end{cases}$$

In this case, since $p \geq 2s$,

$$S \leq 2 \sum_{p \leq r} \frac{\eta(p)}{p} \log p \leq 2s \sum_{p \leq r} \frac{\log p}{p} = O(\log r).$$

Proceeding as we did for the first Kubilius model we obtain the estimate (cf. (1) of the present chapter)

$$|A(k_1, \ldots, k_s)| = \{1 + O(L)\} \cdot \frac{x}{k} \prod_{p | (Q/k)} \left(1 - \frac{s}{p}\right)$$

where $k = k_1 \ldots k_s$, and

$$L = \exp\left(-\frac{1}{8}\frac{\log x}{\log r}\log\left(\frac{\log x}{\log r}\right)\right) + x^{-1/15}.$$

One defines a second measure, μ, on the algebra $\mathfrak{F}$, by

$$\mu A(k_1, \ldots, k_s) = \frac{1}{k}\prod_{p\,|\,(Q/k)}\left(1 - \frac{s}{p}\right),$$

and shows that the estimate

$$\nu A = \mu A + O(L)$$

holds uniformly for all members A of $\mathfrak{F}$.

To complete the proof of lemma (3.9) define the functions

$$\varepsilon_j(p, n) = \begin{cases} k_j(p) & \text{if } p\,|\,(n + a_j), \\ 0 & \text{otherwise} \end{cases}$$

for $j = 1, \ldots, s$, and set

$$V_p = V_p(n) = \sum_{j=1}^{s} \varepsilon_j(p, n).$$

The V_p, $p_0 < p \leq r$, are random variables with respect to both the measures ν and μ, and are independent with respect to μ. Clearly

$$V_p = \begin{cases} l_j(p) & \text{with probability } \dfrac{1}{p}, (j = 1, \ldots, s), \\ 0 & \text{with probability } 1 - \dfrac{s}{p}. \end{cases}$$

We conclude our sketch with the observation that

$$\sum_{p_0 < p \leq r} V_p = \sum_{j=1}^{s} g_j(n + a_j).$$

Multiplicative Functions

The Kubilius model by which we established lemma (3.2) may also be applied directly to the study of multiplicative functions. The following result is typical.

Lemma 3.10. *Let r and x be real numbers, $2 \le r \le x$. Define the (strongly) multiplicative function*

$$g(n) = \prod_{p|n,\, p \le r} f(p),$$

where the $f(p)$ assume real values. Define independent random variables X_p, one for each prime p not exceeding r, by

$$X_p = \begin{cases} f(p) & \text{with probability } \dfrac{1}{p}, \\ 1 & \text{with probability } 1 - \dfrac{1}{p}. \end{cases}$$

Then the estimate

$$\nu_x(n; g(n) \le z) = P\left(\prod_{p \le r} X_p \le z\right) + O(L),$$

where

$$L = \exp\left(-\frac{1}{8}\frac{\log x}{\log r}\log\left(\frac{\log x}{\log r}\right)\right) + x^{-1/15}$$

holds uniformly for all real numbers $f(p)$, z, x (≥ 2), and r ($2 \le r \le x$).

This result often allows one to reduce the study of the value distribution of multiplicative functions to problems concerning products of independent random variables. This last theory is not so well developed as the theory for sums of independent random variables; but see, for example, Zolotarev [1], and lemma (1.45) of Chapter one, where it is shown that an appropriate tool is a Mellin–Stieltjes transformation.

Here we make two remarks. Firstly, if

$$(\text{weak}) \lim_{x \to \infty} \nu_x(n; g(n) \le z)$$

exists, then so does

$$(\text{weak}) \lim_{x \to \infty} \nu_x(n; |g(n)| \le z).$$

Secondly, many multiplicative arithmetic functions assume only non-negative, or, indeed, positive values.

The direct application of lemma (3.10) is not always convenient, since it allows the possibility that a particular random variable X_p may attain the value zero, and so prevents relating the product of random variables to a sum. If $g(n)$ is non-negative for every integer $n \geq 1$, then there is an alternative procedure.

Let $m_1 < m_2 < \ldots$ denote the sequence of positive integers on which $g(n)$ does not vanish, where $g(n)$ is a strongly multiplicative function, defined for $n \geq 1$. Thus $g(n)$ is multiplicative, and satisfies $g(p) = g(p^2) = \ldots$ for for each prime p. Clearly $g(n) = 0$ if and only if n is divisible by a prime p for which $g(p) = 0$. It follows from an application of the sieve of Selberg, as say embodied in estimate (1) of the present chapter, with $k = 1$, that an estimate for the number of integers m which do not exceed x, and which are not divisible by any prime $p \leq r$ for which $g(p) = 0$, is

$$\{1 + O(L)\} \cdot x \prod_{\substack{p \leq r \\ g(p)=0}} \left(1 - \frac{1}{p}\right),$$

where

$$L = \exp\left(-\frac{1}{8}\frac{\log x}{\log r}\log\left(\frac{\log x}{\log r}\right)\right) + x^{-1/15},$$

and provided $2 \leq r \leq x$. Let $M(x)$ denote the number of m_i which do not exceed x. Then it is easy to see that $M(x) = o(x)$ as $x \to \infty$ if and only if the series

$$\sum_{g(p)=0} \frac{1}{p}$$

diverges. In this case

$$\nu_x(n; g(n) = 0) \to 1 \qquad (x \to \infty),$$

and we say that the function $g(n)$ is essentially zero. If $g(n)$ is not essentially zero,

$$\lim_{x\to\infty} \frac{M(x)}{x} = \prod_{g(p)=0} \left(1 - \frac{1}{p}\right) > 0.$$

We now carry out an argument analogous to that used in the construction of the simplest Kubilius model $(\mathfrak{B}, \nu)$ and $(\mathfrak{B}, \mu)$. A direct application of lemma (3.5) does not readily lead to the best results since there is only weak information available concerning the distribution of the sequence $\{m_1, m_2, \ldots\}$ in residue classes.

It is convenient to first concentrate our attention on the sequence of those integers $n_1 < n_2 < \ldots$, which are not divisible by any of the primes $p \le r$ for which $g(p) = 0$, $2 \le r \le x$. In the estimate corresponding to (1) of this chapter we set

$$D = \prod_{p \le r, g(p) \ne 0} p, \qquad Q = \prod_{p \le r} p$$

and have

$$\begin{aligned} &|\{n_j; 1 \le n_j \le x, n_j \equiv 0 (\text{mod } k), (n_j, Dk^{-1}) = 1\}| \\ &\qquad = |\{n; 1 \le n \le x, n \equiv 0 (\text{mod } k), (n, Qk^{-1}) = 1\}| \\ &\qquad = \{1 + O(L)\} \frac{x}{k} \prod_{p | Q/k} \left(1 - \frac{1}{p}\right) \qquad (1 \le k \le x^{1/2}), \end{aligned}$$

with the same value of L as in lemma (3.2).

A direct application of lemma (2.1) shows that $N(x)$, the number of the n_j which do not exceed x, may be estimated by

$$N(x) = \{1 + O(L)\} \cdot x \prod_{p \le r,\, g(p) = 0} \left(1 - \frac{1}{p}\right)$$

From these results we obtain

$$\sum_{\substack{n_j \le x \\ g(n_j) \in F}} 1 = N(x) P\left(\prod_{\substack{p \le r \\ g(p) \ne 0}} X_p \in F \right) + O(xL).$$

Since

$$|N(x) - M(x)| \le \sum_{r < p \le x} \sum_{n \equiv 0 (\text{mod } p)} 1 \le x \sum_{\substack{r < p \le x \\ g(p) = 0}} \frac{1}{p}$$

we arrive at the approximation

$$\frac{1}{M(x)} \sum_{\substack{m_j \le x \\ g(m_j) \in F}} 1 = P\left(\prod_{\substack{p \le r \\ g(p) \ne 0}} X_p \in F \right) + O(L_1) \tag{5}$$

where

$$L_1 = \exp\left(-\frac{1}{8} \frac{\log x}{\log r} \log\left(\frac{\log x}{\log r} \right) \right) + x^{-1/15} + \sum_{\substack{r < p \le x \\ g(p) = 0}} \frac{1}{p}.$$

Although we have established this estimate subject to the condition that $g(n)$ not be essentially zero, this last condition may be removed. It is easy to see that the above construction is perfectly valid, even if the function $g(n)$ is allowed to depend upon the parameters r and x, provided that the inequality

$$\sum_{\substack{r<p\leq x\\ g(p)=0}} \frac{1}{p} \leq c_0$$

holds for a certain positive absolute constant c_0. However, unless $g(n)$ is everywhere zero in the range $1 \leq n \leq x$, the estimate (5) is trivially valid when this inequality fails to be satisfied.

Define independent random variables Z_p, by

$$Z_p = \begin{cases} \log g(p) & \text{with probability } \dfrac{1}{p}, \\ 0 & \text{with probability } 1 - \dfrac{1}{p}, \end{cases}$$

one for each of the primes p, $2 \leq p \leq r$, $g(p) \neq 0$. For each such prime

$$e^{Z_p} = X_p,$$

and

$$P\left(\prod_{p\leq r}{}' X_p \leq e^w\right) = P\left(\sum_{p\leq r}{}' Z_p \leq w\right),$$

where $'$ denotes that summation is restricted to those primes for which $g(p)$ does not vanish. This allows us to apply the well-developed theory of sums of independent random variables, and is of particular advantage when considering theorems of the 'if and only if' kind.

Concluding Remarks

Simple probability spaces, of the kind which appears in the proof of lemma (3.1), were used, independently, and in completely different ways, by Wirsing [5], p. 55, and Elliott [4], pp. 193–194. Very likely such spaces have been used before.

Professor Kubilius informs me that he first gave a detailed proof of a form of lemma (3.2) in his survey paper [1]. We remark here that he later gave greater emphasis to the construction of an underlying finite probability space with associated approximation of measures. See, for example, Chapter 3 of his monograph [5].

Lemma (3.2), and the earlier results of Kubilius in the same direction, give non-trivial information only when $(\log r)/\log n$ is small. This is as it should be. For suppose that one could accurately mimic the behaviour of additive arithmetic functions by sums of independent random variables, as in lemma (3.2), with $\log r$ close to $\log x$ in value. Then let $f(p) \equiv \log p$. For each integer n let n_0 denote the product of the distinct prime divisors of n. We should then have (say),

$$1 \sim v_x(n; \log n_0 \le \log x) \sim P\left(\sum_{p \le x} X_p \le \log x\right).$$

If m is a squarefree integer, the probability that the sum of the variables X_p has the value $\log m$ is

$$\frac{1}{m} \prod_{\substack{p \le x \\ p \nmid m}} \left(1 - \frac{1}{p}\right).$$

Our temporary hypothesis therefore leads to

$$1 \sim \prod_{p \le x} \left(1 - \frac{1}{p}\right) \sum_{m \le x} \frac{\mu^2(m)}{\varphi(m)} \sim \frac{e^{-\gamma}}{\log x} \log x = e^{-\gamma}.$$

Calculations show that $e^{-\gamma} \ne 1$, and so we have reached an impossible situation. Thus lemma (3.2) must fail if $(\log r)/\log x$ is suitably near to one in value.

The application of Selberg's method which underlies the model of Kubilius also cannot be improved. For example, if $k = 1$ then the estimate (1) becomes

$$|E_1| = \{1 + O(L)\} \cdot x \prod_{p \le r} \left(1 - \frac{1}{p}\right).$$

If we set $r = x^{1/2}$ then by the prime number theorem, (see lemma (2.9)),

$$|E_1| = \pi(x) - \pi(x^{1/2}) \sim \frac{x}{\log x} \qquad (x \to \infty).$$

However, by lemma (2.4),

$$x \prod_{p \le x^{1/2}} \left(1 - \frac{1}{p}\right) \sim 2e^{-\gamma} \frac{x}{\log x} \qquad (x \to \infty).$$

Once again an erroneous factor occurs.

These examples show that there is a limitation upon the valid use of the notion, due to Kac, that whether or not an integer is divisible by one prime is independent from it being divisible from another; beautiful and powerful though it is. However, even when this idea is not exactly applicable it remains a very useful heuristic device. For further comments upon this situation we refer to Chapter 12, where the history of the well-known Erdös–Kac theorem is discussed.

In Chapter VIII of his monograph [5], Kubilius points out that his method will also extend to the study of the joint distribution of additive arithmetic functions.

Chapter 4

The Turán–Kubilius Inequality and Its Dual

Let $f(n)$ be a complex-valued additive arithmetic function,

$$f(n) = \sum_{p^k \| n} f(p^k).$$

For real numbers $x > 0$ set

$$E(x) = \sum_{p^k \le x} \frac{f(p^k)}{p^k}\left(1 - \frac{1}{p}\right)$$

$$D(x) = \left(\sum_{p^k \le x} \frac{|f(p^k)|^2}{p^k}\right)^{1/2} \ge 0.$$

Lemma 4.1 (Turán, Kubilius). *The inequality*

$$\sum_{n \le x} |f(n) - E(x)|^2 \le 32xD^2(x)$$

holds uniformly for all additive functions $f(n)$, and positive real numbers x.

We shall show that this inequality has the following as a dual:

Lemma 4.2. *The inequality*

$$\sum_{p^k \le x} p^k \left| \sum_{\substack{n \le x \\ p^k \| n}} a_n - \frac{1}{p^k}\left(1 - \frac{1}{p}\right) \sum_{n \le x} a_n \right|^2 \le 32x \sum_{n \le x} |a_n|^2$$

holds uniformly for all real $x \ge 0$, and complex numbers a_n, $n = 1, \ldots, [x]$.

Remark. We shall show that in place of the factor $32x$ one may obtain the function

$$\Delta = 2x + \left(\sum_{\substack{p^k q^l \le x \\ p \ne q}}\sum p^k q^l\right)^{1/2} + 4\left(\sum_{p^k \le x} \frac{1}{p^k} \sum_{q^l \le x} q^l\right)^{1/2}.$$

As $x \to \infty$ this has the estimate (see lemmas (2.5) and (2.7))

$$\Delta = 2x + O\left(x\left(\frac{\log\log x}{\log x}\right)^{1/2}\right)$$

so that

$$\limsup_{x\to\infty} x^{-1}\Delta \le 2,$$

which improves a result of Kubilius [7], who obtained the upper bound 2.08. The exact value of this upper limit is not known, but in the paper to which we have just referred it is shown, by an example, that it is at least 1.47.

Proof of lemma (4.1). Without loss of generality we may assume that x is an integer, and that $x \ge 2$. Let us denote $E(x)$ and $D(x)$ by E and D respectively.

Assume, for the moment, that the function $f(n)$ is real and non-negative. The sum which we wish to estimate may be written in the form

$$S = \sum_{n\le x} |f(n)|^2 - 2E\sum_{n\le x} f(n) + xE^2.$$

We invert the order of summation in the first of the sums which appear on the right-hand side, and which we shall denote by S_1, to obtain

$$S_1 = \sum_{p^k\le x} f(p^k)^2 \sum_{\substack{n\le x\\ p^k\|n}} 1 + \sum\sum_{\substack{p^kq^l\le x\\ p\ne q}} f(p^k)f(q^l) \sum_{\substack{n\le x\\ p^k\|n,\ q^l\|n}} 1.$$

The number of integers, not exceeding x, for which $p^k\|n$ and $q^l\|n$, is

$$\left[\frac{x}{p^kq^l}\right] - \left[\frac{x}{p^{k+1}q^l}\right] - \left[\frac{x}{p^kq^{l+1}}\right] + \left[\frac{x}{p^{k+1}q^{l+1}}\right]$$
$$= \frac{x}{p^k}\left(1-\frac{1}{p}\right)\frac{1}{q^l}\left(1-\frac{1}{q}\right) + 2\theta,$$

where $|\theta| \le 1$. Hence, since $f(n) \ge 0$,

$$S_1 \le xD^2 + xE^2 + 2\sum_{\substack{p^kq^l\le x\\ p\ne q}} f(p^k)f(q^l).$$

Similarly, the sum

$$S_2 = E\sum_{n\le x} f(n) = E\sum_{p^k\le x} f(p^k) \sum_{\substack{n\le x\\ p^k\|n}} 1$$

is at least as large as

$$E^2 - E \sum_{p^k \le x} f(p^k).$$

Putting these estimates together, the terms involve E^2 cancel, leaving us with the upper bound

$$S = S_1 - 2S_2 + xE^2 \le xD^2 + 2 \sum_{\substack{p^k q^l \le x \\ p \ne q}} \sum f(p^k) f(q^l) + 2E \sum_{p^k \le x} f(p^k).$$

The last two sums which appear in this line may be estimated, in turn, using appropriate applications of the Cauchy–Schwarz inequality:

$$2 \sum_{\substack{p^k q^l \le x \\ p \ne q}} \sum \frac{f(p^k)}{p^{k/2}} \frac{f(q^l)}{q^{l/2}} \cdot p^{k/2} q^{l/2} \le 2D^2 \Bigg(\sum_{\substack{p^k q^l \le x \\ p \ne q}} \sum p^k q^l \Bigg)^{1/2},$$

$$2E \sum_{p^k \le x} f(p^k) p^{-k/2} \cdot p^{k/2} \le 2 \Bigg(D^2 \sum_{p^k \le x} \frac{1}{p^k} \cdot D^2 \sum_{p^k \le x} p^k \Bigg)^{1/2}.$$

We have shown that when $f(n)$ is real and non-negative the inequality of lemma (4.1) is valid with $\Delta/2$ in place of the factor $32x$.

We shall now remove the restrictions upon f.

If f is real, but possibly assuming negative values, we define additive functions $g_j(n)$, $(j = 1, 2)$, by

$$g_1(p^k) = \begin{cases} f(p^k) & \text{if } f(p^k) \ge 0, \\ 0 & \text{if } f(p^k) < 0, \end{cases}$$

$$g_2(p^k) = \begin{cases} 0 & \text{if } f(p^k) \ge 0, \\ -f(p^k) & \text{if } f(p^k) < 0. \end{cases}$$

Corresponding to these functions define

$$E_j = \sum_{p^k \le x} \frac{g_j(p^k)}{p^k} \left(1 - \frac{1}{p}\right) \qquad (j = 1, 2).$$

Then

$$f(n) = g_1(n) - g_2(n),$$

so that, applying the Cauchy–Schwarz inequality once again,

$$|f(n) - E|^2 \leq 2 \sum_{j=1}^{2} |g_j(n) - E_j|^2.$$

Hence

$$\sum_{n \leq x} |f(n) - E|^2 \leq 2 \sum_{j=1}^{2} \sum_{n \leq x} |g_j(n) - E_j|^2$$

$$\leq \Delta \sum_{p^k \leq x} p^{-k} \sum_{j=1}^{2} |g_j(p^k)|^2 = \Delta D^2,$$

and we have established our desired inequality for real functions.

If, now, $f(n)$ is complex-valued, define additive functions

$$g_1(n) = \operatorname{Re} f(n) \qquad g_2(n) = \operatorname{Im} f(n),$$

so that, in an obvious abuse of notation,

$$|f(n) - E|^2 = \sum_{j=1}^{2} |g_j(n) - E_j|^2.$$

Summing over n, $1 \leq n \leq x$, and noting that

$$\sum_{j=1}^{2} \sum_{p^k \leq x} p^{-k} |g_j(k)|^2 = D^2$$

we complete the proof of lemma (4.1); for it is a straightforward calculation to show that Δ does not exceed $32x$ when $x \geq 2$. The details of such a calculation appear in the notes at the end of the present chapter.

A Principle of Duality

Lemma 4.3. *Let c_{ij}, $(i = 1, \ldots, m)$, $(j = 1, \ldots, n)$, be mn complex numbers. Let λ be a real number.*

Then the inequality

$$\sum_{i=1}^{m} \left| \sum_{j=1}^{n} c_{ij} a_j \right|^2 \leq \lambda \sum_{j=1}^{n} |a_j|^2$$

is valid for all complex numbers $a_1, \ldots, a_n$, *if and only if the inequality*

$$\sum_{j=1}^{n} \left| \sum_{i=1}^{m} c_{ij} b_i \right|^2 \leq \lambda \sum_{i=1}^{m} |b_i|^2$$

is valid for all complex numbers $b_1, \ldots, b_m$.

Remark. This result can be viewed from a number of positions. We shall presently consider it in more detail. First we shall prove it, and then use it to establish lemma (4.2).

Proof of lemma (4.3). The proof is a simple exercise in the use of the Cauchy–Schwarz inequality. Consider the matrix

$$\mathbf{C} = (c_{ij}) \qquad 1 \leq i \leq m, \quad 1 \leq j \leq n.$$

We write $\mathbf{a}$ and $\mathbf{b}$ for the (vertical) column vectors with entries $a_1, \ldots, a_n$, and $b_1, \ldots, b_m$, respectively. Then the first of the two inequalities which appear in the statement of lemma (4.3) may be written in the form

$$|\mathbf{Ca}|^2 \leq \lambda |\mathbf{a}|^2 \quad \text{for all } \mathbf{a}.$$

Let us assume it to be valid. Clearly $\lambda \geq 0$, so that (with the interpretation $0^{1/2} = 0$, if necessary),

$$|\mathbf{Ca}| \leq \lambda^{1/2} |\mathbf{a}|.$$

Then, for any (m-dimensional) vector $\mathbf{b}$, whose transpose we shall denote by ^{T}, we have

$$\begin{aligned} |\mathbf{b}^T\mathbf{Ca}| &\leq |\mathbf{b}^T||\mathbf{Ca}| && \text{(by the Cauchy–Schwarz inequality),} \\ &\leq \lambda^{1/2}|\mathbf{b}||\mathbf{a}| && \text{(from the temporary hypothesis).} \end{aligned}$$

We set $\mathbf{a} = \overline{\mathbf{C}}^T\overline{\mathbf{b}}$, so that

$$|\mathbf{b}^T\mathbf{C}|^2 \leq \lambda^{1/2}|\mathbf{b}||\mathbf{b}^T\mathbf{C}|$$

Therefore

$$|\mathbf{b}^T\mathbf{C}| \leq \lambda^{1/2}|\mathbf{b}| = \lambda^{1/2}|\mathbf{b}^T|$$

unless perhaps $\mathbf{b}^T\mathbf{C} = \mathbf{0}^T$, when the inequality holds trivially. This gives the second of the two inequalities in lemma (4.3).

The converse implication is obtained immediately by interchanging the rôles of i and j.

This completes the proof of lemma (4.3).

Proof of lemma (4.2). Let

$$c(p^k, n) = \begin{cases} p^{k/2} - p^{-k/2}(1 - p^{-1}) & \text{if } p^k \| n, \\ -p^{-k/2}(1 - p^{-1}) & \text{otherwise.} \end{cases}$$

Then, if in lemma (4.1) we replace each $f(p^k)$ by $p^{k/2}f(p^k)$, we may write the inequality of that lemma in the form

$$\sum_{n \le x} \left| \sum_{p^k \le x} f(p^k)c(p^k, n) \right|^2 \le 32x \sum_{p^k \le x} |f(p^k)|^2.$$

Since this inequality holds for all complex numbers $f(p^k)$, by lemma (4.3)

$$\sum_{p^k \le n} \left| \sum_{n \le x} c(p^k, n)a_n \right|^2 \le 32x \sum_{n \le x} |a_n|^2$$

holds for all complex numbers a_n, $n = 1, \ldots, [x]$.

It is clear that

$$\sum_{n \le x} c(p^k, n)a_n = p^{k/2}\left\{ \sum_{\substack{n \le x \\ p^k \| n}} a_n - p^{-k}(1 - p^{-1}) \sum_{n \le x} a_n \right\},$$

so that lemma (4.2) is proved.

It is sometimes convenient to use a variant of lemma (4.1). We state two such variants here, together with their duals. The first variant may be deduced from lemma (4.1), and the second can be established along the lines of the proof of lemma (4.1). We may refer to either of these variants as the Turán–Kubilius inequality.

Lemma 4.4. *The inequality*

$$\sum_{n \le x} \left| f(n) - \sum_{p^k \le x} p^{-k} f(p^k) \right|^2 \le 45xD^2(x)$$

holds uniformly for all additive functions $f(n)$, and positive real numbers x.

Define

$$A(x) = \sum_{p \le x} p^{-1} f(p)$$

$$B(x) = \left(\sum_{p \le x} p^{-1} |f(p)|^2 \right)^{1/2} \ge 0.$$

Lemma 4.5. *The inequality*

$$\sum_{n \le x} |f(n) - A(x)|^2 \le 16xB^2(x)$$

holds uniformly for all strongly additive functions $f(n)$, and positive real numbers x.

The respective duals of these last two inequalities are

Lemma 4.6. *The inequality*

$$\sum_{p^k \le x} p^k \left| \sum_{\substack{n \le x \\ p^k \| n}} a_n - p^{-k} \sum_{n \le x} a_n \right|^2 \le 45x \sum_{n \le x} |a_n|^2$$

holds uniformly for all real $x \ge 0$ and all complex numbers a_n, $n = 1, \ldots, [x]$.

Lemma 4.7. *The inequality*

$$\sum_{p \le x} p \left| \sum_{\substack{n \le x \\ n \equiv 0 (\mathrm{mod}\ p)}} a_n - p^{-1} \sum_{n \le x} a_n \right|^2 \le 16x \sum_{n \le x} |a_n|^2$$

holds uniformly for all real numbers $x \ge 0$, and complex numbers a_n, $n = 1, \ldots, [x]$.

The constants 32, 45 etc., which appear in the above inequalities, are far from being the best possible. For example, in a later section of this chapter we shall show that lemmas (4.5) and (4.7) are certainly valid with 16 replaced by 4. Although the value of these constants will not be of importance in the present monograph, there are applications of the above inequalities where a smaller constant improves the final result. We give one example, taken from the author's paper, ref. [11].

The Least Pair of Quadratic Non-Residues (mod p)

Let p be an odd prime. There are many problems which ask for the least residue class representative (mod p) which satisfies a certain property. As one example we note the least positive quadratic non-residue n_2 (mod p).

That is to say, n_2 is the smallest integer greater than zero for which the equation

$$n_2 \equiv y^2 \pmod{p}$$

is not soluble. According to a conjecture of I. M. Vinogradov

$$n_2 = O(p^{\varepsilon})$$

holds for every fixed $\varepsilon > 0$. This much, and more, is certainly derivable from an extended form of the Riemann hypothesis.

The Pólya–Vinogradov inequality asserts, in particular, that the Legendre symbol satisfies

$$\sum_{m \le H} \left(\frac{m}{p}\right) = O(p^{1/2} \log p)$$

uniformly for all real values of H, and primes p. It follows at once that

$$n_2 = O(p^{1/2} \log p).$$

By appealing to the fact that the product of quadratic residues is also a quadratic residue (mod p), Vinogradov improved this estimate to

$$n_2 = O(p^{1/(2\sqrt{e})+\varepsilon})$$

for any fixed $\varepsilon > 0$.

After a period of about forty years, Burgess proved that if $\varepsilon > 0$ is fixed, and $H \ge p^{(1/4)+\varepsilon}$, then there exist positive numbers c_0 and δ, depending upon ε alone, so that

$$\left| \sum_{m \le H} \left(\frac{m}{p}\right) \right| \le c_0 p^{-\delta} H.$$

Once again,

$$n_2 = O(p^{(1/4)+\varepsilon}),$$

and using Vinogradov's device

$$n_2 = O(p^{(1/4\sqrt{e})+\varepsilon}).$$

For a detailed proof of this last result, together with further references, see Burgess [1], and there the matter presently rests.

For each odd prime $p \geq 5$ let $\alpha(p)$ denote the least positive residue class $n \pmod p$ so that both n and $n + 1$ are quadratic non-residues (mod p). It is also likely that $\alpha(p) = O(p^{\varepsilon})$ for any fixed $\varepsilon > 0$, but we cannot prove anything like this at present.

From the estimate

$$\sum_{m \leq H} \left(\frac{m(m+1)}{p}\right) = O(p^{1/2} \log p)$$

which can be deduced from results of A. Weil [1], it follows that

$$\alpha(p) = O(p^{1/2} \log p).$$

Note that an elementary account of the character sum estimates necessary for these last two inequalities may be found in Stepanov [1].

An improvement upon this estimate for $\alpha(p)$ was given in the author's paper, ref. [5], where it was shown that

$$\alpha(p) = O(p^{(1/4)+\varepsilon}).$$

In this problem there does not seem to be a natural structural argument which will play the rôle of Vinogradov's device. We shall show that in problems of such a type a small improvement may be obtained by making use of a form of lemma (4.2).

Theorem 4.8. *Let ε be a fixed positive number. Then, for a suitably chosen constant c, the inequality*

$$\alpha(p) \leq c p^{(1/4)(1-(1/2)e^{-10})+\varepsilon}$$

holds for every prime $p \geq 5$.

Well, it *is* an improvement!

We need a simple preliminary result.

Define the function

$$\sigma(n) = \left(\frac{n}{p}\right) + \left(\frac{2n+1}{p}\right) + \left(\frac{n+1}{p}\right).$$

Lemma 4.9. *If $2(n + 1) < p$ there are two possibilities:*

(i) *One of the pairs $(n, n + 1)$, $(2n, 2n + 1)$, $(2n + 1, 2n + 2)$ consists of quadratic non-residues only.*
(ii) $\sigma(n) \geq 1$.

Proof. Assume that the possibility (i) fails. If both n and $n + 1$ are quadratic residues (mod p) then we already have (ii). Hence, without loss of generality,

$$\left(\frac{2n}{p}\right) + \left(\frac{2n+2}{p}\right) = \left(\frac{2}{p}\right)\left\{\left(\frac{n}{p}\right) + \left(\frac{n+1}{p}\right)\right\} = 0.$$

Since the second and third cases of (i) fail, $2n + 1$ must be a quadratic residue (mod p), and $\sigma(n) = 1$.

Whilst we could now apply lemma (4.2) a better result may be obtained by making use of the following variant of it:

Lemma 4.10. *The inequality*

$$\sum_{\substack{q \le x^{1/2} \\ q \text{ prime}}} q \left| \sum_{\substack{n \le x \\ n \equiv 0 (\text{mod } q)}} a_n - q^{-1} \sum_{n \le x} a_n \right|^2 \le (x + 36x(\log x)^{-1}) \sum_{n \le x} |a_n|^2$$

holds for all real $x \ge 2$, *and complex numbers* a_n, $n = 1, \ldots, [x]$.

Proof. Define

$$g(n) = \sum_{q | n,\, q \le x^{1/2}} f(q) \qquad G(x) = \sum_{q \le x^{1/2}} q^{-1} f(q).$$

Then in much the same manner that we established lemma (4.1) we prove that

$$\sum_{n \le x} |g(n) - G(x)|^2 \le (x + 36x(\log x)^{-1}) \sum_{q \le x} q^{-1} |f(q)|^2.$$

Here we deal with the complex-valued function $g(n)$ directly, without introducing functions corresponding to the $g_j(n)$, $(j = 1, 2)$. That is facilitated by the fact that if q and l are primes which do not exceed $x^{1/2}$, then ql does not exceed x.

Dualising, according to lemma (4.3), we at once obtain the desired result.

We can now prove the theorem.

Set

$$a_n = \begin{cases} 2\left(\dfrac{n/2}{p}\right) & \text{if } n \text{ is even,} \\[2ex] \left(\dfrac{n}{p}\right) & \text{if } n \text{ is odd.} \end{cases}$$

We apply lemma (4.10), in a slight abuse of notation:

$$\sum_{q \le x^{1/2}} q \left| \sum_{\substack{n \le x \\ n \equiv 0 (\mathrm{mod}\ q)}} a_n - q^{-1} \sum_{n \le x} a_n \right|^2 \le (1 + o(1))x \left\{ \sum_{2m \le x} 4 + \sum_{2m+1 \le x} 1 \right\}$$

$$\le (1 + o(1))\tfrac{5}{2}x^2,$$

where q runs through prime numbers.

If $q \ge 3$ and $q | n$ then we have $n = qr$, when r has the same parity as n. Therefore

$$\sum_{\substack{n \le x \\ n \equiv 0 (\mathrm{mod}\ q)}} a_n = \left(\frac{q}{p}\right) \sum_{n \le x} a_n.$$

Let μ be a real number in the range $0 < \mu < 1$, temporarily considered fixed. Then

$$\min_{(\sqrt{x})^\mu < q \le \sqrt{x}} q^2 \left| \sum_{n \le q^{-1}x} a_n - q^{-1}\left(\frac{q}{p}\right) \sum_{n \le x} a_n \right|^2$$

$$\le (1 + o(1))\tfrac{5}{2}x^2 \left\{ \sum_{x^{\mu/2} < q \le x^{1/2}} \frac{1}{q} \right\}^{-1} = (1 + o(1))x^2 5 \Big/ \left(2 \log \frac{1}{\mu}\right).$$

We now introduce $\sigma(n)$. For any $y \ge 2$:

$$\sum_{n \le y} a_n = \sum_{2m \le y} \left\{ 2\left(\frac{m}{p}\right) + \left(\frac{2m+1}{p}\right) \right\} + O(1)$$

$$= \sum_{m \le y/2} \left\{ \left(\frac{m}{p}\right) + \left(\frac{m+1}{p}\right) + \left(\frac{2m+1}{p}\right) \right\} + O(1)$$

$$= \sum_{m \le y/2} \sigma(m) + O(1),$$

where the term $O(1)$ is absolutely bounded.

Let $x = p^{(1/4)+\varepsilon}$, and let q be a prime for which the above minimum is attained, $x^{\mu/2} < q \le x^{1/2}$. From Burgess' character sum estimate

$$\sum_{n \le x} a_n = \sum_{m \le x/2} \sigma(m) + O(1) = O(p^{-\delta}x)$$

so that for our special prime q:

$$\left| \sum_{m \le x/(2q)} \sigma(m) \right| \le O\left(p^{-\delta}\frac{x}{q}\right) + \frac{x}{q}(1 + o(1))\left(\frac{5}{2\log(1/\mu)}\right)^{1/2}.$$

We set $\mu = \exp(-10 - \varepsilon)$, and note that when p (and so x) is sufficiently large, the upper bound in this last inequality is less than $[x/(2q)]$. Then, from lemma (4.9),

$$\alpha(p) \leq 2\left[\frac{x}{2q}\right] + 1 \leq 2x^{1-\mu/2} \leq p^{((1/4)+\varepsilon)(1-(1/2)e^{-10}+O(\varepsilon))}.$$

Since ε may be initially chosen to be arbitrarily small, the proof of theorem (4.8) is complete.

Remarks. It is clear that the result of theorem (4.8) would be improved if one could replace the restriction $p \leq x^{1/2}$ of lemma (4.10) by a weaker one of the form $p \leq x^{\beta}$, for some fixed $\beta > 1/2$.

Consider a sequence of primes p for which the number of distinct prime divisors of $(p-1)$ is uniformly bounded. Suppose that for this sequence

$$4^{\omega(p-1)}\left(\frac{p-1}{\varphi(p-1)}\right)^2 \leq c.$$

Then the method of proof of theorem (4.8) allows one to assert that for such primes, $g(p)$, the least positive primitive root (mod p), satisfies

$$g(p) = O(p^{(1/4)(1-(1/2)e^{-c})+\varepsilon})$$

for each fixed $\varepsilon > 0$, the implied constant depending at most upon ε and c.

This ends the remarks.

Further Inequalities

For use in Chapter 12 we note two further examples of inequalities of the Turán–Kubilius type.

As in Chapter 3, let $h(w)$ be a polynomial in w, with integer coefficients which is positive for $w > w_0(\geq o)$. Let $\rho(p)$ denote the number of distinct residue class solutions to the congruence $h(k) \equiv 0 (\text{mod } p)$, and let $\eta(p)$ denote the number of such solutions which are prime to p.

Lemma 4.11. *Let β be a (fixed) real number, $0 < \beta < 1$. Define*

$$g(n) = \sum_{p|n,\ p\leq x^{\beta}} f(p)$$

$$A(x) = \sum_{p\leq x} \rho(p)\frac{f(p)}{p}$$

$$B(x) = \left(\sum_{p\leq x} \rho(p)\frac{|f(p)|^2}{p}\right)^{1/2} \geq 0.$$

Then there is a positive constant c, depending at most upon β and h(w), so that the inequality

$$\sum_{w_0 < n \leq x} |g(h(n)) - A(x^\beta)|^2 \leq cxB^2(x^\beta)$$

holds uniformly for all $x > w_0$, and complex numbers $f(p)$.

Lemma 4.12. *Let β be a (fixed) real number, $0 \leq \beta < 1$. Define*

$$g(n) = \sum_{p|n,\, p \leq x^\beta} f(p)$$

$$A(x) = \sum_{p \leq x} \eta(p) \frac{f(p)}{\varphi(p)}$$

$$B(x) = \left(\sum_{p \leq x} \eta(p) \frac{|f(p)|^2}{\varphi(p)} \right)^{1/2} \geq 0.$$

Then there is a positive constant c, depending at most upon β and h(w), so that the inequality

$$\sum_{\substack{w_0 < q \leq x, \\ q \text{ prime}}} |g(h(q)) - A(x^\beta)|^2 \leq c \frac{x}{\log x} B^2(x^\beta)$$

holds uniformly for $x > \max(w_0, 2)$, *and complex numbers $f(p)$.*

Proofs. These results may be established in essentially the same way as lemma (4.1). We confine ourselves to a few remarks concerning the proof of lemma (4.12).

Suppose first that $\beta \leq 1/6$. Then we seek an estimate for the expression

$$V = \sum_{w_0 < q \leq x} g^2(h(q)) - 2A(x^\beta) \sum_{w_0 < q \leq x} g(h(q)) + A^2(x^\beta) \sum_{w_0 < q \leq x} 1,$$

where $f(p)$ is real, and non-negative, for each prime p. Typically, we examine the sum

$$S_1 = \sum_{w_0 < q \leq x} g^2(h(q))$$

$$= \sum_{p \leq x^\beta} f^2(p) \sum_{h(q) \equiv 0 (\text{mod } p)} 1 + \sum_{\substack{p, l \leq x^\beta \\ p \neq l}} \sum f(p) f(l) \sum_{h(q) \equiv 0 (\text{mod } pl)} 1.$$

From lemma (2.7), (Brun–Titchmarsh), the double-sum which appears on the right-hand side of this equation does not exceed

$$\frac{c_1 x}{\log x} \sum_{p \le x^\beta} \eta(p) \frac{f^2(p)}{\varphi(p)} = \frac{c_1 x}{\log x} B^2(x^\beta).$$

We estimate the triple sum, on the same side of the equation, by means of lemma (2.10). For $1 \le D \le x$, let

$$E(x, D) = \max_{(l, D) = 1} \left| \pi(x, D, l) - \frac{Li(x)}{\varphi(D)} \right|.$$

Then the triple sum does not exceed

$$Li(x)A^2(x^\beta) + \sum_{\substack{p,l \le x^\beta \\ p \ne l}}\sum f(p)f(l)\eta(pl)\{E(x, pl) + 2w_0\}$$

Note that, applying the Cauchy–Schwarz inequality

$$\sum_{\substack{p,\, l \le x^\beta \\ p \ne l}}\sum f(p)f(l)\left\{\frac{\eta(pl)}{\varphi(pl)}\right\}^{1/2} \{\eta(pl)\varphi(pl)\}^{1/2}\{E(x, pl) + 2w_0\}$$

$$\le B^2(x^\beta)\left(\sum_{\substack{p,\, l \le x^\beta \\ p \ne l}}\sum \eta(pl)\varphi(pl)\{E(x, pl) + 2w_0\}^2\right)^{1/2}.$$

Since each $pl \le x^{2\beta} \le x^{1/3}$, and $\eta(pl) = \eta(p)\eta(l)$ is uniformly bounded independently of x, this last sum, over the pairs of primes p and l, is

$$O\left(\sum_{pl \le x^{1/3}}\sum (E(x, pl) + w_0)x(\log x)^{-1}\right)$$

by the Brun–Titchmarsh lemma,

$$= O(x^2(\log x)^{-2A})$$

by lemma (2.10), for any fixed $A > 0$.

We have shown that

$$S_1 \le Li(x)A^2(x^\beta) + c_2 \frac{x}{\log x} B^2(x^\beta).$$

Arguing similarly for the other terms involved in V, we obtain for it an upper-bound of the form

$$\frac{c_3 x}{\log x} B^2(x^\beta).$$

Following the arguments which were used in the proof of lemma (4.1) we can complete the proof of the present lemma when β does not exceed 1/6.

If $1/6 < \beta < 1$ define

$$g_1(n) = \sum_{p|n,\, p \le x^{1/6}} f(p)$$

$$g_2(n) = \sum_{p|n,\, x^{1/6} < p \le x^\beta} f(p)$$

$$A_2 = \sum_{x^{1/6} < p \le x^\beta} \eta(p) \frac{f(p)}{\varphi(p)}.$$

Then, applying the Cauchy–Schwarz inequality once again,

$$|g(h(q)) - A(x^\beta)|^2 \le 2|g_1(h(q)) - A(x^{1/6})|^2 + 2|g_2(h(q)) - A_2|^2$$

so that the sum we wish to estimate does not exceed

$$2 \sum_{w_0 < q \le x} |g_2(h(q)) - A_2|^2 + O\left(\frac{x}{\log x} B^2(x^{1/6})\right).$$

For primes q in the range $w_0 < q \le x$, the integer $h(q)$ cannot have more than $O(1)$ distinct prime divisors p which exceed $x^{1/6}$. Hence

$$\begin{aligned} 2\sum_q |g_2(h(q)) - A_2|^2 &\le 4 \sum |g_2(h(q))|^2 + 4 \frac{x}{\log x} |A_2|^2 \\ &\le c_4 \sum_q \sum_{\substack{p|h(q) \\ x^{1/6} < p \le x^\beta}} |f(p)|^2 + 4 \frac{x}{\log x} B^2(x^\beta) \sum_{x^{1/6} < p \le x^\beta} \frac{1}{p} \\ &\le c_5 \frac{x}{\log x} B^2(x^\beta). \end{aligned}$$

Here we have made use of the Brun–Titchmarsh lemma in the final step, and the Cauchy–Schwarz inequality in the penultimate step, together with the following estimate, which is derived from lemma (2.5):

$$\sum_{x^{1/6} < p \le x^\beta} \frac{1}{p} = \log\left(\frac{\log x^\beta}{\log x^{1/6}}\right) + O\left(\frac{1}{\log x}\right) = O(1).$$

This completes our sketch of the proof of lemma (4.12).

The proof of lemma (4.11) is similar but simpler. Towards the end of this chapter, in lemma (4.18) we shall establish a form of lemma (4.11) in which there is no truncation condition $p \le x^{\beta}$, but which applies only to (not-necessarily integral) moments of order lower than two.

More on the Duality Principle

Let us consider the situation which is described in lemma (4.3) in more detail.

The matrix $\mathbf{A} = \overline{\mathbf{C}}^T\mathbf{C}$ is Hermitian, of order $n \times n$. It is well known (see, for example, Mirsky [1] Theorem 12.6.1, p. 387), that every Hermitian form may be diagonalised by a suitable unitary transformation. Thus, there is a matrix $\mathbf{U}$, satisfying

$$\mathbf{U}\overline{\mathbf{U}}^T = \mathbf{I},$$

with the appropriate identity matrix $\mathbf{I}$, so that if $\mathbf{x} = \mathbf{U}\mathbf{y}$, then

$$|\mathbf{C}\mathbf{x}|^2 = \overline{\mathbf{x}}^T\mathbf{A}\mathbf{x} = \overline{\mathbf{y}}^T\overline{\mathbf{U}}^T\overline{\mathbf{C}}^T\mathbf{C}\mathbf{U}\mathbf{y} = \sum_{j=1}^{n} \lambda_j |y_j|^2,$$

where $\mathbf{y}^T = (y_1, \ldots, y_n)$. Here the λ_j are the eigenvalues of the non-negative definite matrix $\mathbf{A}$, they are real and non-negative. Without loss of generality we shall therefore assume that

$$\lambda_1 \ge \lambda_2 \ge \cdots \ge \lambda_n \ge 0.$$

In particular

$$|\mathbf{C}\mathbf{x}|^2 \le \lambda_1 |\mathbf{y}|^2 = \lambda_1 |\mathbf{x}|^2,$$

where λ_1 is a maximal eigenvalue. Moreover, it is possible to choose a non-zero vector $\mathbf{x}$ with which to effect equality.

It follows that in lemma (4.3) λ can be taken to have the value λ_1, the greatest eigenvalue (=spectral radius) of the matrix (operator) $\overline{\mathbf{C}}^T\mathbf{C}$, and that if the a_j are to be unrestricted, then this choice is best possible.

The inequality

$$|\mathbf{b}^T\mathbf{C}|^2 \le \lambda |\mathbf{b}^T|^2$$

can be considered in the same way. Here the associated Hermitian matrix is $\mathbf{C}\overline{\mathbf{C}}^T$. It, also, is non-negative definite, with eigenvalues

$$\mu_1 \ge \mu_2 \ge \cdots \ge \mu_m \ge 0,$$

say. The best value for λ in this last inequality, with the vectors $\mathbf{b}$ unrestricted, is $\lambda = \mu_1$.

The result of lemma (4.3) may now be reformulated as $\lambda_1 = \mu_1$. In fact more is true, namely

Lemma 4.13. *Let* $w = \min(m, n)$. *Then* $\lambda_i = \mu_i$ *for* $i = 1, \ldots, w$. *All remaining eigenvalues are zero.*

Remark. Considerations of rank show that when $i > w$, if λ_i or μ_i exist then they have the value zero.

We need the following preliminary result, which is sometimes known as the Courant–Fisher theorem:

Lemma 4.14. *For any vectors* $\mathbf{p}_1, \ldots, \mathbf{p}_s$, $1 \leq s \leq w$, *the inequality*

$$\max|\mathbf{b}^T\mathbf{C}|^2/|\mathbf{b}|^2 \geq \lambda_{s+1}$$

holds when the maximum is taken over the non-zero vectors $\mathbf{b}$ *which satisfy*

$$\mathbf{p}_v^T \cdot \mathbf{b} = 0 \qquad (v = 1, \ldots, s).$$

Moreover, a set of vectors $\mathbf{p}_v$ *exists for which equality is attained.*

Remark. The vectors $\mathbf{p}_v$ and $\mathbf{b}$ are, of course, $1 \times n$.

Proof of lemma (4.12). This result follows easily from our remarks concerning the reduction of the matrix $\mathbf{A}$ $(= \overline{\mathbf{C}}^T\mathbf{C})$ to a diagonal form by means of the transformation $\mathbf{x} = \mathbf{U}\mathbf{y}$.

Let $\mathbf{u}_1^T, \ldots, \mathbf{u}_n^T$ denote the rows of the matrix $\mathbf{U}$. Let s be an integer in the range $0 \leq s \leq n - 1$. Let $\mathbf{p}_1, \ldots, \mathbf{p}_s$ be any s vectors. The dimension of the space spanned by the vectors

$$\mathbf{p}_v, (v = 1, \ldots, s) \qquad \mathbf{u}_k, k = (s + 2, \ldots, n),$$

is at most

$$s + n - (s + 1) = n - 1.$$

Hence we can find a non-zero vector $\mathbf{x}$ which satisfies all of the conditions

$$\mathbf{p}_v \cdot \mathbf{x} = 0 \qquad 1 \leq v \leq s,$$

$$\mathbf{u}_k \cdot \mathbf{x} = 0 \qquad s + 2 \leq k \leq n.$$

Clearly

$$|\mathbf{C}\mathbf{x}|^2 = \sum_{i=1}^{n} \lambda_i |y_i|^2 = \sum_{i=1}^{s+1} \lambda_i |y_i|^2,$$

since $y_i = \mathbf{u}_i \mathbf{x} = \mathbf{0}$ for $i > s + 1$. Hence

$$\max |\mathbf{C}\mathbf{x}|^2/|\mathbf{x}|^2 \geq (\lambda_{s+1}|\mathbf{y}|^2)/|\mathbf{x}|^2 = \lambda_{s+1}.$$

This is one part of lemma (4.14).

One may obtain equality in this last inequality by means of the choice $\mathbf{p}_\nu = \mathbf{u}_\nu$, $(\nu = 1, \ldots, s)$.

This completes the proof of lemma (4.14).

Proof of lemma (4.13). Let s be an integer, $0 \leq s \leq n - 1$. Let $\mathbf{p}_\nu$, $(\nu = 1, \ldots, s)$, be vectors so that equality holds in the result of lemma (4.14). Then for any vectors $\mathbf{a}$ and $\mathbf{b}$

$$\begin{aligned} |\mathbf{b}^T\mathbf{C}\mathbf{a}|^2 &\leq |\mathbf{b}^T\mathbf{C}|^2|\mathbf{a}|^2, \qquad \text{(the Cauchy–Schwarz inequality)} \\ &\leq \lambda_{s+1}|\mathbf{b}|^2|\mathbf{a}|^2, \end{aligned}$$

provided that $\mathbf{p}_\nu \mathbf{b} = 0$, $(\nu = 1, \ldots, s)$. In particular we can set $\mathbf{b} = \mathbf{C}\mathbf{a}$, and deduce that

$$|\mathbf{C}\mathbf{a}|^2/|\mathbf{a}|^2 \leq \lambda_{s+1}$$

holds for all vectors $\mathbf{a}$, different from zero, which belong to the space

$$\mathbf{p}_\nu^T\mathbf{C}\mathbf{a} = 0 \qquad (\nu = 1, \ldots, s).$$

From a second application of lemma (4.14) we see that the maximum of the left-hand side in the last inequality is at least μ_{s+1}, so that $\mu_{s+1} \leq \lambda_{s+1}$.

The converse inequality, and the validity of lemma (4.13), are now apparent.

This last analysis often enables one to give lower bounds beyond which one cannot sharpen the value of the constant λ in lemma (4.3). In fact

$$\sum_{i=1}^{n} \lambda_i = \operatorname{trace} \overline{\mathbf{C}}^T\mathbf{C} = \sum_{i,j}\sum |c_{ij}|^2$$

$$\sum_{i=1}^{m} \mu_i = \operatorname{trace} \mathbf{C}\overline{\mathbf{C}}^T = \sum_{i,j}\sum |c_{ij}|^2$$

so that

$$\lambda \min(m, n) \geq \sum_{i=1}^{m} \sum_{j=1}^{n} |c_{ij}|^2.$$

The exact determination of the eigenvalues of a given matrix (d_{ij}), $1 \leq i, j \leq J$, is often difficult. The best known universal estimate is that each eigenvalue λ must lie in one of the Gershgorin discs

$$|\lambda - d_{ii}| \leq \sum_{\substack{j=1 \\ j \neq i}}^{J} |d_{ij}| \qquad (i = 1, \ldots, k).$$

We shall make use of this estimate in Chapter 22. It may readily be established by noting that if $\mathbf{x}$ is an eigenvector corresponding to the eigenvalue λ, then for each value of i, $(1 \leq i \leq J)$,

$$(\lambda - d_{ii})x_i = \sum_{\substack{j=1 \\ j \neq i}}^{J} d_{ij} x_j.$$

We choose a value of i for which $|x_i|$ is maximal, and therefore positive. For such a value of i

$$|\lambda - d_{ii}| \leq \sum_{j \neq i} |d_{ij}| \left| \frac{x_j}{x_i} \right| \leq \sum_{j \neq i} |d_{ij}|.$$

The Large Sieve

Although we shall not need the results in this monograph, it is interesting to here consider the inequality of the Large Sieve, and its connection with the Turán–Kubilius inequality.

Let $x_j, (j = 1, \ldots, J)$, be real numbers which satisfy

$$\|x_j - x_k\| \geq \delta > 0 \qquad (j \neq k).$$

Here $\|y\|$ denotes the distance of y from the nearest integer. Then

Lemma 4.15 (The Large Sieve). *The inequality*

$$\sum_j \left| \sum_{n=1}^{N} a_n e^{2\pi i n x_j} \right|^2 \leq (N + \delta^{-1}) \sum_{n=1}^{N} |a_n|^2$$

holds uniformly for all complex numbers a_n, $(n = 1, \ldots, N)$, for all positive integers N.

This result is clearly one which falls within the scope of lemma (4.3). The dual assertion is that the inequality

$$\Gamma = \sum_{n=1}^{N} \left| \sum_{j=1}^{J} c_j e^{2\pi i n x_j} \right|^2 \leq (N + \delta^{-1}) \sum_{j=1}^{J} |c_j|^2$$

holds for all complex numbers c_j, and it is this inequality that we shall first establish.

It is convenient to consider the skew-Hermitian form

$$\sum_{\substack{j=1 \\ j \neq k}}^{J} \sum_{k=1}^{J} \bar{u}_j u_k \operatorname{cosec} \pi(x_j - x_k) = \bar{\mathbf{u}}^T \mathbf{D} \mathbf{u},$$

where

$$\mathbf{D} = (D_{jk}) \qquad 1 \leq j, k \leq J,$$

and

$$D_{jk} = \begin{cases} \operatorname{cosec} \pi(x_j - x_k) & \text{if } j \neq k, \\ 0 & \text{if } j = k. \end{cases}$$

The matrix $\mathbf{D}$ may be diagonalized by means of a unitary transformation. (See, for example, Mirsky [1] p. 305). Its eigenvalues are purely imaginary. Indeed, if τ is an eigenvalue of D, with an associated eigenvector $\mathbf{u}$, normalised so that $|\mathbf{u}| = 1$, then

$$\bar{\tau} = \bar{\tau}^T = (\bar{\tau}|\mathbf{u}|^2)^T = (\overline{\bar{\mathbf{u}}^T \mathbf{D} \mathbf{u}})^T = \bar{\mathbf{u}}^T \bar{\mathbf{D}}^T \mathbf{u} = -\bar{\mathbf{u}} \mathbf{D} \mathbf{u} = -\tau.$$

Set

$$\Delta = \max_k 3\pi^{-2} \sum_{\substack{j=1 \\ j \neq k}}^{J} \|x_j - x_k\|^{-2}.$$

Note that from our hypothesis concerning the points x_j,

$$\Delta < \frac{3}{\pi^2} \sum_{m=1}^{\infty} \frac{2}{(m\delta)^2} = \delta^{-2}.$$

Lemma 4.16. *Each eigenvalue τ, of the matrix D, is bounded by*

$$|\tau|^2 \leq \Delta.$$

Remarks. According to the remark which was made immediately preceding this section, each τ must lie in one of the discs

$$\begin{aligned} |\tau| &\leq \sum_{j \neq k} |\operatorname{cosec} \pi(x_j - x_k)| \\ &\leq \sum_{j \neq k} |\operatorname{cosec} \pi\|x_j - x_k\|| < \frac{1}{2} \sum_{j \neq k} \|x_j - x_k\|^{-1} \\ &\leq \sum_{m \leq 1/(2\delta)} (m\delta)^{-1} < \delta^{-1}(1 + \log \delta^{-1}). \end{aligned}$$

This is slightly weaker than the present result $|\tau| \leq \delta^{-1}$. Whilst this is of no consequence in most applications, there is a certain charm in the sharper inequality, and we shall use it to improve slightly the result of lemma (4.7).

For use later in the proof of this lemma we note that

$$3 \sin^3 \theta - \theta^2(1 + 2 \cos \theta) = \frac{3}{2}(1 - \cos 2\theta) - \theta^2(1 + 2 \cos \theta) = \sum_{k=3}^{\infty} a_k \frac{\theta^{2k}}{(2k)!},$$

where

$$a_k = (-1)^{k+1}(3.2^{2k-1} - 4k(2k - 1)).$$

It is readily checked that for $k \geq 3$, $(-1)^{k+1}a_k > 0$, and, with a certain amount of tedium, that for $|\theta| \leq 2$ the terms of the series given here are decreasing in absolute size. Since they alternate in sign the series must be non-negative for $|\theta| \leq 2$. We set $\theta = \pi w$, and deduce that

$$\operatorname{cosec}^2 \pi w + 2|\cot \pi w \operatorname{cosec} \pi w| \leq 3\pi^{-2}\|w\|^{-2}$$

holds for all real w.

This ends our remarks.

Proof of lemma (4.16). Let τ be an eigenvalue of D, and let $\mathbf{u}$ be a corresponding eigenvector, normalised so that $|\mathbf{u}| = 1$. Hence, for $j = 1, \ldots, J$,

$$\sum_{\substack{k=1 \\ k \neq j}}^{J} u_k \operatorname{cosec} \pi(x_j - x_k) = \tau u_j.$$

Applying the Cauchy–Schwarz inequality to the form $\bar{\mathbf{u}}^T\mathbf{D}\mathbf{u}$ we have

$$|\tau|^2 = |\bar{\mathbf{u}}^T\mathbf{D}\mathbf{u}|^2 \le \sum_j |u_j|^2 \cdot \sum_j \left| \sum_{k \neq j} u_k \operatorname{cosec} \pi(x_j - x_k) \right|^2$$
$$= S_1 + S_2$$

where

$$S_1 = \sum_{j=1}^{J} |u_j|^2 \sum_{\substack{k=1 \\ k \neq j}}^{J} \operatorname{cosec}^2 \pi(x_j - x_k)$$

and

$$S_2 = \sum_k \sum_l \bar{u}_k u_l \sum_{j \neq k, l} \operatorname{cosec} \pi(x_j - x_k) \operatorname{cosec} \pi(x_j - x_l)$$

We simplify the sum S_2 by making use of the identity

$$\operatorname{cosec} \pi(x_j - x_k) \operatorname{cosec} \pi(x_j - x_l)$$
$$= \operatorname{cosec} \pi(x_k - x_l)\{\cot \pi(x_j - x_k) - \cot \pi(x_j - x_l)\}.$$

In fact $S_2 = S_3 - S_4 + 2S_5$, where

$$S_3 = \sum_{k \neq l}\sum \bar{u}_k u_l \operatorname{cosec} \pi(x_k - x_l) \sum_{j \neq k} \cot \pi(x_j - x_k),$$

$$S_4 = \sum_{k \neq l}\sum \bar{u}_k u_l \operatorname{cosec} \pi(x_k - x_l) \sum_{j \neq l} \cot \pi(x_j - x_l),$$

$$S_5 = \sum_{k \neq l} \bar{u}_k u_l \operatorname{cosec} \pi(x_k - x_l) \cot \pi(x_k - x_l).$$

Note that in each of the innersums which appear in the representations for S_3 and S_4, the range $j \neq k, l$ has been enlarged by the addition of one term. This enables us to take advantage of the fact that τ is an eigenvalue of the matrix $\mathbf{D}$. We obtain

$$S_3 = \sum_{j \neq k}\sum \bar{u}_k \cot \pi(x_j - x_k) \sum_{\substack{l=1 \\ l \neq k}}^{J} u_l \operatorname{cosec} \pi(x_k - x_l)$$

$$= \tau \sum_{j \neq k}\sum |u_k|^2 \cot \pi(x_j - x_k)$$

and, similarly,

$$S_4 = -\bar{\tau} \sum_{j \neq l} \sum |u_l|^2 \cot \pi(x_j - x_l).$$

Since τ is imaginary

$$S_3 = S_4.$$

Using the fact that

$$2|\bar{u}_k u_l| \leq |u_k|^2 + |u_l|^2$$

we see that

$$|\tau|^2 \leq \sum_{j=1}^{J} |u_j|^2 \sum_{k \neq j} (\operatorname{cosec}^2 \pi(x_j - x_k) + 2|\operatorname{cosec} \pi(x_j - x_k) \cot \pi(x_j - x_k)|)$$

which, by the second of our remarks, does not exceed

$$\sum_{j=1}^{J} |u_j|^2 \sum_{k \neq j} 3\pi^{-2} \|x_j - x_k\|^{-2} \leq \Delta.$$

This completes the proof of lemma (4.16).

Proof of lemma (4.15). This is now straightforward. Expanding the dual form we have

$$\Gamma = \sum_{j=1}^{J} |c_j|^2 N + \sum_{j \neq k} \sum \bar{c}_j c_k \sum_{n=1}^{N} e^{2\pi i n(x_j - x_k)}.$$

The coefficient of each $\bar{c}_j c_k$ is the sum of a geometric progression, which, since $\|x_j - x_k\| \neq 0$, has the value

$$\frac{e^{2\pi i\alpha} - e^{-2\pi i N\alpha}}{1 - e^{2\pi i\alpha}} = (e^{\pi i\alpha} - e^{2\pi i(N + (1/2))\alpha}) \frac{1}{2i} \operatorname{cosec} \alpha,$$

where $\alpha = x_j - x_k$. We apply lemma (4.16) to the forms

$$\frac{i}{2} \sum_{j \neq k} \sum \overline{c_j e^{\pi i x_j}} c_k e^{\pi i x_k} \operatorname{cosec} \pi(x_j - x_k)$$

and

$$\frac{i}{2}\sum\sum_{j\neq k} \overline{c_j \exp\left(2\pi i\left(N+\frac{1}{2}\right)x_j\right)} c_k \exp\left(2\pi i\left(N+\frac{1}{2}\right)x_k\right) \operatorname{cosec} \pi(x_j - x_k)$$

to deduce that

$$\Gamma = (N + \theta\Delta^{1/2}) \sum_{j=1}^{J} |c_j|^2,$$

where $|\theta| \leq 1$, and $\Delta^{1/2} \leq \delta^{-1}$.

Dualising, we complete the proof of lemma (4.15).

Remarks. The above argument clearly shows that every eigenvalue λ of the matrix associated with the form Γ satisfies

$$|\lambda - N| \leq \delta^{-1},$$

so that if $\delta^{-1} < N$ then each such eigenvalue is non-zero. We are not able to deduce from this any asymptotic estimate of the type

$$\sum_j \left| \sum_{n=1}^{N} a_n e^{2\pi i n x_j} \right|^2 = (N + O(\delta^{-1})) \sum_{n=1}^{N} |a_n|^2,$$

however, since it is clear that $J\delta \leq 1$ so that we should have $J \leq \delta^{-1} < N$. Thus, the matrix underlying the above form in the a_n has at least one eigenvalue which is zero.

This ends our remark.

An Application of the Large Sieve

For each real number α set

$$S(\alpha) = \sum_{n=1}^{N} a_n e^{2\pi i n\alpha}.$$

Let Q be a positive integer, and let x_j, $(j = 1, \ldots, J)$, run through the rational numbers of the form bd^{-1}, where $(b, d) = 1$, $1 \leq b < d$, $1 \leq d < Q$. In this case

$$\min\left|\frac{b_1}{d_1} - \frac{b_2}{d_2}\right| \geq \frac{1}{d_1 d_2} \geq \frac{1}{Q^2},$$

where the minimum is taken over pairs of distinct fractions. From lemma (4.15) we deduce that

$$\sum_{d \le Q} \sum_{\substack{b=1 \\ (b,d)=1}}^{d} \left| S\left(\frac{b}{d}\right) \right|^2 \le (N + Q^2) \sum_{n=1}^{N} |a_n|^2.$$

We let $Q = N^{1/2}$ and confine our attention to prime moduli p in place of the general modulus d. Then

$$\sum_{p \le N^{1/2}} \sum_{b=1}^{p-1} \left| S\left(\frac{b}{p}\right) \right|^2 \le 2N \sum_{n=1}^{N} |a_n|^2.$$

If we set

$$Z = \sum_{n=1}^{N} a_n$$

then

$$\sum_{b=1}^{p-1} \left| S\left(\frac{b}{p}\right) \right|^2 = \sum_{m,n} \sum a_m \bar{a}_n \sum_{b=1}^{p-1} \exp\left(2\pi i(m - n)\frac{b}{p}\right).$$

The innermost sum is $p - 1$ when $m \equiv n \pmod{p}$, and -1 otherwise. Hence the triple sum which occurs on the right-hand side of this equation may be written in the form

$$p \sum_{m \equiv n \pmod{p}} \bar{a}_m a_n - |Z|^2 = p \sum_{r=0}^{p-1} \left| \sum_{n \equiv r \pmod{p}} a_n - p^{-1} Z \right|^2$$

and we have established the inequality

$$\sum_{p \le N^{1/2}} p \sum_{r=0}^{p-1} \left| \sum_{\substack{n=1 \\ n \equiv r \pmod{p}}}^{N} a_n - p^{-1} \sum_{n=1}^{N} a_n \right|^2 \le 2N \sum_{n=1}^{N} |a_n|^2.$$

This inequality is of the same form as that in lemma (4.7), save that the range $p \le x$ has been shortened to (essentially) $p \le x^{1/2}$. To compensate for this an enormous improvement is embraced by the addition of the summation over all residue classes $r \pmod{p}$ for each prime modulus p, and at no extra cost on the right-hand side.

We shall not need the result of the large sieve itself, but we shall need the uniformity $p \le x$ of lemma (4.7). To derive a result of the same type as lemma (4.7) from the large sieve the following result is useful.

Lemma 4.17. *The inequality*

$$\sum_{n\le N}\left|\sum_{\substack{p|n\\ p>N^{1/2}}} f(p) - \sum_{N^{1/2}<p\le N} p^{-1}f(p)\right|^2 \le 2N \sum_{N^{1/2}<p\le N} p^{-1}|f(p)|^2$$

holds for all complex numbers $f(p)$, and positive integers N.

Proof. If we follow the method of proof of lemma (4.1) it will suffice to establish the desired inequality for real and non-negative $f(p)$, and with a factor N in place of $2N$ in the upper bound.

In fact, the sum to be estimated then has the value

$$V = \sum_{N^{1/2}<p\le N} f^2(p) \sum_{\substack{n=1\\ n\equiv 0(\text{mod } p)}}^{N} 1 - 2S \sum_{N^{1/2}<p\le N} f(p) \sum_{\substack{n=1\\ n\equiv 0(\text{mod } p)}}^{N} 1 + NS^2,$$

where

$$S = \sum_{N^{1/2}<p\le N} p^{-1}f(p).$$

For, any integer n not exceeding N can be divisible by at most one prime greater than $N^{1/2}$.

Typically,

$$\sum_{n\equiv 0(\text{mod } p)} 1 = \left[\frac{N}{p}\right] \ge \frac{N}{p} - 1 \ge \frac{N}{2p},$$

if p does not exceed $N/2$. However, if $(N/2) < p \le N$, then

$$\left[\frac{N}{p}\right] = 1 \ge \frac{N}{2p},$$

so that in any case

$$-2S \sum_{N^{1/2}<p\le N} f(p) \sum_{\substack{n=1\\ n\equiv 0(\text{mod } p)}}^{N} 1 \le -2S \sum_{N^{1/2}<p\le N} \frac{1}{2} Np^{-1}f(p) = -NS^2.$$

Hence

$$V \le N \sum_{N^{1/2}<p\le N} p^{-1}f^2(p),$$

and the proof of lemma (4.17) is now readily completed.

Dualising the inequality in lemma (4.17) we obtain

$$\sum_{N^{1/2}<p\leq N} p\left|\sum_{n\equiv 0(\mathrm{mod}\ p)} a_n - p^{-1}\sum_{n=1}^{N} a_n\right| \leq 2N\sum_{n=1}^{N}|a_n|^2.$$

This may be combined with the relevant part of the large sieve inequality, as obtained above for primes not exceeding $N^{1/2}$. Dualising one last time leads, (in the notation of lemma (4.5)), to the inequality

$$\sum_{n\leq N}|f(n) - A(N)|^2 \leq 4NB^2(N).$$

We have replaced the constant 16, in lemma (4.5), by 4.

We now establish the modified form of the Turán–Kubilius inequality which was mentioned following lemma (4.12).

Set

$$A(x) = \sum_{p\leq x}\frac{f(p)}{\varphi(p)}$$

$$B(x) = \left(\sum_{p\leq x}\frac{|f(p)|^2}{\varphi(p)}\right)^{1/2}.$$

Lemma 4.18. *Let α be a real number, $0 \leq \alpha < 2$. Then there is a further number $D = D(\alpha)$ so that the inequality*

$$\sum_{p+1\leq x}|f(p+1) - A(x)|^\alpha \leq D\cdot\frac{x}{\log x}\cdot B(x)^\alpha$$

holds uniformly for all strongly-additive functions $f(n)$, and all real $x \geq 2$.

Remarks. If a and b are real non-negative numbers, then

$$(a+b)^\alpha \leq 2^\alpha(a^\alpha + b^\alpha).$$

Define the additive functions

$$g(n) = \sum_{p|n,\ p\leq x^{1/2}} f(p)$$

$$h(n) = \sum_{p|n,\ p>x^{1/2}} f(p).$$

Then the sum which we wish to estimate does not exceed

$$2^\alpha \sum_{p+1\le x} |g(p+1) - A(x^{1/2})|^\alpha + 2^\alpha \sum_{p+1\le x} |h(p+1) - A(x) + A(x^{1/2})|^\alpha.$$

But, for any fixed $a_i \ge 0$, $(i = 1, \ldots, k)$, the expression

$$\left(k^{-1}\sum_{i=1}^{k} a_i^\alpha\right)^{1/\alpha} \qquad (\alpha > 0),$$

is an increasing function of α. Thus

$$\left(\frac{1}{\pi(x)} \sum_{p+1\le x} |g(p+1) - A(x^{1/2})|^\alpha\right)^{1/\alpha} \le \text{(same expression with } \alpha = 2)$$
$$\le \text{constant} \,.\, B(x^{1/2}),$$

this last step by means of lemma (4.12). Moreover, an application of the Cauchy–Schwarz inequality shows that

$$|A(x) - A(x^{1/2})|^2 \le \sum_{x^{1/2}<p\le x} \frac{1}{p-1} \sum_{p\le x} \frac{|f^2(p)|}{\varphi(p)} = O(B(x)).$$

In order to prove lemma (4.18) it will therefore suffice to show that

$$\sum_{p+1\le x} |h(p+1)|^\alpha \le c\,\frac{x}{\log x}\, B(x)^\alpha,$$

which is an analogue of lemma (4.17).

This we shall do.

We need the following result concerning primes in arithmetic progressions. The particular case $m = 2$ was established by Barban, Vinogradov and Levin [1].

Lemma 4.19. *Let m be a non-negative integer, and δ a real number, $0 < \delta \le 1/2$. Then there is a number c, depending upon m but not δ, so that the inequality*

$$\sum_{x^{1-\delta}<p\le x} p^{m-1}\pi(x, -1, p)^m \le \delta c\left(\frac{x}{\log x}\right)^m$$

holds for all sufficiently large values of x.

Remarks. It is interesting to view this inequality in the form

$$\sum_{x^{1-\delta}<p\le x} p^{-1}\left(\frac{\pi(x,-1,p)}{\varphi(p)^{-1}\pi(x)}\right)^m \le \delta c'.$$

It is clear that we may assume that $m \ge 1$, the case $m = 0$ following at once from lemma (2.5).

Proof. For positive integers k_i, $(i = 1, \ldots, m)$, let

$$N(x; k_1, \ldots, k_m) = \sum{}' p^{m-1}$$

where $'$ indicates that summation is confined to those primes p, $x^{1-\delta} < p \le x$, for which each of the $k_i p - 1$ is a prime not exceeding x, $(i = 1, \ldots, m)$.

Inverting the order of summation, we see that the sum (in lemma (4.19)) which we wish to estimate has the alternative representation

$$S = \sum_{k_1}\cdots\sum_{k_m} N(x; k_1, \ldots, k_m).$$

Here we note that if $pk_i - 1 = q$ (a prime) with $q \le x$, then

$$k_i = \frac{q+1}{p} \le 2x^\delta.$$

Hence

$$S \le m! \sum_{k_1\le\cdots\le k_m\le 2x^\delta}\cdots\sum N(x; k_1, \ldots, k_m).$$

We shall consider the contribution towards this last sum which arises from the m-tuples $(k_1, \ldots, k_m)$ which are distinct; and show that it is

$$O\left(\delta\left(\frac{x}{\log x}\right)^m\right).$$

Since allowing two or more of the k_i to coincide amounts to reducing the value of m, the sum S will have a similar upper bound, and lemma (4.19) will be established.

Consider a (typical) innersum when $1 \le k_1 < k_2 < \cdots < k_m \le 2x^\delta$, where for convenience we shall write k in place of k_m.

Let $\rho(d)$ denote the number of residue classes $r \pmod d$ which satisfy

$$r\prod_{i=1}^{m}(k_i r - 1) \equiv 0 (\text{mod } d).$$

Further, let

$$\Lambda = \prod_{i=1}^{m} k_i \prod_{1 \le i < j \le m} (k_j - k_i).$$

Thus, if p is prime, $p \ge m + 1$, $p \nmid \Lambda$, then $\rho(p) = m + 1$.

Let r denote a real number, $2 \le r \le x^{1/4}$, and let Q denote the product of those primes p in the interval $2(m + 1) < p \le r$ which do not divide Λ. If there are no such primes we set $Q = 1$. Set

$$a_n = n \prod_{i=1}^{m} (k_i n - 1).$$

Then

$$N(x; k_1, \ldots, k_m) \le \sum_{\substack{n \le (x+1)/k \\ (a_n, Q) = 1}} n^{m-1}.$$

We shall estimate this last sum by means of the Selberg sieve method. In order to apply such a method we need an estimate for the sum

$$T_d(y) = \sum_{\substack{n \le y \\ n \equiv 0 (\mathrm{mod}\ d)}} n^{m-1} = d^{m-1} \sum_{w \le y/d} w^{m-1} \qquad (y \ge 1).$$

If $m \ge 2$, $t \ge 1$, then, integrating by parts,

$$\sum_{w \le t} w^{m-1} = \int_{1-}^{t} w^{m-1}\, d[w] = t^{m-1}[t] - \int_{1}^{t} (m-1) w^{m-2} [w] dw$$

$$= \frac{t^m}{m} + 4\theta t^{m-1} \qquad |\theta| \le 1,$$

so that

$$T_d(y) = \frac{y^m}{md} + 4\theta' y^{m-1} \qquad |\theta'| \le 1 \qquad (1 \le d \le y).$$

If $m = 1$ this result is trivially valid.

We now apply lemma (2.1), with

$$f(n) = n^{m-1} \qquad X = \frac{1}{m}\left(\frac{x+1}{k}\right)^m \qquad \eta(d) = d^{-1}\rho(d).$$

We set

$$z = (xk^{-1}(\log x)^{-m-1})^{1/4},$$

and since $k \le 2x^{1/2}$, this ensures that $z \ge x^{1/9}$ for all sufficiently large values of x. Putting $\log r = (\log z)/65m$ we see that

$$\sum_{2(m+1) < p \le r} \frac{\eta(p)}{1 - \eta(p)} \log p \le 8m \sum_{p \le r} \frac{\log p}{p} < \frac{1}{8} \log z.$$

With this choice

$$\sum_{d \le z^3} 3^{\omega(d)} \left| T_d\left(\frac{x+1}{k}\right) - \left(\frac{x+1}{k}\right)^m \frac{1}{md} \right|$$

$$\le 4\left(\frac{x+1}{k}\right)^{m-1} \sum_{d \le z^3} 3^{\omega(d)} \le c_0\left(\frac{x}{k}\right)^{m-1} z^4 \le c_1(\log x)^{-1}\left(\frac{x}{k \log x}\right)^m,$$

and we deduce that

$$N(x; k_1, \ldots, k_m) \le c_2\left(\frac{x}{k}\right)^m \prod_{p|Q}\left(1 - \frac{\rho(p)}{p}\right) + 2c_1(\log x)^{-1}\left(\frac{x}{k \log x}\right)^m$$

$$\le c_3(\log x)^{-1}\left(\frac{x}{k \log x}\right)^m \exp\left(\sum_{\substack{p|\Lambda \\ 2(m+1) < p \le r}} \frac{m+1}{p}\right).$$

Define

$$L_k = \sum_{k_1 < k_2 < \cdots < k_{m-1} < k} \cdots \sum \exp\left(\sum_{p|\Lambda} \frac{m+1}{p}\right),$$

then the sum which we wish to estimate does not exceed

$$W = c_3\left(\frac{x}{\log x}\right)^m (\log x)^{-1} \sum_{k \le 2x^{\delta}} k^{-m} L_k.$$

Define, further,

$$\Gamma = \Gamma(k_{m-1}) = k_{m-1} \prod_{i=1}^{m-2} (k_{m-1} - k_i).$$

Clearly

$$L_k \le \sum_{k_1 < \cdots < k_{m-2} < k} \cdots \sum \exp\left(\sum_{p|\Lambda\Gamma_{-1}} \frac{m+1}{p}\right) \sum_{k_{m-1} < k} \exp\left(\sum_{p|\Gamma} \frac{m+1}{p}\right).$$

Let us estimate the innermost sum. For convenience we shall write l in place of k_{m-1}, and define $k_0 = 0$. Since

$$a_1 \dots a_{m-1} \le a_1^{m-1} + \cdots + a_{m-1}^{m-1}$$

for non-negative numbers a_i, the sum does not exceed

$$U = \sum_{i=0}^{m-2} \sum_{l<k} \exp\left(\sum_{p|(l-k_i)} \frac{m^2-1}{p} \right).$$

Moreover, a typical summand is not more than

$$c_4 \prod_{p|(l-k_i)} \left(1 + \frac{m^2}{p}\right) = c_4 \sum_{d|(l-k_i)} \mu^2(d) m^{2\omega(d)} d^{-1}.$$

Therefore

$$U \le \sum_{i=0}^{m-2} c_4 \sum_{d \le k} m^{2\omega(d)} d^{-1} \sum_{\substack{l<k \\ l \equiv k_i (\text{mod } d)}} 1$$

$$\le mc_4 \max_i \sum_{d \le k} m^{2\omega(d)} d^{-1} \cdot \rho(d) 2kd^{-1} \le c_5 k.$$

We proceed similarly, summing over each of the variables $k_{m-2}, \dots, k_1$ in turn, to arrive at

$$L_k \le c_6 k^{m-1} \exp\left(\sum_{p|k} \frac{m+1}{p} \right) \le c_7 k^{m-1} \cdot \sum_{d|k} \frac{(m+1)^{\omega(d)}}{d}.$$

Hence

$$\sum_{k_1 < \cdots < k_m \le 2x^\delta} \cdots \sum N(x; k_1, \dots, k_m) \le W$$

$$\le c_8 \left(\frac{x}{\log x} \right)^m \cdot (\log x)^{-1} \sum_{k \le 2x^\delta} k^{-1} \sum_{d|k} \frac{(m+1)^{\omega(d)}}{d}$$

$$\le c_8 \left(\frac{x}{\log x} \right)^m \cdot (\log x)^{-1} \sum_{d=1}^{\infty} \frac{(m+1)^{\omega(d)}}{d} \sum_{t \le 2x^\delta/d} \frac{1}{dt}$$

$$\le c_9 \delta \left(\frac{x}{\log x} \right)^m.$$

This completes our proof of lemma (4.19).

Proof of lemma (4.18). Let m be a positive integer, to be chosen presently, and set $\beta = 2(1 - m^{-1})$.

Since each integer n in the range $1 \leq n \leq x + 1$ can have at most two prime divisors q which satisfy $q > x^{1/2}$, we have

$$|h(n)|^\beta \leq 2 \sum_{q|n,\, q > x^{1/2}} |f(q)|^\beta.$$

Hence

$$\sum_{p+1 \leq x} |h(p+1)|^\beta \leq 2 \sum_{x^{1/2} < q \leq x} |f(q)|^\beta \pi(x, -1, q).$$

To this last sum we apply Hölder's inequality with exponents $2/\beta$, $2/(2-\beta)$, in the following manner:

$$\sum_{x^{1/2} < q \leq x} |f(q)|^\beta q^{-\beta/2} \cdot q^{\beta/2} \pi(x, -1, q)$$

$$\leq \left(\sum_{x^{1/2} < q \leq x} |f(q)|^2 q^{-1} \right)^{\beta/2} \left(\sum_{x^{1/2} < q \leq x} q^{\beta/(2-\beta)} \pi(x, -1, q)^{2/(2-\beta)} \right)^{(2-\beta)/2}.$$

From our definition of β, $\beta/(2-\beta) = m - 1$, $2/(2-\beta) = m$, and we may apply lemma (4.19) to obtain for this last expression the upper bound

$$B(x)^\beta \cdot c_{10} x (\log x)^{-1}.$$

Given α, $0 \leq \alpha < 2$, we choose for m a value sufficiently large that $\alpha \leq \beta$. Then

$$\left(\pi(x)^{-1} \sum_{p+1 \leq x} |h(p+1)|^\alpha \right)^{1/\alpha} \leq \text{(a similar expression with } \alpha \text{ replaced by } \beta)$$

$$\leq c_{11} B(x),$$

from what we have just shown.

The proof of lemma (4.18) is now complete.

Remark. In the result of lemma (4.18) we may replace $p + 1$ by $p + l$ for any fixed $l \neq 0$; likewise $\pi(x, -l, q)$ may replace $\pi(x, -1, q)$ in lemma (4.19).

Concluding Remarks

Let us show that the function Δ, which is defined in the remark following the statement of lemma (4.2), is at most $32x$, as asserted earlier.

In lemma (3.1), of Chapter 3, it was shown that for $r > 0$

$$\sum_{p \leq r} \log p \leq (4 \log 2) r < 4r.$$

Hence

$$\sum_{p\le r} 1 \le \sum_{p\le r} 1 + \sum_{p\le r} \frac{\log p}{\log r^{1/2}} \le r^{1/2} + \frac{8r}{\log r} \le \frac{10r}{\log r},$$

since $\log r = 2 \log r^{1/2} \le 2r^{1/2}$.

Next, we note that

$$\sum_{\substack{p^k \le r \\ k\ge 2}}\sum 1 \le \sum_{p\le r^{1/2}} 1 + \sum_{p\le r^{1/3}} 1 \cdots \le \frac{r^{1/2}\log r}{\log 2} \le 32\frac{r}{\log r},$$

since $(\log r)^2 = (4 \log r^{1/4})^2 \le 16r^{1/2}$, and $2 \log 2 = \log 2^2 > \log e = 1$.

As 1 is not a power of a prime

$$\sum_{p^k\le r} \frac{1}{p^k} \le \sum_{2\le m\le r} \frac{1}{m} \le \sum_{2\le m\le r} \int_{m-1}^{m} \frac{dy}{y} \le \int_1^r \frac{dy}{y} = \log r.$$

Hence

$$\Delta \le 2x + \left(2\sum_{m\le x} m\right)^{1/2} + 4\left(\log x \cdot \frac{42x}{\log x}\right)^{1/2} \le x(2 + 2 + 4(7)),$$

which justifies our assertion.

It is clear that the constant 32 could be much improved by a recourse to tables.

Lemma (4.1) for the special function $\omega(m)$ was proved by Turán [1], [2]. More exactly, he showed that

$$\sum_{n\le x} (\omega(n) - \log\log x)^2 = O(x \log\log x).$$

A little later [3] he generalised this result to

$$\sum_{n\le x} (f(n) - A(x))^2 = O(xA(x)),$$

under the conditions that $f(p)$ be real, $0 \le f(p) \le K$. The implied constant depended upon K.

Kubilius established lemma (4.5), for real functions and with an unspecified constant, in his survey article, ref. [1]. It was generalized to complex functions, and lemma (4.6) was proved, in his monograph [5].

Except for a number of simplifying devices, all of these proofs, including our present treatment, are along the original lines of Turán [1]. Note that the inequality in Kubilius' form, with no side condition upon the size of $f(p)$, is the one best suited to dualisation.

The dual (or conjugate) of a linear operator is defined as follows. We adopt the convention of Yosida [1].

Let X be a space, linear over the complex numbers, with norm $\| \quad \|$. The space of bounded linear functionals f on X into the complex numbers may be given a norm by

$$\|f\| = \sup_{x \neq 0} |f(x)|/\|x\|,$$

x being taken over the elements of X. We call this space the (strong) dual of X, and denote it by X'. It may be considered a linear space by defining

$$(\alpha f_1 + \beta f_2)(x) = \alpha f_1(x) + \beta f_2(x).$$

Let Y be a further normed linear space, with dual Y'. Let T be a bounded linear map on X into Y. To each functional g of Y' we make correspond the functional f in X' which is defined by

$$f(x) = g(T(x)).$$

The map $T^*: Y' \to X'$ so defined may be viewed as linear, and is said to be the *dual* of T.

Giving T^* the norm

$$\|T^*\| = \sup_{g \neq 0} \|T^*(g)\|/\|g\|$$

it follows from the Hahn–Banach theorem that

$$\|T\| = \|T^*\|.$$

In the particular case that X and Y are finite dimensional vector spaces the map T is represented by a matrix (a_{ij}). A straightforward calculation shows that T^* is represented by the transposed matrix (a_{ji}).

In this we have followed the account of Yosida [1], Chapter VII, pp. 193–197. According to his account a further operator, the *adjoint* of T, may be defined if X and Y are Hilbert spaces. The upshot of this, in the particular case that we have just mentioned, is that the adjoint operator is represented by the complex-conjugate transposed matrix $(\bar{a}_{ji})$. In some accounts (for example, Kantorovich and Akilov [1] Chapter IX, §3, pp. 304–308) the adjoint and the conjugate (it is also called the transpose in this same account) are not distinguished, and both are defined exactly in the manner of T^* given above, save that the space Y' is linearized by the requirement that

$$(\alpha f_1 + \beta f_2)(x) = \bar{\alpha} f_1(x) + \bar{\beta} f_2(x).$$

With this definition the matrix which represents T^* in the special case is once again $(\bar{a}_{ji})$.

For the purposes of the present monograph the distinction between dual and adjoint is irrelevant, save that the former leads to a more euphonius verb.

The result of lemma (4.11) in the case $f(p) = 1$, $h(w) = w^2 + 1$, was already proved by Turán in his Ph.D. thesis (reprinted in ref. [1]). The latter is also interesting in that the asymptotic estimate

$$\sum_{n \le x} 2^{r\omega(n)} = \frac{b(1)}{\Gamma(2^r)} x(\log x)^{2^r - 1} + O(x(\log x)^{2^{r-1}-1})$$

is obtained by means of contour integration. Here the function $b(s)$, for complex values of s, $\mathrm{Re}(s) > 1/2$, is defined by

$$\sum_{n=1}^{\infty} 2^{r\omega(n)} n^{-s} = b(s)\zeta^{2^r}(s),$$

and the above result is uniform in $|r| \le 1/2$. This anticipates a similar result of A. Selberg, ref. [3]. We shall return to consider both these results in Chapter 12, when we discuss the Erdös–Kac theorem.

Lemma (4.11), in the case $f(p^k) \equiv 1$, and with an arbitrary polynomial, was established by Turán [3].

There are inequalities which are dual to those which appear in lemmas (4.11) and (4.12). For example, the dual of lemma (4.12) asserts that

$$\sum_{p \le x^\beta} \frac{\varphi(p)}{\eta(p)} \left| \sum_{\substack{w_0 < q \le x \\ h(q) \equiv 0 (\mathrm{mod}\ p)}} a_q - \frac{\eta(p)}{\varphi(p)} \sum_{w_0 < q \le x} a_q \right|^2 \le \frac{cx}{\log x} \sum_{w_0 < q \le x} |a_q|^2$$

holds for all complex numbers a_q, $w_0 < q \le x$, q prime.

We mention here the following interesting theorem of Wolke, (ref. [2] Satz 3), concerning the Turán–Kubilius inequality.

In the notation of lemma 4.1, *let* $f(n)$ *be real, let*

$$D^2(y) - D^2(y^{2/3}) \le 10^{-4} D^2(y)$$

hold for $y \geq y_0(f)$, *and let*

$$\sum_{\substack{p^k \leq y \\ k \geq 2}} \sum p^{-k} f^2(p^k) \leq 10^{-5} D^2(y)$$

hold for $y \geq y_1(f)$. *Then,*

Theorem (Wolke). *There is an absolute constant C so that the inequality*

$$\sum_{n \leq x} |f(n) - \lambda|^2 \geq 10^{-2} x D^2(x)$$

holds uniformly for all $x \geq \max(y_0, y_1^3, C)$, *and for all real* λ.

Here the sum involving λ will be smallest when

$$\lambda = \frac{1}{[x]} \sum_{n \leq x} f(n) \qquad \left(= E(x) + \frac{28\theta}{\sqrt{\log x}} D(x), \; (|\theta| \leq 1) \right).$$

The method of the Large Sieve was invented by Linnik [1], who also introduced the terminology.

Let

$$V(p) = \sum_{r=0}^{p-1} \left| \sum_{\substack{n=1 \\ n \equiv 0 (\bmod p)}}^{N} a_n - p^{-1} \sum_{n=1}^{N} a_n \right|^2.$$

In the case that a_n assumes values 0 or 1 the inequality

$$\sum_{p \leq (N/12)^{1/3}} pV(p) \leq 2N \sum_{n \leq N} |a_n|^2 \qquad N \geq 12,$$

is contained in Rényi [3], who was responsible for a number of reformulations of the large sieve inequality, in particular in probabilistic terms (see, for example, Rényi [5]).

In fact, a result (essentially) of the form

$$\sum_{p \leq N^\alpha} pV(p) \leq c_1 N \sum_{n=1}^{N} |a_n|^2$$

for some fixed $\alpha > 0$ is already enough to prove that, in a certain quantitative sense, 'most' Dirichlet L-series $L(s, \chi_d)$, formed with respect to moduli d not exceeding N^β, have no zeros in a region of the form

$$1 - \frac{1}{(\log d)^\gamma} \leq \sigma \leq 1 \qquad |\tau| \leq (\log d)^3 \qquad (s = \sigma + i\tau)$$

with certain constants $0 < \gamma < 1$, $\beta > 0$. This was shown by Rényi in 1947, 1948 (ref. [1], [2], respectively), and he was thus able to obtain a weighted (and so slightly weaker) form of lemma (2.10) with the range $D \leq x^{1/2}(\log x)^{-B}$ replaced by $D \leq x^{\delta}$, for some (absolute) positive constant δ. Combining this with Brun's sieve he proved that all sufficiently large even integers may be represented as the sum of a prime, and a number with (absolutely) boundedly many prime factors.

The same arguments, combined with a zero-density estimate, were applied by Barban [1] to establish a form of lemma (2.10) with a range $D \leq x^{\eta}$ for varying values of η, all less than 1/2. See the comments made in Chapter 2 following the statement of lemma (2.10); see also Pan [1].

All of these accounts were largely based upon various ideas of Linnik. (See, for example, Rényi's paper [2]). Broadly speaking, in obtaining information concerning the difference

$$\pi(x, D, l) - \frac{1}{\varphi(D)} Li(x)$$

one needs good information concerning the behaviour of appropriate Dirichlet L-series in the neighbourhood of the lines $\sigma = 1$, $\sigma = 1/2$ ($|\tau|$ not too large). Up until (and including) these last results, the large sieve had only been used to obtain good results in the neighbourhood of $\sigma = 1$.

Without making use of the large sieve, A. I. Vinogradov [1], [2] obtained an 'almost sure' estimate for the size of $|L(\frac{1}{2} + i\tau, \chi)|$. Together with the results of Barban [1] this showed that in lemma (2.10) one could take $D \leq x^{1/2-\varepsilon}$, for any fixed $\varepsilon > 0$.

Meanwhile, K. F. Roth [1] had improved Rényi's result concerning the case $a_n = 0, 1$ to

$$\sum_{p \leq Q} pV(p) \leq c_1 Q^2 \log Q \sum_{n=1}^{N} |a_n|^2.$$

A short while later, Bombieri's paper [3] appeared, in which he established the result

$$\sum_{d \leq Q} \sum_{\substack{b=1 \\ (b,d)=1}}^{d} \left| S\left(\frac{b}{d}\right) \right|^2 \leq c_3(N + Q^2) \sum_{n=1}^{N} |a_n|^2.$$

Apart from the value of c_3, this result cannot be much improved. Moreover, Bombieri introduced a form of the large sieve inequality which involved Dirichlet characters, and which was convenient for the study of Dirichlet L-series. By applying the large sieve method at both $\sigma = 1/2$ and $\sigma = 1$, Bombieri established lemma (2.10), (independently of A. I. Vinogradov), with the range $D \leq x^{1/2}(\log x)^{-B}$.

The large sieve was given a form involving points x_j (mod 1), as well as a new proof, by Davenport and Halberstam [1].

The application of the duality principle to the investigation of inequalities of large-sieve type was noted, independently, explicitly or implicitly, by Bombieri (see Forti and Viola [1]), Elliott [7], Kobayashi [1], and Matthews [1]. The proof of lemma (4.15) which we give here is that of Montgomery and Vaughan [1]. Apparently, the simplifying remark, made during the proof of lemma (4.16), that $S_3 = S_4$ when $\mathbf{u}$ gives rise to an extremal, is due to A. Selberg, (see Montgomery [4]). There is a close relation between the large sieve inequality and the well-known inequality of Hilbert. An explicit connection appears first in Matthews [2], see also Montgomery and Vaughan [1]. This last paper also contains a form of the large-sieve inequality which is more sensitive to the distribution of the points x_j.

An upper bound, of the type usually associated with Selberg's method, may be deduced from the large sieve. A particular result of this type was derived by Bombieri and Davenport [1], a general result by Montgomery [1]. See also: Kobayashi [2], Halberstam and Richert [2] pp. 125–126.

A derivation of the Turán–Kubilius inequality from the large sieve was first given by the author, ref. [6]. The method used was different from that given above; at that time it was apparently not realised that these inequalities are connected by duality. The proof depended upon the representation

$$\sum_{p|n,\ p \le x^{1/2}} f(p) = \sum_{p \le x^{1/2}} p^{-1} \sum_{r=0}^{p-1} e^{(2\pi i r n/p)} f(p).$$

It was shown by Gallagher, [3], that this method can be applied to establish a form of the Turán–Kubilius inequality for polynomials, which is sharper than that which may be obtained by the analogue of the original argument of Turán.

The inequality

$$\sum_{d \le Q} \sum_{\substack{b=1 \\ (b,d)=1}}^{d} \left| S\left(\frac{b}{d}\right) \right|^2 \le (N + Q^2) \sum_{n=1}^{N} |a_n|^2$$

cannot be much improved. The remarks made following the proof of lemma (4.13) show that any number which replaces the factor $(N + Q^2)$ must be at least as large as

$$\max\left(N, \sum_{q \le Q} \varphi(q)\right).$$

It was pointed out by Bombieri (see Forti and Viola [1]) that there is a duality between appropriately defined "Dirichlet-series" operators if they are considered as mapping L_p-spaces into L_{p_1}-spaces. In the notation of lemma (4.3) let p and p_1 be non-negative numbers. Then the inequality

$$\left(\sum_{i=1}^{m}\left|\sum_{j=1}^{n} c_{ij}a_j\right|^p\right)^{1/p} \leq \lambda\left(\sum_{j=1}^{n}|a_j|^{p_1}\right)^{1/p_1}$$

holds for all complex numbers a_j, $j = 1, \ldots, n$, if and only if the inequality

$$\left(\sum_{j=1}^{n}\left|\sum_{i=1}^{m} \bar{c}_{ij}b_i\right|^q\right)^{1/q} \leq \lambda\left(\sum_{i=1}^{m}|b_i|^{q_1}\right)^{1/q_1}$$

holds for all complex numbers b_i, $i = 1, \ldots, m$. Here $p^{-1} + q^{-1} = 1$ and $p_1^{-1} + q_1^{-1} = 1$. This last result may be found as theorem 286, p. 205, in the Hardy, Littlewood and Pólya volume [1] on inequalities.

As an example here let us consider lemma (4.18). An application of Hölder's inequality shows that

$$B(x)^\alpha \leq \pi(x)^{(\alpha/2)-1}\sum_q (q^{-1}|f(q)|^2)^{\alpha/2},$$

and therefore

$$\left(\sum_{p+1\leq x}\left|\sum_{\substack{q|(p+1)\\ x^{1/2}<q\leq x}} q^{1/2}f(q)\right|^\alpha\right)^{1/\alpha} \leq c_0\pi(x)^{1/2}\left(\sum_{x^{1/2}<q\leq x}|f(q)|^\alpha\right)^{1/\alpha}.$$

We can regard this as an upper bound for an L_α-norm. Define β by $\alpha^{-1} + \beta^{-1} = 1$. Then, if $1 < \alpha < 2$ (so that any $\beta > 2$ may be reached) we may dualise this last inequality to obtain

$$\sum_{x^{1/2}<q\leq x}\left|q^{1/2}\sum_{\substack{p+1\leq x\\ p\equiv -1 \pmod q}} d_p\right|^\beta \leq c_1^\beta\pi(x)^{\beta/2}\sum_{p+1\leq x}|d_p|^\beta$$

which holds uniformly for all complex numbers d_p, $x \geq 2$ and $\beta \geq \beta_0$, where β_0 is any fixed number, $\beta_0 > 2$. Here c_1 depends upon β_0.

We remark that the proof of lemma (4.19) may easily be modified to show that

$$\sum_{x^{1/2}<d\leq x} d^{m-1}\pi(x, -1, d)^m \leq c_2\pi(x)^m \log x.$$

and so on.

This completes our preparations, and we may now embark.

Chapter 5

The Erdös–Wintner Theorem

In this chapter we begin the study of the distribution of the values of additive arithmetic functions with two theorems that have become classical. The proofs which we shall give differ considerably from those which were given in the original papers.

Theorem 5.1 (Erdös, Erdös–Wintner). *In order that the additive function $f(n)$ should possess a limiting distribution, it is both necessary and sufficient that the three series*

$$\sum_{|f(p)|>1} \frac{1}{p} \qquad \sum_{|f(p)|\le 1} \frac{f(p)}{p} \qquad \sum_{|f(p)|\le 1} \frac{f^2(p)}{p}$$

converge.

When this condition is satisfied, the characteristic function, $\upsilon(t)$, of the limiting distribution, has the representation

$$\upsilon(t) = \prod_p \left(1 - \frac{1}{p}\right)\left(1 + \sum_{m=1}^{\infty} p^{-m} \exp(itf(p^m))\right),$$

where the product is taken over all prime numbers. The limiting distribution is then of pure type, and will be continuous if and only if the series

$$\sum_{f(p)\neq 0} \frac{1}{p}$$

diverges.

Theorem 5.2 (Erdös). *Let the series*

$$\sum_{|f(p)|>1} \frac{1}{p} \qquad \sum_{|f(p)|\le 1} \frac{f^2(p)}{p}$$

converge. For each integer n define

$$A(n) = \sum_{\substack{p \le n \\ |f(p)| \le 1}} \frac{f(p)}{p}.$$

Then the frequencies

$$\nu_n(m;\, f(m) - A(n) \le z) \qquad (n = 1, 2, \ldots),$$

converge weakly. The characteristic function, $\psi(t)$, of the limiting distribution has the representation

$$\psi(t) = \prod_{|f(p)| > 1} (1 + g(p)) \prod_{|f(p)| \le 1} (1 + g(p)) e^{-itf(p)/p},$$

where

$$g(p) = -\frac{1}{p} + \left(1 - \frac{1}{p}\right) \sum_{m=1}^{\infty} p^{-m} \exp(itf(p^m)),$$

and the products are taken over those primes p for which $|f(p)| > 1$, and $|f(p)| \le 1$, respectively. The limiting distribution is of pure type, and will be continuous if and only if the series

$$\sum_{f(p) \ne 0} \frac{1}{p}$$

diverges.

Before establishing these theorems we give some examples.

Let $\varphi(n)$ denote Euler's totient function. This is a multiplicative function of n, and never vanishes. We may, therefore, apply theorem (5.1) to the function $f(n) = \log(\varphi(n)/n)$. For each prime p

$$0 < |f(p)| = -\log\left(1 - \frac{1}{p}\right) \le \frac{2}{p},$$

and it is immediately clear that the three series of theorem (5.1) converge absolutely. Hence

$$g(z) = \lim_{n \to \infty} n^{-1} \sum_{\substack{m=1 \\ \varphi(m) \le zm}}^{n} 1$$

exists, and is a continuous function of z. This particular result, the first of its type, was obtained by Schoenberg (ref. [1]).

It was proved by Erdös (ref. [8]) that $g(z)$ is singular. It is clear that $g(z) = 0$ if $z \le 0$, and $g(z) = 1$ if $z \ge 1$, but it is difficult to get a picture of the behaviour of $g(z)$ in the range $0 < z < 1$. This proves to be generally true for functions to which theorem (5.1) applies.

As another example, let $f(n) = \log(\sigma(n)n^{-1})$, where $\sigma(n)$ denotes the sum of the divisors of the integer n. The function $\sigma(n)$ is multiplicative and, as in the above argument, we deduce the existence of the continuous limiting-distribution

$$h(z) = \lim_{n\to\infty} n^{-1} \sum_{\substack{m=1 \\ \sigma(m)\le zm}}^{n} 1.$$

This was established, independently, by Behrend, Chowla, Davenport, and Erdös, (see the introduction to this volume and, for example, Behrend [1]). Apparently, interest was sparked in this particular case by the remark, made in a book of Bessel–Hagen, that it was not known whether abundant numbers, those for which $\sigma(n) > 2n$, had an asymptotic density. As we can see, they have the density $1 - h(2)$, which is about $1/4$ in value. We shall return to consider the behaviour of the function $\sigma(n)n^{-1}$, more carefully, toward the end of the present chapter.

It is convenient to first establish theorem (5.2), and then use it in the proof of both the sufficiency and the necessity parts of theorem (5.1).

Proof of theorem 5.2. We shall prove the theorem in essentially two steps.

Step one. For each integer n, $n \ge 4$, let r be the integer $[(\log n)^{1/4}]$. Define the function

$$j(m) = \sum_{\substack{p^k \| m \\ p \le r,\ k < r}} f(p^k).$$

Then the frequencies

$$\nu_n(m; j(m) - A(r) \le z)$$

converge weakly as $n \to \infty$, to a law with characteristic function $\psi(t)$.

According to the representation of lemma (3.1), Chapter 3, with $N = r$, we may write

$$\nu_n(m; j(m) - A(r) \le z) = P\left(\sum_{p \le r} X_p - A(r) \le z\right) + O(n^{-1/2}),$$

where the X_p are independent random variables defined by

$$X_p = \begin{cases} f(p^j) & \text{with probability } \left(1 - \frac{1}{p}\right)\frac{1}{p^j}, j = 1, \ldots, r - 1, \\ 0 & \text{with probability } 1 - \frac{1}{p} + \frac{1}{p^r}, \end{cases}$$

and the estimate is uniform in all real numbers z.

The desired result now follows from lemma (1.20), but in the present case the details are sufficiently light that it is worthwhile to carry out a direct proof.

Since the variables X_p are independent, the characteristic function, $\psi_n(t)$, of the distribution function

$$P\left(\sum_{p \le r} X_p - A(r) \le z\right)$$

has the form

$$\psi_n(t) = \exp(-itA(r)) \prod_{p \le r} (1 + g(p) + \varepsilon_p(r)),$$

where

$$\varepsilon_p(r) = -p^{-r} - (1 - p^{-1}) \sum_{m=r}^{\infty} p^{-m} \exp(itf(p^m)).$$

We next make use of the identity (empty products having the value 1):

$$\prod_{l=1}^{L} (1 + x_l + y_l) = \prod_{l=1}^{L} (1 + x_l) + \sum_{m=1}^{L} y_m \prod_{l=1}^{m-1} (1 + x_l) \prod_{l=m+1}^{L} (1 + x_l + y_l).$$

Bearing in mind that the factors $1 + g(p)$ and $1 + g(p) + \varepsilon_p(r)$ are all characteristic functions, with absolute value not exceeding 1, we obtain

$$\begin{aligned} \left|\psi_n(t) - \exp(-itA(r)) \prod_{p \le r} (1 + g(p))\right| &\le \sum_{p \le r} |\varepsilon_p(r)| \\ &\le 2 \sum_{p \le r} p^{-r} \le 2^{1-(r/2)} \sum_p p^{-2} = o(1), \end{aligned}$$

as n, and so r, becomes unbounded.

It is now convenient to recall the simple inequalities

$$|\log(1 + \alpha) - \alpha| \le |\alpha|^2$$

$$|e^{i\beta} - 1 - i\beta| \le |\beta|^2$$

which are certainly valid if α is complex, $2|\alpha| \le 1$, and the value of the logarithm is the principal one on the real axis; and if β is a real number.

Consider, then, the inequalities

$$\sum_{|f(p)| \le 1} |\log(1 + p^{-1}\{e^{itf(p)} - 1\}) - itf(p)p^{-1}|$$
$$\le \sum_{|f(p)| \le 1} p^{-1}|e^{itf(p)} - 1 - itf(p)| + \sum_{|f(p)| \le 1} p^{-2}|e^{itf(p)} - 1|^2$$
$$\le |t|^2 \sum_{|f(p)| \le 1} p^{-1}f^2(p) + 4\sum p^{-2},$$

and

$$\sum_{|f(p)| > 1} |\log(1 + p^{-1}\{e^{itf(p)} - 1\})| \le \sum_{|f(p)| > 1} 2p^{-1}|e^{itf(p)} - 1| \le 4 \sum_{|f(p)| > 1} p^{-1},$$

which may be obtained by setting

$$\alpha = p^{-1}\{e^{itf(p)} - 1\} \qquad \beta = itf(p),$$

in turn. From the hypotheses of theorem (5.2), we deduce that the series which appear on the extreme left-hand end of each chain of inequalities converge uniformly on every compact set of t-values.

It follows, almost at once, that as n (and so r) $\to \infty$ the product

$$\exp(-itA(r)) \prod_{p \le r} (1 + g(p))$$

converges with the same uniformity, and to the function $\psi(t)$ which was defined in the statement of theorem (5.2). In particular, $\psi(t)$ will be a continuous function of t at the point $t = 0$, and so will be a characteristic function. Let $G(z)$ denote its corresponding distribution function. We shall establish

Step two. As $n \to \infty$

$$v_n(m; f(m) - A(m) \le z) \Rightarrow G(z).$$

Define the functions

$$h(m) = \sum_{\substack{p \| m,\, r < p \le n \\ |f(p)| \le 1}} f(p)$$

$$b(m) = {\sum_{p^k \| m}}' f(p^k),$$

where $'$ indicates that we consider only prime powers p^k for which one of the mutually exclusive conditions

$$p > r \quad \text{and} \quad k \geq 2,$$

$$p > r, k = 1 \quad \text{and} \quad |f(p)| > 1,$$

$$p \leq r \quad \text{and} \quad k \geq r,$$

is satisfied. From our definitions of these functions

$$f(m) - A(n) = j(m) - A(r) + h(m) - \{A(n) - A(r)\} + b(m).$$

We shall prove that for every fixed $\varepsilon > 0$ each of the frequencies

$$L_1 = v_n(m; |h(m) - \{A(n) - A(r)\}| > \varepsilon)$$

and

$$L_2 = v_n(m; |b(m)| > \varepsilon)$$

converge to zero as $n \to \infty$. According to a remark made following lemma (1.7) this will suffice to complete our second step.

We may readily estimate the size of the first of these two frequencies by appealing to the Turán–Kubilius inequality in the form of lemma (4.4). In our present circumstances it becomes

$$\sum_{n=1}^{n} |h(m) - \{A(m) - A(r)\}|^2 \leq 45n \sum_{\substack{r < p \leq n \\ |f(p)| \leq 1}} p^{-1} f^2(p).$$

After the argument of Tchebycheff, and appealing to the appropriate hypothesis in theorem (5.2), we see that

$$L_1 \leq \frac{1}{n\varepsilon^2} \sum_{m=1}^{n} |h(m) - \{A(n) - A(r)\}|^2 = o(1) \qquad (n \to \infty).$$

If an integer m is to be counted in the frequency L_2, then it must satisfy one of a number of divisibility criteria.

First, it may be divisible by the square of some prime $p > r$. The frequency of these integers is at most

$$\frac{1}{n} \sum_{p > r} \left[\frac{n}{p^2}\right] \leq \sum_{p > r} \frac{1}{p^2} = o(1) \qquad (n \to \infty).$$

Next, it may be exactly divisible by a prime in the range $r < p \le n$, for which $|f(p)| > 1$. From an hypothesis of theorem (5.2) we deduce that the frequency of such integers is at most

$$\frac{1}{n} \sum_{\substack{p > r \\ |f(p)| > 1}} \left[\frac{n}{p}\right] \le \sum_{\substack{p > r \\ |f(p)| > 1}} \frac{1}{p} = o(1) \qquad (n \to \infty).$$

Finally, if neither of these situations arise then it must be divisible by a prime-power p^k, with $p \le r$ and $k \ge r$. These integers do not have a frequency greater than

$$\frac{1}{n} \sum_{p} \sum_{k \le r} \left[\frac{n}{p^k}\right] \le \sum_{p} p^{-r}(1 - p^{-1})^{-1} \le 2^{1-(r/2)} \sum p^{-2} = o(1) \qquad (n \to \infty).$$

We have now shown that $L_2 \to 0$, and completed our second step.

To justify the remaining assertions of theorem (5.2), define independent random variables Y_p, one for each prime p, by

$$Y_p = f(p^k) \quad \text{with probability} \left(1 - \frac{1}{p}\right)\frac{1}{p^k} \qquad (k = 0, 1, \ldots).$$

Inspection shows that $G(z)$, the limiting distribution in theorem (5.2), coincides with the infinite convolution of the distributions

$$P(Y_p \le z) \qquad |f(p)| > 1$$

$$P(Y_p - f(p)p^{-1} \le z) \qquad |f(p)| \le 1.$$

The assertions concerning the type, and the continuity, properties of $G(z)$ now follow from the theorems of Jessen and Wintner, and Lévy, parts (ii) and (i), respectively, of lemma (1.22) of Chapter one.

This completes the proof of theorem (5.2).

We obtain, straightaway, as a corollary

Proof of Theorem 5.1 (Sufficiency). Since $f(m) - A(n)$, $1 \le m \le n$, has a limiting distribution, and $\lim A(n)$, $(n \to \infty)$, exists, then $f(m)$ also has a limiting distribution. The limit law is a translation of that which occurs in theorem (5.2), and the remaining assertions of theorem (5.1) follow from their analogues in theorem (5.2).

We must now prove that if the additive function $f(m)$ has a limiting distribution, then the three series of theorem (5.1) are convergent. In our present proof the essential step is embodied in the following result.

Lemma 5.3. *Assume that the frequencies*

$$\nu_n(m; f(m) - \alpha(n) \le z) \qquad (n = 1, 2, \ldots),$$

converge to a limit law with characteristic function $w(t)$. *Assume, further, that for each positive constant* η, $0 < \eta < 1$, *we have* $\alpha(m) - \alpha(n) = o(1)$ *as* $n \to \infty$, *uniformly for* $\eta n \le m \le n$. *Then*

$$|w(t)|^2 \sum_p p^{-1} |e^{itf(p)} - 1|^2 \le 36.$$

Remark. The constant 36 is not best possible, as may be seen from what follows. With a little effort it could be reduced to a number less than 3.

Proof of lemma. Let P and N be positive integers which satisfy $2 \le P \le N$. We apply lemma (4.7), with $a_n = \exp(itf(n))$, to obtain the inequality

$$\sum_{p \le P} p \left| \sum_{m \le Np^{-1}} e^{itf(mp)} - p^{-1} \sum_{n \le N} e^{itf(n)} \right|^2 \le 16N^2.$$

For each prime p not exceeding P

$$\sum_{m \le Np^{-1}} e^{itf(mp)} = e^{itf(p)} \sum_{m \le Np^{-1}} e^{itf(m)} + 2\theta N p^{-2} \qquad (|\theta| \le 1).$$

We temporarily fix P, divide by N^2, and apply the Cauchy–Schwarz inequality, to deduce that as $N \to \infty$

$$|w(t)|^2 \sum_{p \le P} p^{-1} \left| \exp\left(it\left\{ f(p) - \alpha\left(\left[\frac{N}{p}\right]\right) + \alpha(N) \right\}\right) - 1 \right|^2$$
$$\le (2 + o(1))16 + 8 \sum_{p \le P} p^{-3}.$$

Making use of the hypothesis concerning $\alpha(n)$ in the form

$$\sup_{p \le P} \left| \alpha(N) - \alpha\left(\left[\frac{N}{p}\right]\right) \right| \to 0 \qquad (N \to \infty),$$

letting $N \to \infty$, and then $P \to \infty$, we see that

$$|w(t)|^2 \sum_p p^{-1} |e^{itf(p)} - 1|^2 \le 32 + 8 \sum p^{-3} < 36.$$

This proves the lemma.

Proof of theorem 5.1 (Necessity). We begin by noting that for any real number β,

$$|e^{i\beta} - 1|^2 = 4\left|\frac{e^{i\beta/2} - e^{-i\beta/2}}{2i}\right|^2 = 4(\sin \beta/2)^2,$$

and that if $|\beta| \le \pi/2$, then $|\sin \beta| \ge 2|\beta|/\pi$.

Let T be a positive real number, chosen so that $|w(t)| \ge 1/2$ in the interval $|t| \le T$. Set $\delta = \pi/T$. Then, by lemma (5.3),

$$\sum_{|f(p)| \le \delta} \frac{f^2(p)}{p} \le \delta^2 \sum p^{-1}\left(\sin \frac{Tf(p)}{2}\right)^2 \le \frac{33}{4}\delta^2 |w(T)|^{-2} \le 33\delta^2.$$

Moreover, we may integrate over the interval $0 \le t \le T$ to deduce that

$$\frac{1}{2}\left(1 - \frac{1}{\pi}\right) \sum_{|f(p)| > \delta} \frac{1}{p} \le \sum_{|f(p)| > \delta} \frac{1}{2p}\left(1 - \frac{\sin Tf(p)}{Tf(p)}\right)$$

$$\le \sum_p \frac{1}{T}\int_0^T p^{-1}\left(\sin \frac{tf(p)}{2}\right)^2 dt \le \frac{33}{4T}\int_0^T |w(t)|^{-2}\, dt \le 33.$$

These results enable us to prove that both the series

$$\sum_{|f(p)| > 1} \frac{1}{p} \qquad \sum_{|f(p)| \le 1} \frac{f^2(p)}{p}$$

converge.

It follows from theorem (5.2) that the frequencies

$$\nu_n(m;\, f(m) - A(n) \le z) \qquad (m = 1, 2, \ldots),$$

converge. Since, by hypothesis, so do the frequencies

$$\nu_n(m;\, f(m) \le z) \qquad (m = 1, 2, \ldots),$$

we may apply lemma (1.9) to deduce the (finite) existence of $\lim_{n\to\infty} A(n)$.

This completes the proof of theorem (5.1).

Note. At the end of this chapter we give an alternative proof, making no appeal to the result of Lévy, of the condition for the limiting distribution in theorem (5.1) to be continuous.

We continue this chapter with two theorems which refine, in some way, theorems (5.1) and (5.2).

Theorem 5.4. *In order that there should exist a function* $\alpha(x)$ *which satisfies the condition*

$$\alpha(x) - \alpha(\eta x) = o(1) \qquad (x \to \infty),$$

for each fixed real number η, $0 < \eta < 1$, *and is such that the frequencies*

$$\nu_x(n;\ f(n) - \alpha(x) \le z)$$

converge weakly as $x \to \infty$, *it is both necessary and sufficient that the series*

$$\sum_{|f(p)|>1} \frac{1}{p} \qquad \sum_{|f(p)|\le 1} \frac{f^2(p)}{p}$$

converge.

Proof. After our treatment of theorem (5.2), it will suffice to note that if we define

$$\alpha(x) = \sum_{p \le x,\ |f(p)| \le 1} \frac{f(p)}{p},$$

an application of the Cauchy–Schwarz inequality then gives

$$|\alpha(x) - \alpha(\eta x)|^2 \le \sum_{\eta x < p \le x} \frac{1}{p} \sum_{|f(p)|\le 1} \frac{f^2(p)}{p} = o(1) \qquad (x \to \infty).$$

This completes our sketch of the proof of theorem (5.4).

Theorem 5.5. *The additive function* $f(n)$ *possesses a limiting distribution with a finite mean and variance if and only if the series*

$$\sum_{p,\ w \ge 1} p^{-w} f(p^w) \qquad \sum_{p,\ w \ge 1} p^{-w} f^2(p^w)$$

converge.

Moreover, when this condition is satisfied

$$\lim_{x\to\infty} x^{-1} \sum_{n \le x} f(n)$$

and

$$\lim_{x\to\infty} x^{-1} \sum_{n \le x} f^2(n)$$

exist, and are respectively equal to the first and second moment of the limiting distribution.

Remark. There is an interesting criterion which is equivalent to the convergence of the two series over the pairs (p, w), but it is more convenient to delay its consideration to a later occasion, theorem (7.7) in Chapter 7.

Proof. Consider, first, the situation when $f(n)$ is known to possess a limiting distribution. Then, by theorem (5.1), the series

$$\sum_{|f(p)|>1} \frac{1}{p} \qquad \sum_{|f(p)|\le 1} \frac{f(p)}{p} \qquad \sum_{|f(p)|\le 1} \frac{f^2(p)}{p}$$

all converge. Let $p_1 < p_2 < \dots$ denote the sequence of primes for which $|f(p)| > 1$.

Let $F(z)$ denote the limiting distribution of $f(n)$, and let $\pm b$, with $b > 0$, be continuity points of $F(z)$. Define

$$c(b) = \int_{-b}^{b} z^2 \, dF(z).$$

Then, as $n \to \infty$,

$$n^{-1} \sum_{\substack{m=1 \\ |f(m)|\le b}}^{n} f^2(m) \to c(b).$$

We shall now prove that, for a suitable choice of the constant δ, the series

$$\sum_{\substack{p,\, w\ge 1 \\ |f(p^w)|>\delta}} \frac{f^2(p^w)}{p^w}$$

converges. The convergence of the series in the statement of theorem (5.5) then follows, the one directly, and the other after an application of the Cauchy–Schwarz inequality.

Let b be so large that $F(b) - F(-b) \ge 3/4$. Choose an integer l, so large that

$$\sum_{i=l}^{\infty} \frac{1}{p_i} < \frac{1}{8}.$$

Let k run through those positive integers for which $|f(k)| \leq b$, and which are not divisible by any one of the primes p_i with $i \geq l$. The number of such integers k which do not exceed a given bound n is at least

$$(1 + o(1))n\{F(b) - F(-b)\} - \sum_{i=l}^{\infty} np_i^{-1} \geq \frac{1}{2}n \qquad (n \text{ sufficiently large}).$$

Consider now the double sum

$$S = \sum_{p_i^w k \leq n}' \sum f(p_i^w k)^2,$$

where the summation is restricted to those integers k which we have just defined, and those prime-powers p_i^w, with $i \geq l$, which do not exceed $n^{1/2}$, and for which (in an obvious notation) $3c_\infty^{1/2} < |f(p_i^w)| \leq b$. We shall estimate this sum S, both from above and below.

From above

$$S \leq \sum_{\substack{m=1 \\ |f(m)| \leq 2b}}^{n} f^2(m) \leq (1 + o(1))c_{2b}n \qquad (n \to \infty),$$

since there can be at most one representation of any integer m in the form $m = p_i^w k$.

From below

$$S \geq \sum_{p_i^w \leq n^{1/2}}' \sum_{k \leq np_i^{-w}} \{\tfrac{1}{2}f^2(p_i^w) - f^2(k)\}.$$

For all sufficiently large n each inner sum has the lower bound

$$\frac{1}{2}\left[\frac{n}{p_i^w}\right] f^2(p_i^w) - \sum_{\substack{k \leq np_i^{-w} \\ |f(k)| \leq b}} f^2(k)$$

$$\geq \frac{n}{3p_i^w}(f^2(p_i^w) - 6c_b) \geq \frac{n}{3p_i^w}(f^2(p_i^w) - 6c_\infty) \leq \frac{nf^2(p_i^w)}{9p_i^w},$$

and this holds uniformly for $p_i^w \leq n^{1/2}$.

Combining the upper and lower bounds for S, dividing by n, and letting $n \to \infty$, we derive the inequality ($\delta = 3c_\infty^{1/2}$),

$$\sum_{\delta < |f(p_i^w)| \leq b} \frac{f^2(p_i^w)}{p_i^w} \leq 9c_{2b} \leq 9c_\infty.$$

Considering partial sums, and letting $b \to \infty$, we obtain the convergence of the series

$$\sum_{|f(p_i^w)| > \delta} p_i^{-w} f(p_i^w)^2.$$

As indicated earlier, we may derive the convergence of the two series

$$\sum p^{-w} f(p^w) \qquad \sum p^{-w} f(p^w)^2.$$

In the other direction, let us assume the convergence of these two last series. It follows at once from theorem (5.1) that $f(n)$ has a limiting distribution, $F(z)$, say

We apply the Turán–Kubilius inequality in the form of lemma (4.4), noting that the expressions $L(n)$, $D(n)$, where

$$L(n) = \sum_{p^k \leq n} p^{-k} f(p^k) \qquad D^2(n) = \sum_{p^k \leq n} p^{-k} f^2(p^k) \qquad D(n) \geq 0,$$

are uniformly bounded for all positive integers n. We obtain, for $n \geq 1$,

$$\sum_{m=1}^{n} f^2(m) \leq 2 \sum_{m=1}^{n} (f(m) - L(n))^2 + 2nL^2(n) \leq 90n(D^2(n) + L^2(n)),$$

which does not exceed λn for a positive constant λ.

Hence, for any pair of continuity points $\pm b$ of $F(z)$,

$$\int_{-b}^{b} z^2 \, dF(z) = \lim_{n \to \infty} n^{-1} \sum_{\substack{m=1 \\ |f(m)| \leq b}}^{n} f^2(m) \leq \lambda.$$

Since b may be chosen arbitrarily large, and λ is independent of b, this proves that $F(z)$ has a finite second moment and, therefore, a finite mean and variance.

We have now completed that part of theorem (5.4) which involves the equivalence. From now until the end of this proof we shall assume that one, and so both, of these equivalent conditions are satisfied.

Let M_1 and M_2 denote the first and second moments, respectively, of the limiting distribution $F(z)$.

In the notation of the above proof,

$$\frac{1}{n} \sum_{\substack{m=1 \\ |f(m)| \leq b}}^{n} f(m) \to \int_{-b}^{b} z \, dF(z) \qquad (n \to \infty),$$

and

$$\frac{1}{n}\sum_{\substack{m=1\\|f(m)|>b}}^{n} |f(m)| \le \frac{1}{bn}\sum_{m=1}^{n} f^2(m) \le \frac{\lambda}{b} \qquad (n = 1, 2, \ldots).$$

Hence

$$\limsup_{n\to\infty}\left|n^{-1}\sum_{m\le n} f(m) - M_1\right| \le \frac{\lambda}{b} + \int_{|z|>b} z\, dF(z).$$

Since b may be chosen arbitrarily large,

$$\lim_{n\to\infty} n^{-1}\sum_{m=1}^{n} f(m) = M_1.$$

Apparently, a simple argument of this kind cannot be made for the assertion concerning the average of $f^2(m)$.

Define independent random variables X_p, one for each prime p, by

$$X_p = f(p^w) \ \text{ with probability } p^{-w}(1 - p^{-1}) \qquad (w = 0, 1, \ldots).$$

Let $\phi_p(t)$ denote the characteristic function of the variable X_p.

The limiting distribution $F(z)$ is the infinite convolution of the variables X_p, and its characteristic function $v(t)$ has the representation.

$$v(t) = \prod_p \phi_p(t),$$

where the product converges uniformly on any bounded interval of t-values. Since $v(t)$ is continuous, and $v(0) = 1$, there is an interval $|t| \le t_0$, with $t_0 > 0$, on which the inequality $|v(t)| \ge 3/4$ holds. From the uniformity of convergence we deduce that, for a suitably chosen prime P, the inequality $|\phi_p(t) - v(t)| \le 1/4$ holds uniformly for all primes $p > P$, and $|t| \le t_0$. By taking into account that each $\phi_p(t)$, with p not exceeding P, is also a characteristic function, we see that there is an interval $-t_1 \le t \le t_1$, with $t_1 > 0$, on which all of the inequalities $|v(t)| \ge 1/2, |\phi_p(t)| \ge 1/2$, are satisfied.

It follows readily from this that the series

$$\log v(t) = \sum_p \log \phi_p(t),$$

$$\frac{v'}{v} = \sum_p \frac{\phi_p'}{\phi_p},$$

$$\frac{v''}{v} - \frac{(v')^2}{v^2} = \sum_p\left(\frac{\phi_p''}{\phi_p} - \frac{(\phi_p')^2}{\phi_p^2}\right)$$

all converge uniformly on the interval $|t| \le t_1$. Here $'$ denotes differentiation with respect to t. For example,

$$\phi_p'' = \int_{-\infty}^{\infty} (iz)^2 e^{itz}\, dP(X_p \le z),$$

so that

$$|\phi_p''| \le \int_{-\infty}^{\infty} z^2\, dP(X_p \le z) = \left(1 - \frac{1}{p}\right) \sum_{w=1}^{\infty} p^{-w} f(p^w)^2.$$

Similarly,

$$|\phi_p'| \le \left(1 - \frac{1}{p}\right) \sum_{w=1}^{\infty} p^{-w} |f(p^w)|,$$

which, after an application of the Cauchy–Schwarz inequality, yields

$$|\phi_p'|^2 \le \left(1 - \frac{1}{p}\right) \sum_{w=1}^{\infty} p^{-w-1} f(p^w)^2.$$

Together with the upper bound $|\phi_p^{-1}| \le 2$, these inequalities enable us to apply the well-known M-test of Weierstrass to the series which represents $(v''/v) - (v'/v)^2$, and so on.

Setting $t = 0$, we deduce that

$$M_1 = \sum_p \text{mean } X_p = \sum_{p,\, w \ge 1} p^{-w} f(p^w)$$

$$M_2 - M_1^2 = \text{variance of } F(z) = \sum_p \text{var } X_p.$$

We now consider the sum

$$U = \sum_{m=1}^{n} f(m)^2$$

by first principles.

Define

$$g(m) = \sum_{p^k \| m,\, p^k \le n^{1/3}} f(p^k).$$

Then

$$\Gamma = \sum_{m=1}^{n} g(m)^2 = \sum_{p^k \le n^{1/3}} f(p^k)^2 \sum_{\substack{m=1 \\ p^k \| m}}^{n} 1 + \sum_{\substack{p^k,\, q^l \le n^{1/3} \\ p \ne q}} \sum f(p^k) f(q^l) \sum_{\substack{m=1 \\ p^k \| m,\, q^l \| m}}^{n} 1.$$

Typically,

$$\sum_{\substack{m=1\\ p^k\|m}}^{n} 1 = \left[\frac{n}{p^k}\right] - \left[\frac{n}{p^{k+1}}\right] = (1 + O(n^{-2/3}))\frac{n}{p^k}\left(1 - \frac{1}{p}\right),$$

$$\sum_{\substack{m=1\\ p^k\|m,\ q^l\|n}}^{n} 1 = (1 + O(n^{-1/3}))\frac{1}{p^k}\left(1 - \frac{1}{p}\right)\frac{1}{q^l}\left(1 - \frac{1}{q}\right),$$

so that

$$\Gamma = (1 + O(n^{-2/3}))n \sum_{p^k \le n^{1/3}} f(p^k)^2 p^{-k}(1 - p^{-1})$$
$$+ n\left(\sum_{p^k \le n^{1/3}} f(p^k)p^{-k}(1 - p^{-1})\right)^2 - n\sum_p\left(\sum_{w=1}^{\infty} f(p^w)p^{-w}(1 - p^{-1})\right)^2$$
$$+ O\left(n^{2/3}\left(\sum_{p^k \le n^{1/3}} |f(p^k)|p^{-k}\right)^2\right).$$

Applying the Cauchy–Schwarz inequality,

$$n^{-1/3}\left(\sum_{p^k \le n^{1/3}} |f(p^k)|p^{-k}\right)^2 \le n^{-1/3}\sum_{p^k \le n^{1/3}} p^{-k}\sum_{p^k} p^{-k}f(p^k)^2 = o(1) \qquad (n \to \infty),$$

and therefore, as $n \to \infty$,

$$n^{-1}\Gamma \to \sum_p \text{mean } X_p^2 + M_1^2 - \sum_p (\text{mean } X_p)^2 = \sum_p \text{var } X_p + M_1^2 = M_2.$$

Next, at most two prime-powers p^k, $> n^{1/3}$, can divide any particular integer m not exceeding n. Therefore,

$$\sum_{m=1}^{n} |f(m) - g(m)|^2 \le \sum_{m=1}^{n} 2 \sum_{\substack{p^k\|m\\ p^k > n^{1/3}}} |f(p^k)|^2 \le 2n \sum_{p^k > n^{1/3}} |f(p^k)|^2 p^{-k}.$$

Bearing in mind the hypotheses of the present theorem, and making use of the Cauchy–Schwarz inequality yet again, we see that

$$U = \sum_{m=1}^{n} (g(m) + f(m) - g(m))^2$$
$$= \Gamma + 2\sum_{m=1}^{n} g(m)(f(m) - g(m)) + \sum_{m=1}^{n} (f(m) - g(m))^2,$$

$$|U - \Gamma| \le 2\left(\Gamma \sum_{m=1}^{n} |f(m) - g(m)|^2\right)^{1/2} + \sum_{m=1}^{n} |f(m) - g(m)|^2 = o(n),$$

as $n \to \infty$. Hence

$$\lim_{n\to\infty} n^{-1} \sum_{m=1}^{n} f^2(m) = \lim_{n\to\infty} n^{-1} U = M_2,$$

and the proof of theorem (5.5) is completed.

The Function $\sigma(n)$

In this section we consider the additive function $\log(\sigma(n)n^{-1})$ in more detail. The method is a combination of Fourier analysis and elementary number theory.

Theorem 5.6. *There is a distribution function $h(z)$, so that the estimate*

$$\nu_n(m; \sigma(m) \leq zm) = h(z) + O(\log\log n(\log n \log\log\log n)^{-1})$$

holds uniformly for all real numbers z, and all integers $n \geq 4$.

Remarks. A similar result holds with $\sigma(m)$ replaced by Euler's function, $\varphi(m)$.

The existence of a continuous limiting distribution $h(z)$ was already obtained as an application of theorem (5.1). We shall denote the characteristic function of the law $h(e^z)$ by $w(t)$.

Our aim is to apply lemma (1.47), a quantitative form of Fourier inversion, of Esseen type. To this end we shall obtain the following two results

Lemma 5.7. (Faĭnleĭb). *Given any positive number A, there is a further positive number γ, so that the estimates*

$$\sum_{m=1}^{n} \left(\frac{\sigma(m)}{m}\right)^{it} = \begin{cases} nw(t) + O(n(\log n)^{-A}), \\ n + O(|t|n) \end{cases}$$

hold, the first uniformly for $|t| \leq \exp(\gamma \log n \,.\, \log\log\log n/\log n)$, and the second uniformly for $|t| \leq 1/4$.

Lemma 5.8 (Erdös). *There is an absolute constant c_1, so that the inequality*

$$\nu_x\left(m; a < \frac{\sigma(m)}{m} \leq a\left(1 + \frac{1}{t}\right)\right) \leq \frac{c_1}{\log t}$$

holds uniformly for all real numbers $a > 0$, $2 \leq t < x$.

Let us assume, for the moment, that these two results have already been established.

We shall apply lemma (5.7) with $A = 2$, and set

$$T = \exp\left(\frac{\gamma \log n \,.\, \log\log\log n}{\log n}\right)$$

with the appropriate value of γ.

We apply lemma (1.47) with

$$F(z) = \nu_n\left(m; \log \frac{\sigma(m)}{m} \le z\right) \qquad G(z) = h(e^z),$$

so that (in the notation of that lemma)

$$f(t) = n^{-1} \sum_{m=1}^{n} \left(\frac{\sigma(m)}{m}\right)^{it} \qquad g(t) = w(t).$$

According to lemma (5.8), letting $x \to \infty$,

$$G\left(\log a + \log\left(1 + \frac{1}{t}\right)\right) - G(\log a) \le \frac{c_1}{\log t},$$

uniformly for $a > 0$, $t \ge 2$. We deduce that, once again in the notation of lemma (1.47),

$$S_G(h) \le \frac{c_2}{\log(1/h)},$$

uniformly for $0 < h < 1$.

Therefore,

$$\sup_z |F(z) - G(z)| \le c_3 S_G\left(\frac{1}{T}\right) + c_3 \int_{-T}^{T} \left|\frac{f(t) - g(t)}{t}\right| dt$$

$$\le \frac{c_3 c_2}{\log T} + c_3\left(\int_{|t| \le (1/\log n)} + \int_{(1/\log n) < |t| \le T}\right) \left|\frac{f(t) - g(t)}{t}\right| dt.$$

It follows from lemma (5.7) that, for $|t| \le 1/4$,

$$w(t) = 1 + O(|t|).$$

Over the range $|t| \le 1/\log n$ we use the second of the estimates in lemma (5.7), and obtain

$$\int_{|t| \le (1/\log n)} \left|\frac{f(t) - g(t)}{t}\right| dt \le c_4 \int_{|t| \le (1/\log n)} 1 \; dt = \frac{c_4}{\log n},$$

whilst over the range $(\log n)^{-1} < |t| \le T$ we use the first of the two estimates:

$$\int_{(1/\log n)<|t|\le T} \left| \frac{f(t)-g(t)}{t} \right| dt = O\left((\log n)^{-2} \int_{1/\log n}^{T} \frac{dt}{t} \right) = O((\log n)^{-1}).$$

Since

$$(\log \; T)^{-1} = \log \log n (\gamma \log n \,.\, \log \log \log n)^{-1},$$

the proof of theorem (5.6) will be completed.

Proof of lemma (5.7). Let $l(d)$ denote the Möbius inverse of the multiplicative function $(\sigma(m)m^{-1})^{it}$, $(m = 1, 2, \ldots)$. Thus

$$\sum_{d|m} l(d) = \left(\frac{\sigma(m)}{m} \right)^{it}.$$

It is readily checked that for each prime-power p^k $(k = 1, 2, \ldots)$,

$$l(p^k) = \left(1 + \frac{p^k - 1}{p^k(p-1)} \right)^{it} - \left(1 + \frac{p^{k-1}-1}{p^{k-1}(p-1)} \right)^{it}.$$

In particular $|l(p^k)| \le 2$, and

$$|l(p)| = \left| \exp\left(it \log\left(1 + \frac{1}{p} \right) \right) - 1 \right| \le |t| \log\left(1 + \frac{1}{p} \right) \le |t| p^{-1}.$$

The sum which we wish to estimate may be represented in the form

$$\sum_{m=1}^{n} \sum_{d|m} l(d) = \sum_{d=1}^{n} l(d) \sum_{\substack{m=1 \\ m \equiv 0 (\text{mod } d)}}^{n} 1 = \sum_{d=1}^{n} l(d) \left[\frac{n}{d} \right].$$

Removing the square brackets, and extending the sum over d to infinity, we obtain

$$\left| \sum_{m=1}^{n} \left(\frac{\sigma(m)}{m} \right)^{it} - n \sum_{d=1}^{\infty} \frac{l(d)}{d} \right| \le \sum_{d=1}^{n} |l(d)| + n \sum_{d>n} \frac{|l(d)|}{d}.$$

We may estimate these last two sums together, by noting that for each real y, $0 < y < 1/4$, their sum does not exceed

$$\sum_{d=1}^{\infty} |l(d)| \left(\frac{n}{d} \right)^{1-y} + n \sum_{d=1}^{\infty} \frac{|l(d)|}{d} \left(\frac{d}{n} \right)^{y} = 2n^{1-y} \sum_{d=1}^{\infty} \frac{|l(d)|}{d^{1-y}}$$

$$= 2n^{1-y} \prod_{p} \left(1 + \frac{|l(p)|}{p^{1-y}} + \frac{|l(p^2)|}{p^{2(1-y)}} + \cdots \right) \le 2n^{1-y} \exp\left(\sum_{p} \sum_{k=1}^{\infty} \frac{|l(p^k)|}{p^{k(1-y)}} \right).$$

Those terms in the exponential which correspond to powers of primes greater than the first contribute a bounded amount:

$$\sum_p \sum_{k=2}^{\infty} \frac{|l(p^k)|}{p^{k(1-y)}} \le 2 \sum_p \sum_{k=2}^{\infty} p^{-3k/4} = 2 \sum_p p^{-3/2}(1 - p^{-3/4})^{-1}.$$

Next we note that

$$\sum_{p > |t|^2+4} \frac{|l(p)|}{p^{1-y}} \le 2|t| \sum_{p > |t|^2+4} \frac{1}{p^{2-y}} \le 2|t| \int_{|t|^2}^{\infty} w^{-3/2}\, dw = 4.$$

Moreover, since $e^\beta - 1 \le \beta e^\beta$ for real non-negative numbers β,

$$\sum_{p \le |t|^2+4} \frac{|l(p)|}{p^{1-y}} \le 2 \sum_{p \le |t|^2+4} \frac{1}{p}(p^y - 1 + 1) \le 2 \sum_{p \le |t|^2+4} \left(\frac{yp^y \log p}{p} + \frac{1}{p}\right)$$

$$= O(y \log(|t|^2 + 4) . \exp(y \log(|t|^2 + 4))) + O(\log\log(|t|^2 + 4)),$$

the last step by means of lemma (2.5).

With

$$w(t) = \sum_{d=1}^{\infty} \frac{l(d)}{d} \qquad y = 3A \log\log n/\log n,$$

the condition $|t| \le \exp(\gamma \log n . \log\log\log n/\log\log n)$ ensures that neither of these last two error terms exceeds $\frac{1}{3}y \log n$, that $\frac{1}{3}y \log n \ge A \log\log n$, and that

$$\sum_{m=1}^{n} \left(\frac{\sigma(m)}{m}\right)^{it} - nw(t) = O(n(\log n)^{-A}),$$

provided only that γ is given a sufficiently small, but otherwise fixed, value.

This gives the first estimate of lemma (5.7).

As for the second estimate of this lemma, we note that, anyway, $|l(p^k)| \le 4|t|/p$ so that

$$|l(d)| \le \prod_{p|d} \left(\frac{4|t|}{p}\right).$$

Hence, if $4|t| \leq 1$,

$$\left|\sum_{m=1}^{n}\left(\frac{\sigma(m)}{m}\right)^{it} - n\right| \leq n\sum_{d=2}^{n}\frac{|l(d)|}{d} \leq 4|t|n\sum_{d=2}^{\infty}\frac{1}{d}\prod_{p|d}\frac{1}{p}$$

$$= 4|t|n\sum_{r=2}^{\infty}\frac{\mu^2(r)}{r^2}\prod_{p|r}\left(1+\frac{1}{p}+\frac{1}{p^2}+\cdots\right) \leq 4|t|n\sum_{r=2}^{\infty}\frac{\mu^2(r)2^{\omega(r)}}{r^2} = O(|t|n).$$

This completes the proof of lemma (5.7).

Let $F(x; a, b)$ denote the number of integers $n \leq x$ for which the bounds

$$a \leq \frac{\sigma(n)}{n} < b$$

are satisfied.

As a mark of respect to Erdös the mathematician as one of the founders of the probabilistic theory of numbers, we shall next give his original proof, typical in his style, of the proposition that for $x > t$

$$(1) \qquad F\left(x; a, a+\frac{1}{t}\right) < c_1\frac{x}{\log t}.$$

It first appeared in a paper in the Pacific Journal of Mathematics [15], pp. 61–63. In our presentation we shall not change anything, including misprints, except for the three boxed reference-numbers at the end of the proof.

We shall give a commentary immediately following his proof.

Erdös' proof:

To prove (1) denote by $B(x, t)$ the set of integers

$$(4) \qquad 1 \leqq b_1 < \cdots < b_k \leqq x \qquad a \leqq \frac{\sigma(b_i)}{b_i} < a+\frac{1}{t}.$$

We have to show that for $x > t$

$$(5) \qquad k < c_1 x/\log t.$$

To prove (5) we show that if we neglect $o(x/\log t)$ of the integers b we can assume that the b's have various properties which make the estimation of their number easier.

First of all we can assume that no b is divisible by a power of a prime p^α, $\alpha > 1$ which is greater than $(\log t)^2$. This is clear since the number of such integers $\leqq x$ is less than

$$\sum_{\substack{p^\alpha > (\log t)^2 \\ \alpha > 1}} \frac{x}{p^2} < c_2 x/\log t. \tag{6}$$

Write now

$$b_i = u_i v_i w_i \tag{7}$$

where all prime factors of u_i are $< \log t$, all prime factors of v_i are in $(\log t, t^{1/2})$ and all prime factors of w_i are $\geqq t^{1/2}$.

Now we show that we can assume

$$u_i < t^{1/10}. \tag{8}$$

For if (8) does not hold then u_i must have at least r distinct prime factors $< \log t$ where $(\log t)^r > t^{1/10}$ or $r > \log t/20 \log \log t$. Thus by a simple computation the number of b's not satisfying (8) is less than

$$x\left(\sum_{p < \log t} \frac{1}{P}\right)\frac{r}{r!} < x\frac{(2 \log \log t)^r}{r!} < \frac{c_3 x}{\log t}.$$

Now we consider the b's with $v_j > 1$, i.e., we consider the b's which have at least one prime factor in $(\log t, t^{1/2})$. Let $p_i | b_i$ be such a prime factor, then we must have $p_i^2 \nmid b_i$. Now we show that the integers b_i/p_i are all distinct, thus the number of these b's is less than $x/\log t$.

To see this assume $b_i/p_i = b_j/p_j$, $p_j > p_i$. But then

$$\frac{\sigma(b_i/p_i)}{b_i/p_i} = \frac{\sigma(b_j/p_j)}{b_j/p_j} \quad \text{or} \quad \frac{\sigma(b_i)b_j}{b_i\sigma(b_j)} = \frac{(p_i + 1)p_j}{p_i(p_j + 1)}. \tag{10}$$

But $a \leqq \sigma(b)/b < a + 1/t$, $p_i < t^{1/2}$, $p_j < t^{1/2}$. Thus

$$1 \leqq \frac{\sigma(b_i)b_j}{b_i\sigma(b_j)} < 1 + \frac{1}{at} \quad \text{and} \quad \frac{(p_i + 1)p_j}{p_i(p_j + 1)} \geqq 1 + \frac{1}{t} \tag{11}$$

(10) and (11) clearly contradict each other. Thus we can henceforth assume that our b's have no prime factor in $(\log t, t^{1/2})$. Thus finally we can restrict ourselves to the b's of the form

$$b_i = u_i w_i$$

where all prime factors of u_i are $< \log t$ and $u_i < t^{1/10}$ and all prime factors of $w_i \geqq t^{1/2}$.

Next we show that we can restrict ourselves to the b's for which

$$\frac{\sigma(w_i)}{w_i} < 1 + \frac{10}{t^{1/2}}. \tag{12}$$

Consider first the b's which for some $r = 0, 1, \ldots$ have two or more prime factors in $(2^n t^{1/2}, 2^{n+1} t^{1/2})$. The number of these b's is clearly less than (in Σ_r the summation is extended over the primes in $(2^r t^{1/2}, 2^{r+1} t^{1/2})$)

$$x \sum_{r=0}^{\infty} \left(\sum_r \frac{1}{p} \right)^2 < x \sum_{r=0}^{\infty} \frac{1}{(\log 2^r t^{1/2})^2} = x \sum_{r=0}^{\infty} \frac{1}{(r \log 2 + \log t)^2} < \frac{c_1 x}{\log t}.$$

For the b's which have only one prime factor in $(2^r t^{1/2}, 2^{r+1} t^{1/2})$, $r = 0, 1, \ldots$ we evidently have

$$\frac{\sigma(w_i)}{w_i} < \prod_{r=0}^{\infty} \left(1 + \frac{1}{2^r t^{1/2}}\right) < 1 + \frac{10}{t^{1/2}}$$

for $t > t_0$. Thus henceforth we can assume that (12) holds.

Thus we obtained that if we neglect $cx/\log t$ integers than all our integers $b_i < x$ satisfying

$$a \leqq \frac{\sigma(b_i)}{b_i} < a + \frac{1}{t}$$

have the following properties. All their prime factors p^α, $\alpha > 1$ satisfy $p^\alpha < (\log t)^2$, they have no factor in $(\log t, t^{1/2})$ and if we put $b_i = u_i w_i$ where all prime factors of u_i are $\leqq \log t$ then $u_i < t^{1/10}$ and

$$\frac{\sigma(w_i)}{w_i} < 1 + \frac{10}{t^{1/2}}.$$

Now observe that for all the b's which remain we must have constant value $\sigma(u_i)/u_i = \alpha$. To see this assume that, say, $\sigma(u_1)/u_1 > \sigma(u_2)/u_2$ then we have

$$\frac{\sigma(u_1)}{u_1} - \frac{\sigma(u_2)}{u_2} \geqq \frac{1}{u_1 u_2} > \frac{1}{t^{1/5}} \tag{13}$$

or by (13)

$$\frac{\sigma(u_2)}{u_2} < a + \frac{1}{t} - \frac{1}{t^{1/5}} \tag{14}$$

but then by (12) and (14) for $t > t_0$

$$\frac{\sigma(b_2)}{b_2} < \left(a + \frac{1}{t} - \frac{1}{t^{1/5}}\right)\left(1 + \frac{10}{t^{1/2}}\right) < a$$

an evident contradiction.

In view of what we just proved all the b's (neglecting perhaps $cx/\log t$ of them) are of the form

$$u_i w_i, \frac{\sigma(u_i)}{u_i} = \alpha_1 \qquad u_i < t^{1/2},$$

where all prime factors of u_i are $\leqq \log t$ and all prime factors of w_i are $\geqq t^{1/2}$.

In a previous paper [14] I proved that there is an absolute constant C so that

$$\sum_{\sigma(u)/u=\alpha} \frac{1}{u} \leqq C. \tag{15}$$

In fact with more trouble we can show $C = 1$ [11], [14].

Now we can complete the estimation of the number of b's not exceeding x.

For fixed u_i the number of w_i for which $u_i w_i$ can be a b is less than the number of integers $\leqq x/u_i$ all whose prime factors are $\geqq t^{1/2}$.

Thus by Brun's method that number is less than

$$\frac{cx}{u_i \log t}$$

summing for u_i we obtain our statement from (15). The restriction $t > t_0$ is clearly irrelevant.

This ends Erdös' proof.

Commentary

On (6): In the estimate (6), the contribution of those numbers $\leq x$ which are divisible by the square of a prime greater than $\log t$, is at most

$$\sum_{p>\log t} \frac{x}{p^2} = O(x/\log t)$$

If p is a prime which does not exceed $\log t$, and for which $p^\alpha > (\log t)^2$, then $\alpha > \alpha_p = (2 \log\log t)/\log p$. The number of integers $\leq x$ which are divisible by such a prime power, is at most

$$\sum_{p\leq\log t}\sum_{\alpha>\alpha_p} \frac{x}{p^\alpha} \leq \sum_{p\leq\log t} \frac{2x}{p^{\alpha_p}} = \frac{2x}{(\log t)^2}\sum_{p\leq\log t} 1 \leq \frac{2x}{\log t}.$$

On (8): If an integer u_i has ω distinct prime factors, and exceeds $t^{1/10}$, then

$$\tfrac{1}{10}\log t < \log u_i = \sum_{p^\alpha \| u_i} \log p^\alpha \leq \omega \log(\log t)^2.$$

Set

$$\eta = \frac{\log t}{20 \log\log t}.$$

Then the number of integers (and so the number of the b's) not exceeding x, and which are divisible by such a u_i, is at most

$$\sum_{u > t^{1/10}} \frac{x}{u} \leq x \sum \frac{2^{\omega(u)-\eta}}{u} \leq x 2^{-\eta} \prod_{p \leq \log t} \left(1 + \frac{1}{p} + \frac{1}{p^2} + \dots\right)$$
$$= O(x 2^{-\eta} \log t) = O(x/\log t).$$

Here the dummy u runs over those integers which are composed entirely of prime factors not exceeding $\log t$, and we have made an appeal to lemma (2.5). It is simpler to use this argument in place of the one given by Erdös, which, incidentally, contains two misprints.

On (11): Note that $a \geq 1$ may be assumed without loss of generality. Since $p_j - p_i \geq 2$, we have

$$\frac{(p_i+1)p_j}{p_i(p_j+1)} = 1 + \frac{p_j - p_i}{p_i(p_j+1)} \geq 1 + \frac{2}{t^{1/2}(t^{1/2}+1)} \geq 1 + \frac{1}{t}$$

if $t \geq 1$. This is a surprising step. It appears, already, in Erdös' early paper [1].

On (12): In the argument which follows the inequality (12) there is a misprint of n in place of r. Note that according to lemma (2.5)

$$\sum_r \frac{1}{p} = \log\left(\frac{\log 2 + \log 2^r t^{1/2}}{\log 2^r t^{1/2}}\right) + O\left(\frac{1}{\log 2^r t^{1/2}}\right)$$
$$= O\left(\frac{1}{\log 2^r t^{1/2}}\right)$$

an estimate which leads to a satisfactory result in Erdös' argument. The sharper upper bound for Σ_r, to which he appeals, may be deduced from any form of prime number theorem, such as lemma (2.6), in which the error term is sufficiently sharp.

On (14): The inequality

$$\left(a + \frac{1}{t} - \frac{1}{t^{1/5}}\right)\left(1 + \frac{10}{t^{1/2}}\right) < a$$

is, in fact, only valid if

$$a < \frac{(t^{4/5} - 1)}{t} \cdot \left(1 + \frac{t^{1/2}}{10}\right).$$

For $30a < t^{3/10}$ we can use the proof given by Erdös. When this last condition fails the following, alternative argument,

$$\begin{aligned} B(x, t) &\leq \frac{1}{a} \sum_{b \leq x} \frac{\sigma(b)}{b} = \frac{1}{a} \sum_{b \leq x} \frac{1}{d} \sum_{d \mid b} \frac{b}{d} \\ &= \frac{1}{a} \sum_{d \leq x} \frac{1}{d} \sum_{\substack{b \leq x \\ b \equiv 0 (\mathrm{mod}\ d)}} 1 \leq \frac{x}{a} \sum_{d=1}^{\infty} \frac{1}{d^2} < \frac{60x}{t^{3/10}}, \end{aligned}$$

leads to the desired result.

On (15): Consider, first, the squarefree solutions n to the equation $\sigma(n) = n\alpha$ which lie in the interval $2^k < n \leq 2^{k+1}$. Let m and n be two such solutions, and let l be the greatest prime divisor of m. Then $m\sigma(n) = n\sigma(m)$ and, since l cannot divide $\sigma(m)$, we must have $l \mid n$.

If there is a solution of the above type, with $l > k^3$, then

$$S_k = \sum_{\substack{2^k < n \leq 2^{k+1} \\ \sigma(n) = \alpha n}} \frac{\mu^2(n)}{n} \leq \sum_{r \leq 2^{k+1}} \frac{1}{lr} = O\left(\frac{1}{k^2}\right).$$

If there is no such solution, then all solutions n are made up of primes not exceeding k^3. Therefore

$$\begin{aligned} S_k &\leq \frac{1}{2^{k/4}} \sum \frac{\mu^2(n)}{n^{3/4}} = \frac{1}{2^{k/4}} \prod_p \left(1 + \frac{1}{p^{3/4}}\right) \\ &\leq 2^{-k/4} \exp\left(\sum_{p \leq k^3} p^{-3/4}\right) = \exp\left(-k \frac{\log 2}{4} + ck^{3/4}\right) = O(k^{-2}). \end{aligned}$$

Summing over $k = 1, 2, \ldots$ we see that

$$\sum_{\sigma(n) = \alpha n} \frac{\mu^2(n)}{n} \leq c_0\left(1 + \frac{1}{2} + \sum_{k=1}^{\infty} k^{-2}\right) < 4c_0$$

for a certain absolute constant c_0.

Each positive integer n may be uniquely expressed in the form $n = n_1 n_2$, where n_1 contains those prime divisors of n which appear to the first power only. Clearly n_1 is squarefree. Moreover, $(n_1, n_2) = 1$, so that if $\sigma(n) = \alpha n$, then $\sigma(n_1)/n_1 = \alpha n_2/\sigma(n_2)$.

We enumerate the (general) solutions to the equation $\sigma(n) = \alpha n$ by first collecting together those for which n_2 has a fixed value. In this way see that

$$\sum_{\sigma(n)=\alpha m} \frac{1}{n} = \sum_{n} \frac{1}{n_2} \sum_{\sigma(n_1)=\beta n_1} \frac{1}{n_1} \qquad (\beta = \alpha n_2/\sigma(n_2)),$$

$$< 4c_0 \prod_p \left(1 + \frac{1}{p^2} + \frac{1}{p^3} + \dots\right) = C.$$

On Brun's method. In place of Brun's method (note Erdös' characteristic reference to it!), we can apply Selberg's sieve method, lemma (2.1). The details of the present application closely follow those of lemma (2.7), the Brun–Titchmarsh lemma. In the notation of that lemma we set $D = 1$, $z = t^{1/4}$, and $r = t^\delta$ for a suitably small, fixed, positive number δ.

This completes our commentary upon Erdös' proof.

Proof of lemma (5.8). Clear.!

As Erdös remarked in his original paper from which the above proof was taken, it is easy to modify his argument to establish the bound of lemma (5.8) with $\sigma(n)$ replaced by $\varphi(n)$ (and possibly a new value for c_1). With slight modifications the above comments will still apply.

This completes our consideration of lemma (5.8).

It is well known that Erdös is, by choice, an itinerant mathematician, often not spending more than a few days in a place at a time. Towards the end of his life, the late Professor Mordell liked to think of himself something of a rival in this respect. On several occasions I heard him recount that:

"When I first saw a paper by Erdös, I said at once, 'this is clearly a man of parts'."

Concluding Remarks

As we mentioned in the introduction, the sufficiency of the convergence of the three series, in order that an additive function should possess a limiting distribution, was first established by Erdös. This was done in a number of refining steps [2], [5], [7] and, in particular, improved upon the earlier results of Schoenberg [1], [2] and [3]. His method was elementary, that of Schoenberg depended upon Fourier analysis. It is interesting that in his pioneering paper [1], §18, Schoenberg recounts that in his lectures of 1923–24 J. Schur demonstrated that for each complex number s

$$\lim_{n\to\infty} n^{-1} \sum_{m=1}^{n} \left(\frac{\varphi(m)}{m}\right)^{s} = \Phi(s),$$

where $\Phi(s)$ is an integral function, with the product representation

$$\Phi(s) = \prod_{p} \left(1 - \frac{1}{p} + \frac{1}{p}\left(1 - \frac{1}{p}\right)^{s}\right).$$

Here the product is taken over all primes p, and converges absolutely and uniformly on each bounded subset of the s-plane.

In order to deduce the necessity of the conditions in theorem (5.1), Erdös and Wintner [1], made use of a result of Erdös and Kac [1], [2], concerning the approximately normal distribution of certain additive functions. We shall consider the latter theorem in a subsequent chapter, (Theorem (12.3) of Chapter 12).

Much of the fundament of the proofs of Erdös [5], [7], and Erdös and Wintner [1], was provided by a theorem of Sperner, concerning collections of finite sets, no one of which contains another, lemma (1.15). The first person to apply Sperner's lemma to number theory was apparently Behrend [2], in a paper concerning sequences of integers, no one of which divides another. Such an application was independently realised by Erdös; as he writes in 1976:

"In 1934, end of September, I left Hungary for the first time; my mother accompanied me until Vienna. I was a bit despondent on the train to Paris, but soon proved that if $a_1 < \cdots < a_k \le x$, $a_i \nmid a_j$, $i \ne j$, then

$$\sum_{a_i < x} \frac{1}{a_i} < \frac{c \log x}{\sqrt{\log \log x}},$$

i.e. I rediscovered Behrend's proof which he found a few months earlier. Davenport and Rado met me at Cambridge on October 1st. I told them my proof and learned that I was anticipated. (I was rather lucky during my long life this did not happen often to me)."

Erdös adapted this argument to the study of additive functions [5], [7]. It is worthwhile, both from a mathematical and an historical standpoint, to give their basic argument.

Lemma 5.9. *There is an absolute constant c with the following property:*

Let x be a real number, $x \ge 2$. Let $m_1 < m_2 < \cdots < m_r$ be a sequence of squarefree integers, and $q_1 < q_2 < \cdots < q_w$ be a sequence of primes, none of *these exceeding x. Set*

$$S = \sum_{i=1}^{w} \frac{1}{q_i}.$$

Suppose that there are no solutions to the equation $m_j = m_k \lambda, j \neq k$, where the integer λ is composed entirely of the primes q_i.

Then

$$S^{1/2} \sum_{j=1}^{r} \frac{1}{m_j} \leq c \log x.$$

Proof. For each positive integer n, not exceeding x, let $\Delta(n)$ denote the number of divisors of n which can be found amongst the m_j. Define

$$D = \sum_{n \leq x} \Delta(n).$$

Clearly

$$D = \sum_{m_j \leq x} \sum_{\substack{n \leq x \\ n \equiv 0 (\mathrm{mod}\ m_j)}} 1 = \sum_{m_j \leq x} \left[\frac{x}{m_j}\right] \geq x \sum_{j=1}^{r} \frac{1}{m_j} - x.$$

We shall now obtain an upper bound for the sum D.

Let g denote a positive real number. Consider those integers n which are divisible by at most g of the primes q_i. For such integers we have

$$\Delta(n) \leq 2^g \sum_{d|n}{}' \mu^2(d),$$

where $'$ indicates that summation is confined to those d which do not have a q_i amongst their prime factors. The contribution towards D which arises from these integers n does not exceed

$$2^g \sum_{d \leq x}{}' \mu^2(d) \left[\frac{x}{d}\right] \leq 2^g x \cdot \frac{\prod_{p \leq x} \left(1 + \frac{1}{p}\right)}{\prod_{i=1}^{w} \left(1 + \frac{1}{q_i}\right)} \leq c_1 2^g e^{-S} x \log x.$$

If, now, n is an integer which has more than g prime factors amongst the q_i, we write $n = n'n''$, where no q_i divides n'. Each (unrestricted) divisor d, of n, has a corresponding decomposition $d = d'd''$. Consider those divisors

m_j, of n, which have a particular common value for m'_j. Clearly we cannot have $m''_j | m''_k$ for any two such divisors, otherwise

$$m_j = m_k \lambda,$$

where $\lambda = m''_k / m''_j$, contradicting an hypothesis of the lemma.

Let $l_1, \ldots, l_k$ denote the prime divisors of the (squarefree) integer n'. Then to each divisor d, which divides some m_j, which divides n, there corresponds a collection of subsets of the suffices $\{1, \ldots, k\}$, no one of which is contained in another. By Sperner's lemma, any such collection can contain at most

$$\binom{k}{\left[\frac{k}{2}\right]} \le \frac{c_2 2^k}{k^{1/2}}$$

members.

Hence

$$\Delta(n) \le \frac{c_2 2^k}{k^{1/2}} \cdot 2^{\omega(n) - k} \le \frac{c_2 \tau(n)}{g^{1/2}},$$

and these integers n contribute towards D at most

$$c_2 g^{-1/2} \sum_{n \le x} \tau(n) \le c_3 g^{-1/2} x \log x.$$

Altogether, therefore,

$$\sum_{j=1}^{r} \frac{1}{m_j} \le \{c_1 2^g e^{-S} + c_3 g^{-1/2} + (\log x)^{-1}\} \log x.$$

Choosing $g = \max(S, 1)$ we complete the proof of lemma (5.9).

As an example in the use of this lemma, let $f(n)$ have a limiting distribution. Then it is not difficult to show, it needs only an integration by parts, that for suitably chosen positive constants δ and η the inequality

$$\sum_{\substack{m \le x \\ |f(m)| \le \delta}} \frac{\mu^2(m)}{m} > \eta \log x$$

holds for all sufficiently large values of x. We apply lemma (5.9) with those integers m which are counted in this sum, and with the q_i running through the primes q, not exceeding x, for which $f(q) > 2\delta$. Then the equation $m_j = m_k\lambda, j \neq k$, is not soluble, otherwise we shall have

$$2\delta \geq f(m_j) - f(m_k) = f(\lambda) > 2\delta,$$

which is impossible.

We conclude from lemma (5.9) that

$$\sum_{f(p) > 2\delta} \frac{1}{p} \leq (c\eta^{-1})^2 < \infty.$$

A similar argument may be made for those primes p with $f(p) < -2\delta$.

For another application, we refer to the author's paper [25]. For an additive function $f(n)$, define

$$M(x) = \max_{n \leq x} |f(n)| \qquad E(x) = \max_{p \leq x} |f(p+1)|,$$

where, in the definition of $E(x)$, p runs over prime numbers. In that paper it is proved that there are positive absolute constants, A and B, so that the inequality

$$M(x) \leq AE(x^B) + AM((\log x)^B)$$

holds for all $x \geq 2$. We do not consider this result further, since its proof demands a large number of auxiliary facts concerning the distribution of prime numbers in arithmetic progressions of short length.

There is an analogy between the proof of the above lemma (5.9) and the proof of the concentration function estimate, lemma (1.14). For example, the separate consideration, in the proof of lemma (5.9), of those integers n which have at least g prime factors amongst the q_i, corresponds to the conditioning, in the proof of lemma (1.14), that more than $s/4$ of the variables ξ_i satisfy one of the inequalities $\xi_i \leq \varepsilon_i$ and $\xi_i > 1 - \varepsilon_i$. This is not entirely an accident.

Early estimates for concentration functions, following Lévy [2], were derived mainly from a suitable form of the Central Limit Theorem, such as an earlier version of lemma 1.48 (Berry–Esseen) due to Liapounoff. See, for example, Lévy [2], Doeblin [1]; see also the later papers of Kolmogorov [2], [3].

Already in 1943, in the third of three papers dealing with the distribution of the roots of algebraic equations with random coefficients, Littlewood and Offord [1], considered, in particular, the following related problem:

Given real numbers a_j, $|a_j| > 1$, $j = 1, \ldots, n$, how many of the 2^n sums $\varepsilon_1 a_1 + \cdots + \varepsilon_n a_n$, with each ε_j having one of the values ± 1, lie in an interval of length one?

By applying the Central Limit Theorem, with a Liapounoff remainder, they showed that there are $O(2^n n^{-1/2} \log n)$ such sums.

In response to this paper, Erdös [9] pointed out that without loss of generality $1 < a_1 \le a_2 \cdots \le a_n$. By considering the set of suffices j, $1 \le j \le n$, for which $\varepsilon_j = 1$, one could apply Sperner's lemma to deduce that the number of such sums is at most

$$\binom{n}{\left[\frac{n}{2}\right]} \le \frac{c_2 2^n}{n^{1/2}}.$$

In two short notes, [1] and [2], Rogozin showed that one could combine the ideas of Erdös and Kolmogorov to obtain the result of lemma (1.14). The proof of this last result which is given in Chapter one is, in all essentials, his.

For two recent papers concerning refinements of lemma (1.14) see Esseen [3], and Kesten [1]. The first of these two papers uses only methods of analysis. There is also a survey of certain applications of concentration functions in Kesten [2]. See, also, Hengartner and Theodorescu [1].

Theorem (5.2) appears in Erdös' paper [10], actually stated without proof.

The present method of treating theorems (5.1) and (5.2) was outlined in the author's talk [12], at the St. Louis conference on Analytic Number Theory, 1971. It benefits from historical hindsight, see, for example, Erdös' original paper [7]. Inequalities of the large sieve type, as used in the proof of lemma (5.3), were introduced (implicitly) by Linnik [1], in 1941, and were not available when theorems (5.1) and (5.2) were originally established.

A number of other proofs of theorem (5.1) are known. Mention should be made of a method of Delange [4], who makes use of a theorem of his own, concerning the existence of non-zero mean-values for sums

$$n^{-1} \sum_{m=1}^{n} g(m)$$

of complex-valued multiplicative functions $g(m)$ which satisfy $|g(m)| \le 1$, $m = 1, 2, \ldots$. We shall consider theorems of this type in Chapter 6.

A fine alternative proof of the sufficiency part of Delange's theorem and so, indirectly, of the sufficiency part of theorem (5.1), was given by Rényi [6]. Rényi's proof makes use only of the Turán–Kubilius inequality. His method was adapted by Galambos [2], [1], [4], to deal with abstract situations involving weakly dependent random variables.

To some extent, one can generalise theorems (5.1) and (5.2) to cover the behaviour of additive functions on sets of integers other than the set of all positive integers. For example, one may consider integers of the type $k(m)$, $m = 1, 2, \ldots$, where $k(\ )$ is an integral valued polynomial; or $p + 1$, where $p = 2, 3, \ldots$ runs through the prime numbers. We shall consider some of these generalisations in the later Chapter 12.

A theorem, similar to theorem (5.4) but with the condition on $\alpha(x)$ strengthened to

$$\alpha(x) - \alpha(x^{\delta}) = o(1) \qquad (x \to \infty),$$

for any fixed real number δ, $0 < \delta < 1$, was proved by Delange [6]. In a certain sense, the condition on $\alpha(x)$ which appears in the present theorem (5.4) cannot be weakened. This may be seen from theorem (7.1), of Chapter 7.

The content of theorem (5.5) is largely contained in the paper [2] of Erdös and Wintner. See also the later, connected theorem (7.7) due to the author, [13]. This last reference also illustrates a connection between additive functions and the theory of Lambert series.

Our treatment of lemma (5.7) is a sharpening of that given by Faĭnleĭb [1], who considered the function $\varphi(m)/m$. However, a few years later, [2], Faĭnleĭb obtained a result of the same strength as lemma (5.7), but by another method. From a value-distribution point of view, the behaviour of the functions $m/\sigma(m)$ and $\varphi(m)/m$ is similar.

In his paper of 1974, from which our treatment of lemma (5.8) is taken, Erdös asserts that by combining his result with the method of Diamond [1], one can establish theorem (5.6) with an error term which is $O((\log n)^{-1})$; a result which would be, in a certain sense, best possible. This stronger result does not seem to be immediately available in such a manner, so that his assertion remains an interesting open conjecture.

If one applies the concentration function estimate of lemma (1.14) directly to the limiting law $h(e^z)$, one obtains the upper bound

$$\limsup_{x \to \infty} \nu_x\left(m; a < \frac{\sigma(m)}{m} \leq a\left(1 + \frac{1}{t}\right)\right) \leq \frac{c_1}{\log \log t},$$

uniformly for all $a > 0$, $t \geq 2$. An estimate of this type which, so to speak, does not take advantage of the fact that the function $\sigma(m)$ assumes integral values was, apparently, first obtained by Tjan [1].

Problems. A favourite problem of Erdös is to give a condition for the absolute continuity of the limit law in theorem (5.1), and presumably in theorem (5.2) also, that is as satisfactory and elegant as Lévy's criterion for continuity.

To ask for a classification of the limit laws in theorems (5.1) and (5.2) is perhaps to bay at the moon. It is known that they are of pure type, and that every type, discrete, absolutely continuous, singular, can actually occur, Erdös [8]. In this same paper he furnishes the example

$$f(p) = \begin{cases} \dfrac{(-1)^{(p-1)/2}}{(\log \log p)^{3/4}}, & p > e^e, \\ 0 & \text{otherwise}, \end{cases}$$

of a strongly additive arithmetic function which has an absolutely continuous limiting distribution.

Let $f(n)$ be an additive function with a limiting distribution $F(z)$, whose corresponding characteristic function is $v(t)$. As was promised earlier, we conclude the present chapter with an

Alternative Proof of the Continuity of the Limit Law

It is convenient to note the following result.

Lemma 5.10. *Let $q_1 < q_2 < \ldots$ be a sequence of primes for which the series*

$$\sum_{j=1}^{\infty} q_j^{-1}$$

converges. Then those squarefree integers which are not divisible by any of the primes q_j have the asymptotic density

$$\frac{6}{\pi^2} \prod_{j=1}^{\infty} \left(1 + \frac{1}{q_j}\right)^{-1}.$$

Proof. We do not give a detailed proof of this result, but note that by means of the representation

$$\mu^2(m) = \sum_{r^2|m} \mu(r)$$

it is straightforward to establish the estimate

$$\sum_{\substack{n \le x \\ n \equiv 0 (\mathrm{mod}\ d)}} \mu^2(n) = \frac{6}{\pi^2} x\eta(d) + O(x^{1/2} d^{-1/2} 2^{\omega(d)})$$

where $\eta(d)$ is the multiplicative function defined by

$$\eta(d) = \frac{\varphi(d)}{d^2} \prod_{p|d} \left(1 - \frac{1}{p^2}\right)^{-1},$$

and where the result is uniform in all squarefree integers d, and real numbers $x \geq d$.

The proof of lemma (5.10) may then be completed by an appeal to lemma (2.1), or, indeed, by the use of a simple Eratosthenian sieve method.

This completes our remarks concerning the proof of lemma (5.10).

Alternative proof (*necessity*). Assume that the series

$$\sum_{f(p)\neq 0} \frac{1}{p} \tag{16}$$

converges. It follows from lemma (5.10) that those integers n which are squarefree and not divisible by any of the primes p for which $f(p) \neq 0$, have the positive asymptotic density

$$\frac{6}{\pi^2} \prod_{f(p)\neq 0} \left(1 + \frac{1}{p}\right)^{-1}.$$

But for each of these integers we have $f(n) = 0$. Thus the limit law $F(z)$ will have a discontinuity at the point $z = 0$.

In order that $F(z)$ be continuous it is therefore necessary that the series (16) diverges.

Alternative proof (*sufficiency*). Assume, now, that the series (16) diverges.

We consider the behaviour, as $T \to \infty$, of the integral

$$J(T) = \frac{1}{2T}\int_{-T}^{T} |\phi(t)|^2\, dt = \frac{1}{2\pi}\int_{-\pi}^{\pi} \left|\phi\left(\frac{uT}{\pi}\right)\right|^2 du.$$

Let w be a positive real number, chosen so large that

$$L(w) = \sum_{\substack{p \leq w \\ f(p)\neq 0}} \frac{1}{p} > 0.$$

According to theorem (5.1) there is a number c so that the characteristic function $\phi(t)$ satisfies the inequality

$$|\phi(t)| \leq c \exp\left(-\sum_{p\leq w} p^{-1}(1 - \cos tf(p))\right)$$

uniformly for all real numbers t and w.

Let m be a further positive real number, and let $S(m)$ be the set of real numbers u, which lie in the interval $-\pi \le u \le \pi$, for which

$$\sum_{p \le w} p^{-1}\left(1 - \cos \frac{uT}{\pi} f(p)\right) \le m.$$

Let $|S(m)|$ denote its Lebesgue measure.

By making use of the second of the two integral representations for $J(T)$ we see that

$$\begin{aligned} J(T) &\le \int_{S(m)} |v(t)|^2 \, dt + \int_{\substack{-\pi \\ \text{not } S(m)}}^{\pi} c^2 \exp\left(-2 \sum_{p \le w} p^{-1}\left(1 - \cos \frac{uT}{\pi} f(p)\right)\right) du \\ &\le |S(m)| + c^2 e^{-2m}. \end{aligned}$$

We shall now show that if T is chosen from a suitable unbounded sequence of real numbers, then

$$(17) \qquad |S(m)| < 16\pi^2 \left(\frac{m}{L(w)}\right)^{1/2}.$$

We note that for all sufficiently large values of T

$$(18) \qquad \frac{1}{2} L(w) \le \frac{1}{2\pi} \int_{-\pi}^{\pi} \sum_{p \le w} p^{-1}\left(1 - \cos \frac{uT}{\pi} f(p)\right) du.$$

According to Dirichlet's theorem on the approximation of irrationals by rationals, for any $\varepsilon > 0$ there is a number T, and integers $r(p)$, so that the inequality

$$\left| \frac{T}{\pi} f(p) - r(p) \right| < \varepsilon$$

holds for each prime p in the range $2 \le p \le w$. It is clear, moreover, that we may choose ε to be so small that

$$(19) \qquad \left| \sum_{p \le w} p^{-1}\left(1 - \cos \frac{uT}{\pi} f(p)\right) - \sum_{p \le w} p^{-1}(1 - \cos ur(p)) \right| < m$$

holds uniformly for all real numbers u in the interval $-\pi \le u \le \pi$. *In general the values of the integers $r(p)$ will depend upon T and ε, but this will not affect the argument which follows.*

Let T (and ε) be chosen so that the requirements (18) and (19) are met simultaneously.

We define the sum

$$D(u) = \sum_{p \le w} p^{-1}(1 - \cos ur(p)),$$

noting that when u belongs to $S(m)$ it satisfies the inequality

$$D(u) \le 2m.$$

If $|S(m)| = 0$ then inequality (17) is trivially valid; otherwise define the integer k by

$$k = \left[\frac{4\pi}{|S(m)|}\right].$$

It is clear that

$$k|S(m)| > 2\pi.$$

Moreover, $S(m)$ is closed and symmetric with respect to the origin. According to lemma (1.6) we may therefore conclude that each u with $|u| \le \pi$, differs (mod 2π) from the sum of at most k members of the set $S(m)$. Typically, we shall have

$$u \equiv u_1 + \cdots + u_l \,(\text{mod } 2\pi),$$

where $1 \le l \le k$. In view of the inequalities

$$|\sin(a + b)| \le |\sin a| + |\sin b|,$$

$$|\sin(u_1 + \cdots + u_l)| \le (|\sin u_1| + \cdots + |\sin u_l|)^2 \le l \sum_{j=1}^{l} \sin^2 u_j,$$

we see that

$$D(u) = 2 \sum_{p \le w} p^{-1} \sin^2 ur(p) = D(u_1 + \cdots + u_l)$$

$$\le l \sum_{j=1}^{l} D(u_j) \le 2l^2 m \le 2k^2 m.$$

Note that we have here made essential use of the fact that the $r(p)$ are integers.

Therefore

$$\int_{-\pi}^{\pi} D(u)du \le 4\pi k^2 m.$$

However, from our construction of the $r(p)$, including (18),

$$\int_{-\pi}^{\pi} D(u)du \geq \int_{-\pi}^{\pi} \sum_{p \leq w} p^{-1}\left(1 - \cos \frac{uT}{\pi} f(p)\right)du - 2\pi m$$

$$\geq \frac{1}{2} L(w) - 2\pi m,$$

so that altogether we have

$$L(w) \leq 4\pi m + 8\pi k^2 m \leq 12\pi k^2 m \leq 12\pi\left(\frac{4\pi}{|S(m)|}\right)^2 m.$$

This gives the inequality (17) at once.

Since the above argument may be carried out for an unbounded sequence of T-values

$$\liminf_{T \to \infty} \frac{1}{2T} \int_{-T}^{T} |v(t)|^2 \, dt \leq 16\pi^2\left(\frac{m}{L(w)}\right)^{1/2} + c^2 e^{-2m}.$$

We let $w \to \infty$, and then $m \to \infty$, and obtain the continuity of the limit law $F(z)$ by appealing to lemma (1.23).

This completes our alternative proof. A similar proof may be given concerning the continuity of the limit law in theorem (5.2).

For a discussion of Dirichlet's theorem on simultaneous approximation we refer to Hardy and Wright [1], Chapter IX, Theorem 200, p. 170.

When the non-zero values of $f(p)$ are distinct, a short proof that the divergence of the series (16) ensures the continuity of the limit law was given by Szüsz [1]. At the end of his proof there is a remark to the effect that this extra condition may be dispensed with. It is not altogether clear how one is to carry out the modification suggested. In our present proof we follow him in considering the integrals $J(T)$ but, by means of a suitable application of Dirichlet's theorem on simultaneous approximation, continue with an argument adapted from that used by Halász, [5], in the study of the local behaviour of additive functions. (See Theorem (21.5)).

As we shall show in Chapter 8, when we consider additive functions (mod 1), this alternative proof lends itself to generalisation.

Chapter 6

Theorems of Delange, Wirsing, and Halász

In this chapter we study the mean-value behaviour of those complex-valued multiplicative functions $g(n)$ which satisfy $|g(n)| \leq 1$ for every positive integer n. For the whole of the present chapter $g(n)$ will denote such a function.

With each such function $g(n)$ we associate the Dirichlet series

$$G(s) = \sum_{n=1}^{\infty} g(n)n^{-s} \qquad s = \sigma + i\tau.$$

This series is uniformly absolutely convergent in any half-plane of the form $\sigma \geq \delta$, when $\delta > 1$. It therefore defines a function, $G(s)$, which is analytic in the half-plane $\sigma > 1$.

Theorem 6.1 (Halász). *Let $g(n)$ be a multiplicative function, $|g(n)| \leq 1$, $(n = 1, 2, \ldots)$.*

Then there are constants C, α, and a slowly-oscillating function $L(u)$ with $|L(u)| = 1$, so that

$$(1) \qquad \sum_{n \leq x} g(n) = \frac{x^{1+i\alpha}}{1 + i\alpha} CL(\log x) + o(x) \qquad (x \to \infty),$$

if and only if, in the same notation, the asymptotic estimate

$$(2) \qquad G(s) = \frac{C}{s - (1 + i\alpha)} L\left(\frac{1}{\sigma - 1}\right) + o\left(\frac{1}{\sigma - 1}\right) \qquad (\sigma \to 1+),$$

holds uniformly on each bounded interval $-M \leq \tau \leq M$.

Theorem 6.2 (Wirsing, Halász). *Let $g(n)$ be a multiplicative function, $|g(n)| \leq 1$, $(n = 1, 2, \ldots)$.*

If there is a real number τ so that the series

$$\sum p^{-1}(1 - \operatorname{Re} g(p)p^{-i\tau})$$

converges, then as $x \to \infty$

$$\sum_{n \le x} g(n) = \frac{x^{1+i\tau}}{1+i\tau} \prod_{p \le x} \left(1 - \frac{1}{p}\right)\left(1 + \sum_{m=1}^{\infty} p^{-m(1+i\tau)} g(p^m)\right) + o(x).$$

On the other hand, if there is no such number τ, *then*

$$x^{-1} \sum_{n \le x} g(n) \to 0 \qquad (x \to \infty).$$

In either case there are constants D, α, *and a slowly-oscillating function* $L(u)$ *with* $|L(u)| = 1$, *so that as* $x \to \infty$

$$\sum_{n \le x} g(n) = Dx^{1+i\alpha} L(\log x) + o(x).$$

This may seem an odd way to state these two results, since the result of theorem (6.2) shows that the conditions of theorem (6.1) are always satisfied if D, α and $L(u)$ are suitably defined. However, each of these theorems is interesting in its own right. In particular, theorem (6.1) allows one to assert, (Halász [1]), that the mean-value

$$A = \lim_{n \to \infty} n^{-1} \sum_{m=1}^{n} g(m)$$

exists if and only if

$$G(s) = \frac{A}{s-1} + o\left(\frac{1}{\sigma - 1}\right)$$

as $\sigma \to 1+$, uniformly for $|\tau| \le M$.

The following, related results will be of great use in the study of translated additive functions, and the distribution of the values of multiplicative functions.

Theorem 6.3 (Delange, Wirsing, Halász). *The mean-value*

$$A = \lim_{n \to \infty} n^{-1} \sum_{m=1}^{n} g(m)$$

exists and is non-zero if and only if

(i) *There is at least one positive integer k so that $g(2^k) \neq -1$, and*
(ii) *The series*

$$\sum p^{-1}(1 - g(p))$$

converges.
When these conditions are satisfied the limit has the value

$$\prod_p \left(1 - \frac{1}{p}\right)\left(1 + \sum_{m=1}^{\infty} p^{-m} g(p^m)\right).$$

The mean-value A exists and is zero if and only if either

(iii) *There is a real number τ, so that for each positive integer k, $g(2^k) = -2^{ik\tau}$, moreover the series*

$$\sum p^{-1}(1 - \operatorname{Re} g(p)p^{-i\tau})$$

converges; or
(iv) *The series*

$$\sum p^{-1}(1 - \operatorname{Re} g(p)p^{-i\tau})$$

diverges for every real number τ.

As a particular case, we note the following fine result, first proved by Wirsing [4].

Theorem 6.4 (Wirsing). *Let $g(n)$ be a real-valued, multiplicative function, for which $|g(n)| \leq 1$, $(n = 1, 2, \ldots)$. Then the mean-value*

$$\lim_{n \to \infty} n^{-1} \sum_{m=1}^{n} g(m)$$

always exists. It has the value

$$\prod_p \left(1 - \frac{1}{p}\right)\left(1 + \sum_{m=1}^{\infty} p^{-m} g(p^m)\right)$$

where the product is considered to be zero if the series

$$\sum p^{-1}(1 - g(p))$$

diverges.

Before giving the proofs of these theorems, we outline an argument which will be useful several times in this and in subsequent chapters.

Application of Parseval's Formula

For non-negative real numbers u, define the function

$$H(u) = \sum_{n \leq u} g(n). \tag{3}$$

Then, integration by parts shows that for $\sigma > 1$

$$s^{-1}G(s) = s^{-1}\int_{1-}^{\infty} u^{-s}\, dH(u) = \int_{1}^{\infty} u^{-s-1}H(u)du. \tag{4}$$

We effect the change of variable $u = e^w$ in this last integral, to obtain

$$s^{-1}G(s) = \int_{0}^{\infty} H(e^w)e^{-w\sigma} \cdot e^{-iw\tau}\, dw.$$

It follows that

$$s^{-1}G(s) \text{ as a function of } \tau,$$

and

$$H(e^w)e^{-w\sigma} \text{ as a function of } w,$$

are Fourier transforms. As a function of τ, $s^{-1}G(s)$ is $O(|\tau|^{-1})$ when $|\tau| \to \infty$, so that both of these functions belong to the Lebesgue class $L^2(-\infty, \infty)$. By Parseval's theorem

$$\int_{-\infty}^{\infty} \left|\frac{G(s)}{s}\right|^2 d\tau = 2\pi \int_{0}^{\infty} |H(e^w)e^{-w\sigma}|^2\, dw. \tag{5}$$

But $|H(e^w)| \leq e^w$, so that the integral on the right hand side of this equation does not exceed

$$2\pi \int_{0}^{\infty} e^{-2w(\sigma-1)}\, dw = \frac{\pi}{(\sigma - 1)}.$$

We deduce that for $1 < \sigma \leq 3$,

$$\int_{-1}^{1} |G(s)|^2\, d\tau \leq 10 \int_{-\infty}^{\infty} \left|\frac{G(s)}{s}\right|^2 d\tau \leq \frac{10\pi}{\sigma - 1}. \tag{6}$$

Let ρ be a real number. Then we may apply the same argument with $g(n)$ replaced by $g(n)n^{-i\rho}$. This shows that the bound

$$\int_{|\tau-\rho|\le 1} |G(s)|^2\, d\tau \le \frac{10\pi}{\sigma - 1} \tag{7}$$

holds uniformly for all real numbers ρ.

An alternative approach to inequalities of this type was suggested by Montgomery [3], lemma II.3, p. 158. His method depends upon the fact that if $y > 0$, and by definition if $y = 0$,

$$\int_{-1}^{1} (1 - |t|)e^{ity}\, dt = \left(\frac{\sin y/2}{y/2}\right)^2 \ge 0.$$

By means of this relation he proves:

Lemma 6.5 (Montgomery). *Let the series*

$$A(s) = \sum_{n=1}^{\infty} a_n n^{-s} \qquad B(s) = \sum_{n=1}^{\infty} b_n n^{-s}$$

converge absolutely for $\mathrm{Re}(s) = \sigma$. *Let* $|a_n| \le b_n$ *hold for all positive integers* n.

Then for any $T \ge 0$

$$\int_{-T}^{T} |A(s)|^2\, d\tau \le 2\int_{-2T}^{2T} |B(s)|^2\, d\tau.$$

Proof. The proof is an example in the use of a Fejér kernel. In fact

$$\int_{-T}^{T} |A(s)|^2\, d\tau \le 2\int_{-2T}^{2T}\left(1 - \frac{|\tau|}{2T}\right)|A(s)|^2\, d\tau$$

$$= 2\sum_{m=1}^{\infty}\sum_{n=1}^{\infty} a_m \bar{a}_n m^{-\sigma} n^{-\sigma}\int_{-2T}^{2T}\left(1 - \frac{|\tau|}{2T}\right)(mn^{-1})^{-i\tau}\, d\tau.$$

According to our earlier remark, each of these last integrals is real and non-negative, so that this expression does not exceed

$$2\sum_{m=1}^{\infty}\sum_{n=1}^{\infty} b_m b_n m^{-\sigma} n^{-\sigma}\int_{-2T}^{2T}\left(1 - \frac{|\tau|}{2T}\right)(mn^{-1})^{-i\tau}\, d\tau$$

$$= 2\int_{-2T}^{2T}\left(1 - \frac{|\tau|}{2T}\right)|B(s)|^2\, d\tau \le 2\int_{-2T}^{2T} |B(s)|^2\, d\tau.$$

This completes the proof of lemma (6.5).

It is also convenient to establish a further preliminary result.

Lemma 6.6. *Let $g(n)$ be a multiplicative function, $|g(n)| \leq 1$ for $n = 1, 2, \ldots$. For each prime p define*

$$h(p) = \sum_{k=1}^{\infty} g(p^k)p^{-ks}.$$

Then there is a representation

$$G(s) = (1 + h(2))\exp\left(\sum_{p \leq 3} g(p)p^{-s}\right)G_1(s),$$

valid in the half-plane $\sigma > 1$. Moreover, $G_1(s)$ is analytic in the half-plane $\sigma > 1/2$, and is bounded by

$$e^{-5} \leq |G_1(s)| \leq e^5 \qquad |G_1'(s)| \leq e^{11}$$

in the half-plane $\sigma \geq 1$.

Remark. The function $h(p)$ is well defined if $\sigma > 0$.

The precise values of the constants $e^{\pm 5}$, e^{11} in this lemma are not important.

Proof of lemma (6.6). From the hypotheses of the lemma we see that for each prime p

$$|h(p)| \leq \sum_{k=1}^{\infty} p^{-k\sigma} = p^{-\sigma}(1 - p^{-\sigma})^{-1} \leq \begin{cases} 4p^{-\sigma} & \text{if } \sigma \geq 1/2, \\ 2p^{-1} & \text{if } \sigma \geq 1. \end{cases}$$

Since $g(n)$ is multiplicative, and the series

$$\sum_{n=1}^{\infty} |g(n)|n^{-\sigma}$$

converges for $\sigma > 1$, we see from lemma (2.13) that in this same half-plane $G(s)$ has the Euler product representation

$$G(s) = \prod_{p \geq 2} (1 + h(p)).$$

If we define

$$G_1(s) = \prod_{p \geq 3} (1 + h(p))\exp(-g(p)p^{-s})$$

then we obtain the desired representation, and we need only demonstrate that this last function has the properties asserted in the statement of the lemma.

Assume first that $\sigma > 1/2$. Then for all sufficiently large primes p we have $|h(p)| \le 2/3$. By making use of the estimate

$$|\log(1 + w) - w| \le 2|w|^2,$$

which is certainly valid for the principal value of the logarithm, and all complex numbers w for which $|w| \le 2/3$, we see that

$$|\log(1 + h(p)) - g(p)p^{-s}| \le |\log(1 + h(p)) - h(p)| + |h(p) - g(p)p^{-s}|$$

$$\le 2|h(p)|^2 + \sum_{k=2}^{\infty} |g(p^k)|p^{-k\sigma} \le 36p^{-2\sigma}.$$

It follows that $G_1(s)$ is defined by a product which converges uniformly absolutely in any half-plane $\sigma \ge \sigma_0$, with $\sigma_0 > 1/2$. Hence $G_1(s)$ is analytic in the region $\sigma > 1/2$.

If now $\sigma \ge 1$, then $|h(p)| \le 2/3$ for every odd prime p, so that

$$\log|G_1(s)| \le \sum_{p \ge 3} 2|h(p)|^2 + \sum_{p \ge 3} \sum_{k=2}^{\infty} |g(p^k)|p^{-k}$$

$$\le 10 \sum_{p \ge 3} p^{-2} < 10 \sum_{n=3}^{\infty} \frac{1}{n(n-1)} = 5.$$

To complete the proof of lemma (6.6) we note that $2^{3/4} > 5/3$, and that if $p \ge 11$, $\sigma \ge 3/4$, then $|h(p)| \le 5/(2p^\sigma) < 2/3$. Therefore, the upper bound

$$|G_1(s)| \le e^5 \prod_{p \le 7} (1 + |h(p)|) < e^5(5/3)^4$$

holds uniformly in the half-plane $\sigma \ge 3/4$. Applying Cauchy's theorem, we deduce that for $\sigma \ge 1$,

$$|G_1'(s)| = \left| \frac{1}{2\pi i} \int_{|s-z|=1/4} \frac{G_1(z)}{(z-s)^2} dz \right| < e^{11}.$$

Proof of theorem (6.1), (Necessity). In this part of the proof we shall assume the existence of α, C, and $L(u)$ so that the asymptotic relation (1) holds, and prove that (2) must be satisfied.

We adopt the notation of (4), and integrate by parts to obtain the representation

$$s^{-1}G(s) = \int_1^{\infty} u^{-s-1}H(u)du.$$

Let ε be a real number, $0 < \varepsilon < 1$. Then for all sufficiently large values of u, say for $u \geq u_0$ (>1),

$$|H(u) - \lambda u^{1+i\alpha} L(\log u)| < \varepsilon u \qquad \lambda = C(1 + i\alpha)^{-1}.$$

Thus

$$\left| \int_{u_0}^{\infty} u^{-s-1} H(u) du - \lambda \int_{u_0}^{\infty} u^{-s+i\alpha} L(\log u) du \right| \leq \varepsilon \int_{u_0}^{\infty} u^{-\sigma}\, du < \frac{\varepsilon}{\sigma - 1}$$

and it follows readily that as $\sigma \to 1+$

$$s^{-1} G(s) = \lambda \int_{1}^{\infty} u^{-s+i\alpha}\, L(\log u) du + o\left(\frac{1}{\sigma - 1}\right).$$

Moreover, this result holds uniformly for all real values of τ.

It is now convenient to introduce a positive number Δ, temporarily regarded as fixed, and two further numbers, $y_1 = \exp(\Delta^{-1}(\sigma - 1)^{-1})$, and $y_2 = \exp(\Delta(\sigma - 1)^{-1})$.

From the slowly-oscillating nature of the function $L(u)$, the estimate

$$L(\log u) = L\left(\frac{1}{\sigma - 1}\right) + o(1) \qquad (\sigma \to 1+),$$

holds uniformly for $y_1 \leq u \leq y_2$. Hence

$$\int_{y_1}^{y_2} u^{-s+i\alpha} L(\log u) du = \int_{y_1}^{y_2} u^{-s+i\alpha} L\left(\frac{1}{\sigma - 1}\right) du + o\left(\frac{1}{\sigma - 1}\right),$$

and

$$\left| s^{-1} G(s) - \lambda L\left(\frac{1}{\sigma - 1}\right) \frac{1}{s - (1 + i\alpha)} \right| \leq 2|\lambda| \left\{ \int_{1}^{y_1} + \int_{y_2}^{\infty} \right\} u^{-\sigma}\, du + o\left(\frac{1}{\sigma - 1}\right)$$

$$\leq \frac{1}{\sigma - 1} (2|\lambda| \Delta^{-1} + 2|\lambda| e^{-\Delta} + o(1)).$$

Allowing first that σ approach 1, and then that Δ become large, we see that, uniformly for all real values of τ,

$$\frac{G(s)}{s} = \frac{\lambda}{s - (1 + i\alpha)} L\left(\frac{1}{\sigma - 1}\right) + o\left(\frac{1}{\sigma - 1}\right).$$

It is now straightforward to deduce (6.2), making use of the identity

$$\frac{s\lambda}{s-(1+i\alpha)} = \lambda + \frac{C}{s-(1+i\alpha)},$$

and the fact that $|L(u)| = 1$.

This completes the easier part of theorem (6.1). To go in the opposite direction we shall use a method of Halász [1]. We shall assume $\alpha = 0$ to begin with.

Proof of theorem (6.1), (Sufficiency). Let us assume that the estimate (2) is valid with $\alpha = 0$.

We begin with the integral, defined for all positive real numbers x and σ:

$$\frac{1}{2\pi i}\int_{\sigma-i\infty}^{\sigma+i\infty} \frac{x^s}{s^2}\,ds = \begin{cases} \log x & \text{if } x > 1, \\ 0 & \text{if } 0 < x \le 1. \end{cases}$$

This integral is readily evaluated by taking a (large) semicircle on the line-segment $\operatorname{Re} s = \sigma$, $|\tau| \le T$, as diameter, and to the left or right according as to whether $x > 1$ or not; and then letting $T \to \infty$.

Since

$$G'(s) = -\sum_{n=1}^{\infty} g(n)\log n\, n^{-s} \qquad (\sigma > 1),$$

interchanging integration and summation shows that

$$-\sum_{n\le x} g(n)\log n \log\frac{x}{n} = \frac{1}{2\pi i}\int_{\sigma-i\infty}^{\sigma+i\infty} \frac{x^s}{s^2}\,G'(s)ds$$

provided that $\sigma > 1$. We shall ultimately choose $\sigma = \sigma_0 = 1 + (\log x)^{-1}$. If $x \ge 2$ then this last condition will be satisfied.

Let K and M be positive real numbers, $K \ge 2$, $M \ge 2$, temporarily regarded as fixed. We divide the range of integration in this last integral into the three parts

$$|\tau| \le K(\sigma-1) \qquad K(\sigma-1) < |\tau| \le M \qquad M < |\tau| < \infty.$$

We shall show that the main contribution towards the integral comes from the first of these parts. Different, but related, arguments will show that the remaining values of τ do not contribute an appreciable amount. This choice of subdivision proves to be appropriate many times, in subsequent chapters, when we consider similar contour integrals; although substantial auxiliary arguments are usually required.

The range $|\tau| > M$.

Clearly

$$\left| \frac{1}{2\pi i} \int_{|\tau|>M} \frac{x^2 G'(s)}{s^2} \, ds \right| \le \frac{x^\sigma}{2\pi} \int_{|\tau|>M} \left| \frac{G'(s)}{s^2} \right| d\tau,$$

and we shall show that

$$\frac{1}{2\pi} \int_{|\tau|>M} \left| \frac{G'(s)}{s^2} \right| d\tau < \frac{e^{20}}{M^{1/2}(\sigma - 1)}$$

provided only that $1 < \sigma \le 2$.

We make use of the factorisation

$$\frac{G'(s)}{s^2} = \frac{G'(s)}{sG(s)} \cdot \frac{G(s)}{s}$$

and apply the Cauchy–Schwarz inequality in the form

$$\left(\int_{|\tau|>M} \left| \frac{G'(s)}{s^2} \right| d\tau \right)^2 \le \int_{|\tau|>M} \left| \frac{G'(s)}{sG(s)} \right|^2 d\tau \cdot \int_{|\tau|>M} \left| \frac{G(s)}{s} \right|^2 d\tau.$$

According to the inequality (7), for each integer m

$$\int_{|\tau - m| \le 1} |G(s)|^2 \, d\tau < \frac{10\pi}{\sigma - 1},$$

so that

$$(8) \quad \int_{|\tau|>M} \left| \frac{G(s)}{s} \right|^2 d\tau \le \sum_{|m|>M} \int_{|\tau - m| \le 1} \left| \frac{G(s)}{s} \right|^2 d\tau$$

$$\le 4 \sum_{|m|>M} m^{-2} \int_{|\tau - m| \le 1} |G(s)|^2 \, d\tau < \frac{160\pi}{M(\sigma - 1)}.$$

We use a modified form of this argument to deal with the integral involving $G'(s)/G(s)$.

Differentiating logarithmically the representation for $G(s)$ which is given in lemma (6.6), we see that, for $\sigma > 1$,

$$\frac{G'(s)}{G(s)} = \frac{h'(2)}{1 + h(2)} - \sum_{p \ge 3} g(p) p^{-s} \log p + \frac{G_1'(s)}{G_1(s)}.$$

According to this same lemma,

$$\int_{-\infty}^{\infty}\left|\frac{G_1'(s)}{sG_1(s)}\right|^2 d\tau \le e^{32}\int_{\sigma-i\infty}^{\sigma+i\infty}|s|^{-2}\,d\tau < \pi e^{32}.$$

Define the function

$$J(u) = \sum_{p\le u} g(p)\log p.$$

By making use of the Tchebycheff inequality which was established at the end of the proof of lemma (3.1), we see that

$$|J(u)| \le \sum_{p\le u}\log p \le 4u \qquad (u > 0).$$

We form the appropriate analogue of the Parseval relation (5), and deduce that

$$\int_{-\infty}^{\infty}\left|s^{-1}\sum_{p\ge 3} g(p)p^{-s}\log p\right|^2 d\tau = 2\pi\int_0^{\infty}|e^{-w\sigma}J(e^w)|^2\,dw$$

$$\le 32\pi\int_0^{\infty} e^{-2w(\sigma-1)}\,dw \le \frac{16\pi}{\sigma-1}.$$

Consider next, the function

$$\{1 + h(2)\}^{-1} = 1 + \sum_{m=1}^{\infty}(-h(2))^m \qquad (\sigma > 1).$$

Viewed as a Dirichlet series, each coefficient of 2^{-ls}, $(l = 1, 2, \ldots)$, in the series which appears on the right-hand side, does not exceed in absolute value the corresponding coefficient in the series

$$1 + \sum_{m=1}^{\infty}\left(\sum_{k=1}^{\infty}2^{-ks}\right)^m = \frac{2^s - 1}{2^s - 2}.$$

We may now apply Montgomery's lemma, lemma (6.5), with $T = 1$,

$$\int_{-1}^{1}\frac{d\tau}{|1 + h(2)|^2} \le 2\int_{-2}^{2}\left|\frac{2^s - 1}{2^s - 2}\right|^2 d\tau < 2^{2\sigma+3}\int_{-2}^{2}\frac{d\tau}{|2^s - 2|^2}.$$

So far we have only used the condition that $\sigma > 1$. We now take advantage of the further restriction $\sigma \le 2$, and apply Parseval's relation, (5), once again, to deduce that

$$\frac{1}{2}\int_{-2}^{2}\frac{d\tau}{|2^s - 2|^2} \le \int_{-2}^{2}|s|^{-2}\left|\sum_{l=1}^{\infty}2^l\cdot 2^{-ls}\right|^2 d\tau \le \int_{-\infty}^{\infty}|s|^{-2}\left|\sum_{l=1}^{\infty}2^l\cdot 2^{-ls}\right|^2 d\tau$$

$$= 2\pi\int_0^{\infty}\left|e^{-w\sigma}\sum_{2^l\le e^w}2^l\right|^2 dw \le 8\pi\int_0^{\infty}e^{-2w(\sigma-1)}dw \le \frac{4\pi}{\sigma-1}.$$

Since

$$|h'(2)| = \left|\sum_{k=1}^{\infty}-g(2^k)2^{-ks}\log 2^k\right| \le \sum_{k=1}^{\infty}2^{-k}k\log 2 < 2 \qquad (\sigma > 1),$$

we have proved that

$$\int_{-1}^{1}\left|\frac{h'(2)}{1+h(2)}\right|^2 d\tau < \frac{2^{12}\pi}{\sigma-1}.$$

All of the preceding argument will remain valid if s is replaced by $s + im$, which enables us to assert that

$$\int_{-\infty}^{\infty}\left|\frac{h'(2)}{s(1+h(2))}\right|^2 d\tau \le \int_{-1}^{1}\left|\frac{h'(2)}{1+h(2)}\right|^2 d\tau$$

$$+ \sum_{|m|\le 2}\frac{4}{m^2}\int_{|\tau-m|\le 1}\left|\frac{h'(2)}{1+h(2)}\right|^2 d\tau < \frac{2^{16}\pi}{\sigma-1}.$$

Putting these results together, and appealing to a simple case of Hölder's inequality, yields

$$\int_{-\infty}^{\infty}\left|\frac{G'(s)}{sG(s)}\right|^2 d\tau \le 3\pi(2^{16} + 16 + e^{32})\frac{1}{(\sigma-1)} < \frac{\pi e^{36}}{(\sigma-1)}.$$

With what we have established in (8), we obtain the upper bound

$$\frac{1}{2\pi}\int_{|\tau|>M}\left|\frac{G'(s)}{s^2}\right|d\tau < \frac{e^{20}}{M^{1/2}(\sigma-1)}.$$

The range $K(\sigma - 1) < |\tau| \le M$

For convenience of exposition let us denote this range by Γ.

We begin, as in the previous section, with the inequality

$$\left|\frac{1}{2\pi i}\int_{\Gamma}\frac{x^s G'(s)}{s^2}ds\right| \le \frac{x^\sigma}{2\pi}\int_{\Gamma}\left|\frac{G'(s)}{s^2}\right|d\tau.$$

We shall show that regarding K and M as fixed, there is a constant c_1, depending at most upon the constant C in the hypothesis of theorem (6.1), so that the inequality

$$\int_\Gamma \left|\frac{G'(s)}{s^2}\right| d\tau < \frac{c_1}{K^{1/4}(\sigma - 1)}$$

is satisfied for all sufficiently large values of x.

We follow the argument of the previous section, from which we also adopt the result

$$\frac{1}{2\pi}\int_{-\infty}^{\infty} \left|\frac{G'(s)}{sG(s)}\right|^2 d\tau < \frac{e^{20}}{(\sigma - 1)},$$

but we treat the integral involving $G(s)$ alone, a little differently.

Clearly

$$\int_\Gamma \left|\frac{G(s)}{s}\right|^2 d\tau \le \max_\Gamma |G(s)|^{1/2} \int_\Gamma |s|^{-2}|G(s)|^{3/2}\, d\tau.$$

According to our present hypothesis (2), if x is sufficiently large then

$$\max_\Gamma |G(s)|^{1/2} \le \left(\frac{|C| + 1}{K(\sigma - 1)}\right)^{1/2}.$$

From lemma (6.6), if $\sigma > 1$ then

$$|G(s)| \le e^5 \exp\left(\operatorname{Re} \sum_{p \ge 3} g(p)p^{-s}\right),$$

so that

$$|G(s)|^{3/4} \le e^5 \exp\left(\operatorname{Re} \sum_{p \ge 3} 3g(p)/(4p^s)\right)$$

which, by another application of lemma (6.6), does not exceed

$$e^{12}\left|\sum_{(n,\,2)=1} \left(\frac{3}{4}\right)^{\Omega(n)} g(n)n^{-s}\right|.$$

Here $\Omega(n)$ denotes the number of prime divisors of the integer n, counted with multiplicity. Note that since $\sigma > 1$,

$$\left|1 + \sum_{k=1}^{\infty} \left(\frac{3}{4}\right)^{\Omega(2^k)} g(2^k)2^{-ks}\right| \ge 1 - \sum_{k=1}^{\infty} \left(\frac{3}{8}\right)^k = \frac{2}{5}.$$

Therefore

$$\int_{-1}^{1} |G(s)|^{3/2}\, d\tau \le e^{24} \int_{-1}^{1} \left| \sum_{(n,2)=1} \left(\frac{3}{4}\right)^{\Omega(n)} g(n) n^{-s} \right|^2 d\tau$$

$$\le 2e^{24} \int_{-2}^{2} \left| \sum_{n=1}^{\infty} \left(\frac{3}{4}\right)^{\Omega(n)} n^{-s} \right|^2 d\tau,$$

this last step by lemma (6.5).

However, applying lemma (6.6) yet again, the integrand in this last integral does not exceed

$$\left| e^5 \exp\left(\sum_{p \ge 3} \frac{3}{4} p^{-s} \right) \right|^2 \le e^{24} |\zeta(s)|^{3/2}.$$

Since $\zeta(s)$ has a simple pole at the point $s = 1$, and is otherwise regular in the complex plane, there are absolute constants c_2 and c_3 so that

$$\int_{-1}^{1} |G(s)|^{3/2}\, d\tau \le c_2 \int_{-2}^{2} \left(\frac{1}{|s-1|^{3/2}} + 1 \right) d\tau < c_3 (\sigma - 1)^{-1/2}.$$

The same argument shows that

$$\int_{|\tau - m| \le 1} |G(s)|^{3/2}\, d\tau < c_3 (\sigma - 1)^{-1/2} \qquad (m = \pm 1, \pm 2, \ldots),$$

from which we deduce that

$$\int_{-\infty}^{\infty} |s|^{-2} |G(s)|^{3/2}\, d\tau \le \int_{-1}^{1} |G(s)|^{3/2}\, d\tau$$

$$+ 4 \sum_{|m| \ge 2} m^{-2} \int_{|\tau - m| \le 1} |G(s)|^{3/2}\, d\tau < 9 c_3 (\sigma - 1)^{-1/2}.$$

We have now proved that

$$\int_{\Gamma} \left| \frac{G'(s)}{s^2} \right| d\tau < \frac{c_1}{K^{1/4}(\sigma - 1)},$$

with

$$c_1 = 6\pi e^{10} (|C| + 1)^{1/4} c_3^{1/2},$$

provided only that x is sufficiently large in terms of K and M.

The range $|\tau| \le K(\sigma - 1)$.

In this section we set $\sigma = \sigma_0 = 1 + (\log x)^{-1}$.

We first need an estimate for $G'(s)$ which is valid on the line segment $\Lambda = \{s; \sigma = \sigma_0, |\tau| \le K(\sigma_0 - 1)\}$.

From Cauchy's theorem

$$G'(s) = \frac{1}{2\pi i}\int_\gamma \frac{G(z)}{(s-z)^2}\,dz,$$

where γ denotes the contour $\{z; |z - s| = (\sigma_0 - 1)/2\}$. For each point z which lies on this contour

$$\frac{2}{3(\sigma_0 - 1)} \le \frac{1}{\operatorname{Re} z - 1} \le \frac{2}{\sigma_0 - 1} \qquad |\operatorname{Im} z| \le |\tau| + (\sigma_0 - 1)/2.$$

From the hypothesis (6.2), and the slowly oscillating nature of $L(u)$, the estimates

$$\begin{aligned} G(z) &= \frac{C}{z-1} L\left(\frac{1}{\operatorname{Re} z - 1}\right) + o\left(\frac{1}{\operatorname{Re} z - 1}\right) \\ &= \frac{C}{z-1} L\left(\frac{1}{\sigma_0 - 1}\right) + o\left(\frac{1}{\sigma_0 - 1}\right) \qquad (x \to \infty), \end{aligned}$$

hold uniformly for z on γ, for all points s on Λ.

Thus, uniformly on Λ,

$$G'(s) = -\frac{C}{(s-1)^2} L\left(\frac{1}{\sigma_0 - 1}\right) + o\left(\frac{1}{(\sigma_0 - 1)^2}\right),$$

so that as $x \to \infty$,

$$\frac{1}{2\pi i}\int_\Lambda \frac{x^s G'(s)}{s^2}\,ds = -\frac{C}{2\pi i} L\left(\frac{1}{\sigma_0 - 1}\right)\int_\Lambda \frac{x^s}{s^2(s-1)}\,ds + o(x \log x).$$

We form a contour by adjoining to Λ the semicircle, on Λ as a diameter, which encloses the point $s = 1$. The contribution of the integral over this added semicircle is $O(K^{-1} x \log x)$. Taking into account the residue at the double pole at $s = 1$, we obtain the asymptotic estimate

$$\frac{1}{2\pi i}\int_\Lambda \frac{x^s G'(s)}{s^2}\,ds = -Cx \log x \,.\, L(\log x) + O(K^{-1} x \log x),$$

where the implied constant is absolute if x is sufficiently large.

By making use of our results concerning the previous two ranges of τ, choosing $\sigma = \sigma_0$, we see that for all sufficiently large values of x

$$\left| \sum_{n \le x} g(n) \log n \log \frac{x}{n} - Cx \log x \,.\, L(\log x) \right| \le c_4(M^{-1/2} + K^{-1/4}) x \log x.$$

We divide by $x \log x$, let $x \to \infty$, $M \to \infty$, and then $K \to \infty$, to obtain

$$\sum_{n \le x} g(n) \log n \log \frac{x}{n} = Cx \log x \,.\, L(\log x) + o(x \log x) \qquad (x \to \infty).$$

Let ε be a temporarily fixed real number, $0 < \varepsilon < 1/2$. Applying this last estimate with x replaced by $x(1 + \varepsilon)$, and subtracting, we see that

$$\sum_{n \le x} g(n) \log n \,.\, \log(1 + \varepsilon) + \sum_{x < n \le x(1+\varepsilon)} g(n) \log n \log \frac{x}{n}$$
$$= C\varepsilon x \log x \,.\, L(\log x) + o(x \log x).$$

Since $\log(1 + \varepsilon) = \varepsilon + O(\varepsilon^2)$, we may divide by ε and conclude that

$$\limsup_{x \to \infty} \left| (x \log x)^{-1} \sum_{n \le x} g(n) \log n - CL(\log x) \right| \le c_5 \varepsilon.$$

But ε may be chosen arbitrarily small, and therefore

$$\sum_{n \le x} g(n) \log n = Cx \log x \,.\, L(\log x) + o(x \log x) \qquad (x \to \infty).$$

Since

$$\sum_{x/\log x < n \le x} g(n) \log n = \sum_{x/\log x < n \le x} g(n) \{\log x + O(\log \log x)\}$$
$$= \log x \,.\, \sum_{n \le x} g(n) + O(x (\log x)^{1/2}),$$

and

$$\sum_{n \le x/\log x} g(n) \log n = O(x),$$

we deduce that

$$\sum_{n \le x} g(n) = CxL(\log x) + o(x) \qquad (x \to \infty),$$

and we have proved theorem (6.1) in the case that $\alpha = 0$.

If we do not assume that $\alpha = 0$, then we may apply the above analysis to the function $g(n)n^{-i\alpha}$, and so establish the estimate

$$D(x) = \sum_{n \le x} g(n)n^{-i\alpha} = CxL(\log x) + o(x) \qquad (x \to \infty).$$

Integrating by parts we have

$$\sum_{n \le x} g(n) = \int_{1-}^{x} y^{i\alpha}\, dD(y) = x^{i\alpha}D(x) - i\alpha \int_{1}^{x} y^{i\alpha - 1}D(y)dy.$$

Let ε be a temporarily fixed real number, $0 < \varepsilon < 1/2$. Then in the range $\varepsilon x < y \le x$,

$$D(y) = CyL(\log x) + o(x),$$

so that

$$\begin{aligned} i\alpha \int_{\varepsilon x}^{x} y^{i\alpha - 1}D(y)dy &= i\alpha CL(\log x) \int_{\varepsilon x}^{x} y^{i\alpha}\, dy + o(x) \\ &= \frac{i\alpha}{1 + i\alpha} Cx^{1 + i\alpha}L(\log x) + O(\varepsilon x). \end{aligned}$$

Estimating trivially the integral over the range $1 \le y < \varepsilon x$, we arrive at

$$\sum_{n \le x} g(n) = \frac{C}{1 + i\alpha} x^{1 + i\alpha}L(\log x) + O(\varepsilon x).$$

Since ε may be chosen arbitrarily small, we have completed the proof of theorem (6.1).

In order to prove theorem (6.2) we shall show that condition (2) of theorem (6.1) is always satisfied for suitably chosen C, α and $L(u)$.

We need a number of preliminary results.

Lemma 6.7. *Let $l_n(w)$, $(n = 1, 2, \ldots)$, be a sequence of real functions which are defined and continuous on the interval $-M \le w \le M$. For each value of w in this interval, let $l_n(w)$ decrease (increase) monotonically to the value of a continuous function $l(w)$.*

Then this convergence is uniform on the whole interval.

Proof. If the convergence to $l(w)$ is not uniform then there is a number $\varepsilon > 0$, a sequence of integers $n_1 < n_2 < \ldots$, and a corresponding set of points $w_1, w_2, \ldots$, all lying in the interval $[-M, M]$, so that

$$l_{n_j}(w_j) - l(w_j) \ge \varepsilon \qquad (j = 1, 2, \ldots).$$

If j is any positive integer, and $m \geq n_1$, then

$$l_m(w_j) - l(w_j) \geq \varepsilon.$$

Since the points w_j all lie in a bounded interval, there is a subsequence of the w_j which converges to a number z for which $|z| \leq M$. Without loss of generality we shall therefore assume that $w_j \to z$ as $j \to \infty$.

We deduce that if m is any (fixed) positive integer and j is sufficiently large, then

$$l_m(z) - l(z) \geq \varepsilon$$

holds for all $m \geq 1$, which contradicts an hypothesis of the lemma.

Hence, the convergence is uniform for $|w| \leq M$, and lemma (6.7) is established.

Remark. The result of lemma (6.7) remains valid if $l_n(w) \to \infty$ as $n \to \infty$. One need only consider the function $\min(1, 1/l_n(w))$, so to speak.

At this point it is convenient to note that if z is a complex number, and $|z| \leq 1$, then $|1 - z|^2 \leq 2(1 - \text{Re } z)$, with equality if and only if $|z| = 1$.

Lemma 6.8. *Let K be a real number, $K \geq 1$. Let the series*

$$\sum_p p^{-1}(1 - \text{Re } g(p))$$

converge. Then

$$\sum_p \left| \frac{1}{p^u} - \frac{1}{p^s} \right| |1 - g(p)| \to 0 \qquad (u \to 1+),$$

uniformly for $|\tau| \leq K(u - 1)$, $(u - 1)/2 \leq \sigma - 1 \leq u - 1$.

Proof. For each prime p

$$\begin{aligned} \left| \frac{1}{p^u} - \frac{1}{p^s} \right|^2 &= p^{-2\sigma} \left| 1 - \frac{1}{p^{u-s}} \right|^2 \\ &\leq p^{-2\sigma} 2(1 - \text{Re} \exp(-(u - s)\log p)) \\ &\leq 2p^{-2\sigma}|u - s|\log p \leq 2(1 + 2K)p^{-2\sigma}(\sigma - 1)\log p, \end{aligned}$$

so that, for a suitable constant c_6,

$$\sum_p p \left| \frac{1}{p^u} - \frac{1}{p^s} \right|^2 \leq 6K(\sigma - 1) \sum p^{-(2\sigma-1)} \log p$$

$$\leq 6K(\sigma - 1)\left\{ -\frac{\zeta'(2\sigma - 1)}{\zeta(2\sigma - 1)} + O(1) \right\} \leq c_6 K.$$

Let ε be a real number, $0 < \varepsilon < 1$, and let N be chosen so large that

$$\sum_{p>N} p^{-1} |1 - g(p)|^2 < \varepsilon^2.$$

That this is possible follows from the hypothesis of this lemma, and the remark immediately preceding its statement. Then, by the Cauchy–Schwarz inequality,

$$\sum_{p>N} \left| \frac{1}{p^u} - \frac{1}{p^s} \right| |1 - g(p)| \leq \left\{ \sum p \left| \frac{1}{p^u} - \frac{1}{p^s} \right|^2 \cdot \sum_{p>N} p^{-1} |1 - g(p)|^2 \right\}^{1/2}$$

$$\leq (c_6 K)^{1/2} \varepsilon.$$

Moreover, for each (fixed) prime not exceeding N,

$$\left| \frac{1}{p^u} - \frac{1}{p^s} \right| |1 - g(p)| \to 0 \qquad (u \to 1+).$$

The sum in the lemma is therefore $o(1) + O(K^{1/2}\varepsilon)$ as $u \to 1+$, and, since ε may be chosen arbitrarily small, lemma (6.8) is proved.

Lemma 6.9. *Let the hypothesis of lemma* (6.8) *be in force, and let M be a further real number, $M \geq 1$. Then there is a constant c_7, depending at most upon the function g, so that the inequality*

$$|G(s)| \leq c_7 K^{-1/2} (\sigma - 1)^{-1}$$

is valid for $K(\sigma - 1) \leq |\tau| \leq M$, for all σ sufficiently near to 1, $\sigma > 1$.

Proof. For each prime p, and real number τ,

$$|1 - p^{i\tau}|^2 \leq 2|1 - g(p)|^2 + 2|g(p) - p^{i\tau}|^2$$

so that

$$2 \sum_p p^{-\sigma}(1 - \operatorname{Re} p^{i\tau}) \leq 2 \sum_p p^{-\sigma} |1 - g(p)|^2 + 4 \sum_p p^{-\sigma}(1 - \operatorname{Re} p^{-i\tau} g(p)).$$

By lemma (6.6),

$$\exp\left(\operatorname{Re}\sum_p p^{-\sigma}(1 - p^{i\tau})\right) \geq \zeta(\sigma)/(4e^{10}|\zeta(\sigma - i\tau)|)$$

and

$$\exp\left(\operatorname{Re}\sum_p p^{-\sigma}(1 - p^{-i\tau}g(p))\right) \leq 3e^{10}\zeta(\sigma)/|G(\sigma + i\tau)|,$$

whilst, by hypothesis

$$\sum p^{-\sigma}|1 - g(p)|^2 \leq 2\sum p^{-1}(1 - \operatorname{Re} g(p)) \leq c_8.$$

Therefore

$$|G(s)| \leq c_9(\zeta(\sigma)|\zeta(\sigma - i\tau)|)^{1/2} \leq c_7 K^{-1/2}(\sigma - 1)^{-1/2},$$

since $\zeta(s)$ has a simple pole at the point $s = 1$, but is otherwise analytic in the disc $|s - 1| < 2M$.

This completes the proof of lemma (6.9).

Proof of theorem (6.2). As remarked earlier, we shall show that the hypothesis (2) of theorem (6.1) is satisfied. There are two cases.

Suppose first that for every real number τ the series

$$\sum p^{-1}(1 - \operatorname{Re} g(p)p^{-i\tau})$$

diverges. Then, by lemma (6.7),

$$\sum_p p^{-\sigma}(1 - \operatorname{Re} g(p)p^{-i\tau}) \to \infty \qquad (\sigma \to 1+),$$

uniformly for all values of τ belonging to a given bounded interval, $|\tau| \leq M$. It follows from lemma (6.6) that with the same uniformity

$$|G(s)/\zeta(s)| \leq 4e^{10} \exp\left(-\sum_{p \geq 3} p^{-\sigma}(1 - \operatorname{Re} g(p)p^{-i\tau})\right) \to 0,$$

as $\sigma \to 1+$. The hypothesis (2) of theorem (6.1) is therefore satisfied with $C = 0 = \alpha$, $L \equiv 1$. In particular, the result

$$x^{-1}\sum_{n \leq x} g(n) \to 0 \qquad (x \to \infty)$$

is obtained.

Suppose now that there is a value of τ, say $\tau = \alpha$, for which the series

$$\sum p^{-1}(1 - \operatorname{Re} g(p)p^{-i\tau})$$

converges. According to the remarks concerning the function $D(x)$ which were made at the end of the proof of theorem (6.1), it will suffice to establish the desired asymptotic estimate for $G(s)$ under the additional assumption that $\alpha = 0$. This we shall do.

Let K be a real number, $K \geq 1$. Consider the functions $G(s)$ and $\zeta(s)$ when $\sigma > 1$ and $|\tau| \leq K(\sigma - 1)$. According to lemma (6.6) there is a representation

$$G(s)/\zeta(s) = Q(s)\exp\left(-\sum_{p \geq 3} p^{-s}(1 - g(p))\right),$$

where the function $Q(s)$ is continuous in every rectangle $1 \leq \sigma \leq 2, |\tau| \leq M$. We define

$$L\left(\frac{1}{\sigma - 1}\right) = \exp\left(i \operatorname{Im} \sum_{p \geq 3} p^{-\sigma}g(p)\right)$$

and

$$C = Q(1)\exp\left(-\sum_{p \geq 3} p^{-1}(1 - \operatorname{Re} g(p))\right).$$

It is clear that $|L| \equiv 1$. In order to prove that the function $L(w)$, $w(\sigma - 1) = 1$, is a slowly oscillating function of w, we note that if $(\sigma - 1)/2 \leq u - 1 \leq \sigma - 1$, then by lemma (6.8),

$$\begin{aligned} L\left(\frac{1}{u - 1}\right)\Big/L\left(\frac{1}{\sigma - 1}\right) &= \exp\left(-i \operatorname{Im} \sum_{p \geq 3} \{p^{-u} - p^{-\sigma}\}\{1 - g(p)\}\right) \\ &= \exp(o(1)) \qquad (\sigma \to 1+), \end{aligned}$$

since $\operatorname{Im}(1 - g(p)) = -\operatorname{Im}(g(p))$.

Moreover, by the same lemma, applying a well known argument of Abel,

$$\begin{aligned} \frac{G(s)}{L\left(\frac{1}{\sigma - 1}\right)\zeta(s)} &= Q(s)\exp\left(-\sum_{p \geq 3} \left\{\frac{1}{p^s} - \frac{1}{p^\sigma}\right\}\{1 - g(p)\}\right. \\ &\qquad \left. - \sum_{p \geq 3} p^{-\sigma}\{1 - \operatorname{Re} g(p)\}\right) \\ &= (C + o(1))\exp(o(1) + o(1)) \to C \qquad (\sigma \to 1+), \end{aligned}$$

uniformly for $|\tau| \le K(\sigma - 1)$. Bearing in mind that $\zeta(s)$ has the residue 1 at its pole at $s = 1$ we may conclude that

$$G(s) - \frac{CL\left(\dfrac{1}{\sigma - 1}\right)}{s - 1} = O\left(\frac{1}{K^{1/2}(\sigma - 1)}\right)$$

uniformly for $|\tau| \le K(\sigma - 1)$, for all σ sufficiently near to 1 in value.

However, if $K(\sigma - 1) \le |\tau| \le M$ for a given number M, then this same estimate is still valid for the trivial reason, supported by lemma (6.9), that both sides are $O(K^{-1/2}(\sigma - 1)^{-1})$.

Therefore

$$\limsup_{\sigma \to 1+} \sup_{|\tau| \le M} (\sigma - 1)\left| G(s) - \frac{CL\left(\dfrac{1}{\sigma - 1}\right)}{s - 1}\right| \le \frac{c_{10}}{K^{1/2}},$$

and since K may be chosen arbitrarily large, a suitable form of the condition (2) of theorem (6.1) is satisfied.

Our proof of theorem (6.2) will now be complete if we obtain the approximate representation, valid in the present case $\alpha = 0$,

$$CL(\log x) = \prod_{p \le x}\left(1 - \frac{1}{p}\right)\left(1 + \sum_{m=1}^{\infty} p^{-m} g(p^m)\right) + o(1) \qquad (x \to \infty).$$

In fact, if we set $\sigma = 1 + (\log x)^{-1}$ then

$$\begin{aligned} CL(\log x) &= \{G(\sigma)/\zeta(\sigma)\} + o(1) \\ &= Q(\sigma)\exp\left(-\sum_{p \ge 3} p^{-\sigma}(1 - g(p))\right) + o(1). \end{aligned}$$

Let ε be a temporarily fixed real number, $0 < \varepsilon \le 1$. Then applying the Cauchy–Schwarz inequality we see that

$$\sum_{p > x^\varepsilon} p^{-\sigma}|1 - g(p)| \le \left\{\sum_{p > x^\varepsilon} p^{-\sigma}\right\}^{1/2}\left\{\sum_{p > x^\varepsilon} p^{-\sigma}|1 - g(p)|^2\right\}^{1/2}.$$

The first of the sums which appear on the right-hand side of this inequality does not exceed

$$\begin{aligned} \int_{x^\varepsilon +}^{\infty} y^{-\sigma}\, d\pi(y) &= -x^{-\varepsilon\sigma}\pi(x^\varepsilon +) + \int_{x^\varepsilon}^{\infty} \sigma y^{-\sigma - 1}\pi(y) dy \\ &= O(1) + O\left(\int_{x^\varepsilon}^{\infty} \frac{\log x}{y \log^2 y}\, dy\right) = O(1), \end{aligned}$$

since

$$\pi(y) = \sum_{p \le y} 1 = O\left(\frac{y}{\log y}\right) \qquad y^{\sigma - 1} = \exp\left(\frac{\log y}{\log x}\right) > \frac{\log y}{\log x}.$$

Moreover, from our hypothesis

$$\sum_{p > x^{\varepsilon}} p^{-\sigma}|1 - g(p)|^2 \le 2 \sum_{p > x^{\varepsilon}} p^{-1}(1 - \operatorname{Re} g(p)) = o(1),$$

as $x \to \infty$. Therefore

$$\sum_{p > x^{\varepsilon}} p^{-\sigma}(1 - g(p)) = o(1) \qquad (x \to \infty).$$

In a similar but simpler way we can prove that

$$\sum_{x^{\varepsilon} < p \le x} p^{-1}(1 - g(p)) = o(1) \qquad (x \to \infty).$$

We next note that

$$\sum_{p \le x^{\varepsilon}} (p^{-1} - p^{-\sigma})|1 - g(p)| \le 2 \sum_{p \le x^{\varepsilon}} p^{-1}\left(1 - \exp\left(-\frac{\log p}{\log x}\right)\right)$$

$$\le \frac{2}{\log x} \sum_{p \le x^{\varepsilon}} \frac{\log p}{p} \le N\varepsilon,$$

for a constant N and all sufficiently large values of x.

Altogether these estimates show that

$$\limsup_{x \to \infty} \left| \sum p^{-\sigma}(1 - g(p)) - \sum_{p \le x} p^{-1}(1 - g(p)) \right| \le N\varepsilon.$$

Since ε may be chosen arbitrarily small

$$\sum p^{-\sigma}(1 - g(p)) = \sum_{p \le x} p^{-1}(1 - g(p)) + o(1)$$

as $x \to \infty$.

The number $Q(1)$ may be written as the quotient of two convergent infinite products taken over the prime numbers (see lemma (6.6)), and the above asymptotic estimate for the function $CL(\log x)$ follows easily.

This completes the proof of theorem (6.2).

Remarks. Note that in the above proof the constant C will have the value zero if and only if $Q(1) = 0$. In other words, remembering that $\alpha = 0$ is assumed,

$$1 + \sum_{k=1}^{\infty} g(2^k)2^{-k} = 0.$$

Note also, that it follows from the argument given in this same proof of theorem (6.2) that there can be at most one real number τ for which the series

$$\sum p^{-1}(1 - \operatorname{Re} g(p)p^{-i\tau})$$

is convergent. Indeed, assume that to the contrary there are two distinct values, τ_1 and τ_2 say, for which this series converges. Define the multiplicative function $h(n)$ by $h(p) = g(p)$ for all odd primes p, and $h(p^k) = 0$ for all remaining prime-powers, including powers of 2. This definition ensures that

$$1 + \sum_{k=1}^{\infty} h(2^k)2^{-k} \neq 0.$$

Then there are non-zero constants C_j, and slowly-oscillating functions L_j with $|L_j(x)| = 1$, so that for $j = 1, 2$,

$$\sum_{n=1}^{\infty} h(n)n^{-s} = \frac{C_j L_j\left(\frac{1}{\sigma - 1}\right)}{s - (1 + i\tau_j)} + o\left(\frac{1}{\sigma - 1}\right) \qquad (\sigma \to 1+),$$

uniformly for $|\tau| \leq \max(|\tau_1|, |\tau_2|)$. In particular,

$$\frac{|C_1|}{|s - (1 + i\tau_1)|} = \frac{|C_2|}{|s - (1 + i\tau_2)|} + o\left(\frac{1}{\sigma - 1}\right),$$

and the choice $s = \sigma + i\tau_1$ leads to an impossible situation as $\sigma \to 1+$.

This ends our remarks. Using the notation of the preceding proof we may now readily establish theorem (6.3), and from it deduce theorem (6.4).

Proof of theorem (6.3). If conditions (i) and (ii) of theorem (6.3) hold then, in the above notation, $Q(1) \neq 0$, and $L(1/(\sigma - 1))$ approaches a finite limit as $\sigma \to 1+$. Hence $g(n)$ has a non-zero finite limit, namely

$$A = Q(1)\exp\left(-\sum_{p \geq 3} p^{-1}(1 - g(p))\right).$$

If condition (iv) holds then $g(n)$ has the mean-value zero, whilst if condition (iii) holds then

$$1 + \sum_{k=1}^{\infty} g(2^k)2^{-k(1+i\tau)} = 0,$$

and according to the first of the above two remarks, $g(n)n^{-i\tau}$ has the mean-value zero. Integration by parts shows that $g(n)$ then also has mean-value zero.

Conversely, assume that $g(n)$ has the mean-value zero. Then either the series

$$\sum p^{-1}(1 - \operatorname{Re} g(p)p^{-i\tau})$$

diverges for each real τ, which is condition (iv), or there is a number τ so that this series converges. Integration by parts shows that

$$x^{-1} \sum_{n \le x} g(n)n^{-i\tau} \to 0 \qquad (x \to \infty),$$

and we are in the same situation as at the end of the proof of theorem (6.2), but with $g(n)$ replaced by $g(n)n^{-i\tau}$. Thus $g(n)n^{-i\tau}$ can have the limiting mean-value zero only if $Q(1) = 0$. In other words

$$1 + \sum_{k=1}^{\infty} g(2^k)2^{-k(1+i\tau)} = 0,$$

and we have established condition (iii).

Last, let $g(n)$ have the non-zero mean-value A. Then according to theorem (6.1),

$$G(s) = \frac{A}{s-1} + o\left(\frac{1}{\sigma - 1}\right) \qquad (\sigma \to 1+),$$

uniformly for $|\tau| \le 1$, say. However,

$$G(s)/\zeta(s) = Q(s)\exp\left(-\sum_{p \ge 3} p^{-s}(1 - g(p))\right)$$

and $\zeta(s) \sim (s-1)^{-1}$ as $s \to 1$, so that

$$Q(\sigma)\exp\left(-\sum_{p \ge 3} p^{-\sigma}(1 - g(p))\right) \to A \qquad (\sigma \to 1+).$$

This will not be possible unless $Q(1) \neq 0$, a condition which leads at once to (i) of theorem (6.3). It is also clear that the series

$$\sum p^{-1}(1 - \operatorname{Re} g(p))$$

must converge. This enables us to apply theorem (6.2) to show that as $x \to \infty$,

$$\prod_{p \leq x} \left(1 - \frac{1}{p}\right)\left(1 + \sum_{m=1}^{\infty} p^{-m} g(p^m)\right) \to A.$$

We readily deduce the convergence of the series (ii), and theorem (6.3) is proven.

Proof of theorem (6.4). Suppose that the function $g(n)$ does not have the mean-value zero. Then from theorem (6.3), parts (iii) and (iv), there is a real number τ so that the series

$$\sum p^{-1}(1 - \operatorname{Re} g(p)p^{-i\tau})$$

converges, whilst

$$1 + \sum_{k=1}^{\infty} g(2^k)2^{-k(1+i\tau)} \neq 0.$$

According to parts (i) and (ii) of that same theorem applied to the function $g(n)n^{-i\tau}$, there is a non-zero number A so that as $x \to \infty$,

$$x^{-1} \sum_{n \leq x} g(n)n^{-i\tau} \to A,$$

and, therefore,

$$x^{-1} \sum_{n \leq x} g(n) = Ax^{i\tau}(1 + i\tau)^{-1} + o(1) \qquad (x \to \infty).$$

Since we may take complex conjugates, if necessary, without loss of generality we may assume that $\tau \geq 0$.

Suppose further, that $\tau > 0$. We set $x = \exp(2\pi m \tau^{-1})$ and let m run through the increasing sequence of all positive integers, to deduce that the number $A(1 + i\tau)^{-1}$ is real. However, if we set $x = \exp(2\pi(m + 1/2)\tau^{-1})$ and let $m \to \infty$ as before, we see that the number $iA(1 + i\tau)^{-1}$ is also real. This is not possible, since $A \neq 0$.

We have now reached the situation that $\tau = 0$, so that conditions (i) and (ii) of theorem (6.3) are satisfied, and $g(n)$ has a non-zero mean value. In this case the value of the limit is

$$\prod \left(1 - \frac{1}{p}\right)\left(1 + \sum_{m=1}^{\infty} p^{-m} g(p^m)\right).$$

We may also use this representation for the limit when $g(n)$ has the mean value zero, for if the series

$$\sum p^{-1}(1 - g(p))$$

converges then condition (iii) of Theorem (6.3) must be satisfied with $\tau = 0$.

This completes the proof of theorem (6.4).

Remark. The above proof shows that the mean-value is non-zero if and only if conditions (i) and (ii) of theorem (6.3) are satisfied. In particular, the Möbius function $\mu(n)$ satisfies

$$x^{-1} \sum_{n \le x} \mu(n) \to 0 \qquad (x \to \infty),$$

a result which leads shortly to the prime number theorem. (See, for example, Chapter 19).

As another example, let $f(n)$ be a real-valued additive arithmetic function for which the three series of theorem (5.1) converge. Let t be a real number, and define $g(n) = \exp(itf(n))$. Then

$$\sum_{|f(p)| \le 1} p^{-1}\{g(p) - 1\} = it \sum_{|f(p)| \le 1} p^{-1} f(p) + O\left(\sum_{|f(p)| \le 1} p^{-1} |tf(p)|^2\right),$$

and

$$\sum_{|f(p)| > 1} p^{-1} |g(p) - 1| \le 2 \sum_{|f(p)| > 1} p^{-1},$$

so that from theorem (6.3)

$$\lim_{x \to \infty} x^{-1} \sum_{n \le x} e^{itf(n)}$$

exists and is continuous. In this way we may establish the existence of a limiting distribution for $f(n)$.

This ends our remarks.

In Chapter 21 we shall consider the local behaviour of additive functions, and it is convenient to establish here a quantitative form of part (iv) of theorem (6.3).

Lemma 6.10. *There is an absolute constant c with the following property:*

Let $g(n)$ be a multiplicative arithmetic function, $|g(n)| \le 1$ for all positive n. Let x and T be a real numbers, $x \ge 2$, $T \ge 2$. Define

$$m(T) = m(x, T) = \min_{|\tau| \le T} \sum_{p \le x} p^{-1}(1 - \operatorname{Re} g(p)p^{-i\tau}).$$

Then the inequality,

$$\left| \sum_{n \le x} g(n) \right| \le cx\{\exp(-\tfrac{1}{4}m(T)) + T^{-1/4}\}$$

is satisfied.

Remark. A form of this lemma was first established by Halász [5]. The exponent 1/4, which appears twice in this formulation, is not best possible. For further remarks concerning this aspect of the result we refer to the aforementioned paper of Halász.

Proof. We sketch a proof, making use of results which were obtained during the proof of theorem (6.1).

Set $\sigma = \sigma_0 = 1 + (\log x)^{-1}$.

For numbers w in the range $1 \le w \le 2x$, we make use of the formula

$$\sum_{n \le w} g(n) \log n \log \frac{w}{n} = -\frac{1}{2\pi i} \int_{\sigma - i\infty}^{\sigma + i\infty} \frac{w^s G'(s)}{s^2} \, ds,$$

from which we deduce that $R(w)$, the sum which appears on the left-hand side of this equation, satisfies

$$w^{-1}|R(w)| \le \frac{e^2}{\pi} \int_{|\tau| \le T} \left| \frac{G'(s)}{s^2} \right| d\tau + \frac{e^2}{\pi} \int_{|\tau| > T} \left| \frac{G'(s)}{s^2} \right| d\tau.$$

The second of these two integrals does not exceed $e^{22}T^{-1/2}(\sigma - 1)^{-1}$. As for the first integral, it does not exceed $\lambda^{1/2} 9c_3(\sigma - 1)^{-1}$, where λ is defined by

$$\lambda = \max_{|\tau| \le T} (\sigma - 1)|G(s)|.$$

We obtain an upper bound for λ in terms of $m(T)$. In fact, from lemma (6.6), there is a positive absolute constant c_0 so that

$$|G(s)|/\zeta(\sigma) \le c_0 \exp\left(-\sum_{p\ge 3} p^{-\sigma}(1 - \operatorname{Re} g(p)p^{-i\tau})\right),$$

whilst towards the end of the proof of theorem (6.2) it was proved that

$$\sum_{p>x} p^{-\sigma} = -x^{-\sigma}\pi(x+) + \sigma\int_x^\infty \pi(y)y^{-\sigma-1}\,dy \le c_{11},$$

and

$$\sum_{p\le x}(p^{-1} - p^{-\sigma}) = \sum_{p\le x} p^{-1}\{1 - \exp(-(\sigma - 1)\log p)\} \le c_{12}.$$

Therefore

$$\lambda \le (\sigma - 1)\zeta(\sigma)c_0 \exp(-m(T) + O(1)) = O(\exp(-m(T))),$$

and we have proved that $R(w) = O(\rho w \log x)$, where

$$\rho = \exp(-m(T)/2) + T^{-1/2},$$

this result holding uniformly for $1 \le w \le 2x$.

Following another argument taken from the proof of theorem (6.1), let $0 < \varepsilon < 1/2$, and apply this estimate with w and $w(1+\varepsilon)$. Then, uniformly for $1 \le w \le x$,

$$\sum_{n\le w} g(n)\log n \log(1+\varepsilon) + \sum_{w<n\le w(1+\varepsilon)} g(n)\log n \log\frac{w}{n}$$
$$= R(w(1+\varepsilon)) - R(w) = O(\rho w \log x).$$

Using the fact that $\log(1+\varepsilon) = \varepsilon + O(\varepsilon^2)$ it is easy to deduce that

$$\sum_{n\le w} g(n)\log n = O(\varepsilon w \log w + \varepsilon^{-1}\rho w \log x)$$
$$= O(\rho^{1/2} w \log x)$$

provided that we choose $\varepsilon = \rho^{1/2}/4$.

An integration by parts now yields the desired result.

This completes our sketch of the proof of lemma (6.10).

Concluding Remarks

The problem of establishing the existence of mean-values

$$A = \lim_{n \to \infty} x^{-1} \sum_{n \leq x} g(n)$$

for multiplicative functions $g(n)$ was considered by Wintner [3], in his book on Eratosthenian Averages. In particular, he asserted that if $g(n)$ assumed only values ± 1, then the limit A always existed. The sketch of a proof which he gave could not be substantiated however, and the problem remained open for a considerable time.

Those multiplicative functions $g(n)$ which satisfy $|g(n)| \leq 1$ for all positive n, and for which a *non-zero* mean value exists were characterized by Delange [4], using, in particular, the Hardy–Littlewood tauberian theorem. This left for consideration the possibility that $A = 0$.

To put this remaining case into its proper perspective, we note that if we set $g(n) = \mu(n)$, the Möbius function, then we are precisely concerned with the case when the series

$$\sum p^{-1}(1 - g(p))$$

diverges. Moreover, the validity of the assertion

$$x^{-1} \sum_{n \leq x} \mu(n) \to 0 \qquad (x \to \infty),$$

was shown by Landau [2], to be essentially as difficult to obtain as the proof of the Prime Number Theorem, as we have already mentioned.

As is his wont, when giving lectures during the course of his travels, Professor Erdös will advise that certain of the problems which he has posed or recorded, carry prizes. Usually they lie in the range U.S. \$50 to \$500, but in the case of the Erdös–Wintner conjecture, fearing that he would not see a solution before he left, he offered a prize of $\$10^{10}$!.

In his paper [4], of 1967, Wirsing constructed an original method of dealing with delicate questions concerning convolutions of real-valued functions, which he applied to the study of certain approximate integral equations. These he brought to bear upon the problem of Wintner, and presently he solved it. As a gentleman, he forgot the prize. Unfortunately his method was best suited to the consideration of real-valued functions, and though it could be modified to deal with a wide class of complex-valued functions, in the end he could only conjecture the full validity of theorem (6.2). (See p. 422 of his paper [4]). It was his belief, in my personal recollection, that an analytic method should be found.

The treatment of the present chapter is based upon such an analytic method, as found by G. Halász, and exposed by him in his fine paper [1]. Both the method of Wirsing, and the method of Halász, are founded essentially upon the use of the identity, assuming $g(n)$ to be completely multiplicative:

$$\sum_{n \le x} g(n) \log n = \sum_{p \le x} g(p) \log p \sum_{m \le x/p} g(m).$$

Wirsing made direct use of this relation, and the relevance of convolutions is clear. In his paper Halász used this identity in the form

$$G'(s) = \frac{G'(s)}{G(s)} \cdot G(s).$$

Theorems (6.1) and (6.2) are due to Halász, being theorems 2 and 3 of his paper [1]. The details of his proofs differ slightly from those which are given here. Whatever presentation one uses, some kind of a technical detour seems necessary in order to take care of the fact that

$$\lim_{s \to 1} \left\{ 1 + \sum_{k=1}^{\infty} g(2^k) 2^{-ks} \right\} = 0$$

is possible.

In his paper [4], Wirsing adopts a more general formulation than that which is considered in the present chapter. We give two of his results which are typical.

Satz 1.1 (Wirsing). *Let $\lambda(n)$ be a multiplicative function which assumes real non-negative values only. Let*

$$\sum_{p \le x} \frac{\log p}{p} \lambda(p) \sim \tau \log x \qquad (x \to \infty),$$

hold with a constant $\tau > 0$. Furthermore, let $\lambda(p) \le G$ hold for some positive constant G, for all primes p, and let

$$\sum_{p, k \le 2} p^{-k} \lambda(p^k) < \infty.$$

Besides this, if $\tau \le 1$ then let

$$\sum_{p^k \le x,\, k \ge 2} \lambda(p^k) = O(x(\log\ x)^{-1}).$$

Then as $x \to \infty$,

$$\sum_{n \le x} \lambda(n) \sim \frac{e^{-\gamma\tau}}{\Gamma(\tau)} \frac{x}{\log x} \prod_{p \le x} \left(1 + \frac{\lambda(p)}{p} + \frac{\lambda(p^2)}{p^2} + \cdots\right)$$

Here γ *denotes Euler's constant.*

Satz 1.2.2 (Wirsing). *Let* $\lambda(n)$ *satisfy the conditions of Satz* (1.1), *and let* $g(n)$ *be a real-valued multiplicative function which satisfies* $|g(n)| \le \lambda(n)$ *for every positive integer* n. *Then*

$$\lim_{x \to \infty} \left\{\sum_{n \le x} g(n)\right\}\left\{\sum_{n \le x} \lambda(n)\right\}^{-1} = \prod_p \left(1 + \sum_{k=1}^{\infty} p^{-k} g(p^k)\right)\left(1 + \sum_{k=1}^{\infty} p^{-k}\lambda(p^k)\right)^{-1},$$

where the product either converges properly to a non-zero limit, or improperly to zero.

In our present circumstances we have restricted ourselves to the case when $\lambda \equiv 1$, so that $\tau = 1$; this being the case which pertains to the probabilistic theory of numbers. Although we do not pursue this point here, we remark that in his paper [1] Halász also develops a method which is in certain respects more powerful than that of the present chapter, and which is more suitable for application to functions which may sometimes exceed one in value. Since we shall have need of it later on, a discussion of this method will be given in Chapter 9.

A more elaborate discussion of the methods of Delange and Wirsing, and their relation to the Erdös–Selberg elementary proof of the Prime Number Theorem is given at the end of Chapter 19.

Chapter 7

Translates of Additive and Multiplicative Functions

In this chapter we apply some of the results of Chapter six. In particular we shall characterize all additive functions which, after a suitable translation, possess a limiting distribution.

Theorem 7.1 (Elliott and Ryavec; Levin and Timofeev). *Let $f(n)$ be a real-valued additive arithmetic function. In order that there exist a real function $\alpha(x)$, defined for $x \geq 1$, for which the frequencies*

$$v_x(n; f(n) - \alpha(x) \leq z)$$

converge to a limiting distribution as $x \to \infty$, it is both necessary and sufficient that there be a constant c so that $f(n) = c \log n + h(n)$, where the series

$$\sum_{|h(p)|>1} \frac{1}{p} \qquad \sum_{|h(p)|\leq 1} \frac{h^2(p)}{p}.$$

are convergent.

Moreover, when this condition is satisfied a suitable function $\alpha(x)$ may be defined by

$$\alpha(x) = c \log x + \sum_{p \leq x,\ |h(p)| \leq 1} \frac{h(p)}{p}.$$

With this choice the limiting distribution has the characteristic function

$$\frac{1}{1 + ict} \prod_{|h(p)|>1} w_p(t) \prod_{|h(p)|\leq 1} w_p(t) \exp(-ith(p)p^{-1}),$$

where

$$w_p(t) = \left(1 - \frac{1}{p}\right)\left(1 + \sum_{k=1}^{\infty} p^{-k} \exp(ith(p^k))\right).$$

It is of pure type, and is continuous if and only if the series

$$\sum_{f(p)\neq 0} \frac{1}{p}$$

diverges.

Remarks. Theorem (7.1) was established, independently, by Elliott and Ryavec [1], and Levin and Timofeev [1]. Apparently, unpublished proofs were also given by J. Kubilius, and by H. Delange.

In fact the limit law is absolutely continuous if $c \neq 0$.

Before giving a proof of theorem (7.1) it is convenient to introduce the following concept of Erdös:

As additive function $f(n)$ is said to be *finitely distributed* if there are positive constants c_1 and c_2, and an unbounded sequence of real numbers $x_1 < x_2 < \ldots$, so that for each x_j at least k positive integers $a_1 < a_2 < \cdots < a_k \leq x_j$ may be found, with $k \geq c_1 x_j$, so that

$$|f(a_m) - f(a_n)| \leq c_2 \qquad 1 \leq m \leq n \leq k.$$

Although their definition appears unwieldy, these functions are very convenient to use, because of the characterization embodied in the following theorem.

Theorem 7.2 (Erdös). *An additive function $f(n)$ is finitely distributed if and only if there is a constant c so that*

$$f(n) = c \log n + h(n),$$

where both the series

$$\sum_{|h(p)|>1} \frac{1}{p} \qquad \sum_{|h(p)|\leq 1} \frac{h^2(p)}{p}$$

converge.

Remark. It follows from theorem (6.2) of Chapter six that

$$\lim_{x\to\infty} x^{-1} \left| \sum_{n\leq x} \exp(itf(n)) \right|$$

always exists. It will transpire that $f(n)$ is finitely distributed if and only if there is a set of real t-values, of positive Lebesgue measure, for which the value of this limit is not zero.

It is convenient to give a cyclic proof of theorems (7.1) and (7.2) by establishing the following three propositions.

(i) *If, for some* $\alpha(x)$,

$$\text{(weak)} \lim_{x\to\infty} \nu_x(n; f(n) - \alpha(x) \le z)$$

exists, then $f(n)$ *is finitely distributed.*

(ii) *If* $f(n)$ *is finitely distributed then it has a decomposition* $f(n) = c \log n + h(n)$ *of the type considered in the statement of theorem* (7.2).

(iii) *If* $f(n)$ *has a representation* $c \log n + h(n)$, *where the series*

$$\sum_{|h(p)|>1} \frac{1}{p} \qquad \sum_{|h(p)|\le 1} \frac{h^2(p)}{p}$$

converge, and if we define

$$\alpha(x) = c \log x + \sum_{p\le x,\ |h(p)|\le 1} \frac{h(p)}{p} \qquad (x \ge 1),$$

then

$$\text{(weak)} \lim_{x\to\infty} \nu_x(n; f(n) - \alpha(x) \le z)$$

exists.

We shall establish the results concerning the nature of the limiting distribution in theorem (7.1) during the proof of proposition (iii).

Proof of proposition (i). If the number w is chosen sufficiently large, and such that $\pm w$ are continuity points of the limiting distribution of $f(n) - \alpha(x)$, then

$$\lim_{x\to\infty} \nu_x(n; |f(n) - \alpha(x)| \le w) > \tfrac{1}{2}.$$

Moreover, if m and n are any two integers which are counted in a typical frequency,

$$|f(m) - f(n)| \le |f(m) - \alpha(x)| + |\alpha(x) - f(n)| \le 2w,$$

from which it is clear that $f(n)$ is finitely distributed.

Proof of proposition (ii). For real numbers t, define the functions

$$H(x, t) = \sum_{n \le x} \exp(itf(n)),$$

$$l(t) = \lim_{x \to \infty} x^{-1}|H(x, t)|.$$

The existence of this limit is guaranteed by theorem (6.2).

Further, define the function

$$D(\theta) = \begin{cases} \left(\dfrac{\sin \pi\theta}{\pi\theta}\right)^2 & \text{if } \theta \neq 0, \\ 1 & \text{if } \theta = 0. \end{cases}$$

Then, for each real number y,

$$\int_{-\infty}^{\infty} e^{2\pi i\theta y} D(\theta) d\theta = \begin{cases} 1 - |y| & \text{if } |y| \le 1, \\ 0 & \text{otherwise.} \end{cases}$$

Interchanging summation and integration shows that for positive λ

$$\int_{-\infty}^{\infty} \lambda |H(x, t)|^2 D(\lambda t) dt = \sum_{\substack{n_1, n_2 \le x \\ |f(n_1) - f(n_2)| \le \lambda}} (1 - \lambda^{-1}|f(n_1) - f(n_2)|).$$

We divide by x, let $x \to \infty$, and apply Lebesgue's theorem on dominated convergence. The hypothesis of (ii) ensures that if λ is sufficiently large then

$$\int_{-\infty}^{\infty} \lambda l(t)^2 D(\lambda t) dt > 0.$$

More exactly, if $f(n)$ satisfies the condition given in the definition of finitely distributed additive functions, and if $\lambda \ge 2c_2$, then the value of this integral is at least as large as $c_1^2/2$.

It follows that there is a set E, of positive Lebesgue measure, on which $l(t) > 0$.

If, for some value of t, we have $l(t) > 0$, then according to the remarks preceding the proof of theorem (6.3) there is a *unique* real number $\tau = \tau(t)$, so that the series

$$\sum_{p \ge 2} p^{-1}(1 - \operatorname{Re} e^{itf(p)} p^{-i\tau})$$

converges. The convergence of this series is equivalent to that of the series

$$L(t, \tau) = \sum p^{-1} \sin^2(\tfrac{1}{2}tf(p) - \tfrac{1}{2}\tau \log p).$$

Such a number τ may be found for each member t of E. Indeed, there is a number K, and a subset F of E, of positive measure, so that whenever t belongs to F the inequality $L(t, \tau) \le K$ is satisfied.

We now appeal to lemma (1.1). This result of Steinhaus shows that there is a proper interval about the origin, $(-2\delta, 2\delta)$ say, each point w of which has a representation $w = t_1 - t_2$, with both t_1 and t_2 belonging to the set F. Let $\tau_j = \tau(t_j)$, $(j = 1, 2)$. In view of the inequality

$$\sin^2(x \pm y) \le 2 \sin^2 x + 2 \sin^2 y,$$

which is valid for all real numbers x and y, we see that

$$L(w, \tau_1 - \tau_2) \le 2L(t_1, \tau_1) + 2L(t_2, \tau_2) \le 4K.$$

In particular $\tau(w)$ exists, and has the value

$$\tau(w) = \tau(t_1 - t_2) = \tau(t_1) - \tau(t_2).$$

A simple extension of this argument shows that $L(t, \tau)$ is defined (and finite) for every real number t, and that for every rational number r, the relation $\tau(rt) = r\tau(t)$ is satisfied.

For our next step we shall need the inequality

$$\frac{1}{k} \sum_{j=1}^{k} (1 - \cos jy) \ge \frac{1}{2}$$

which is certainly valid when k is an integer, $k \ge 2$, and y is a real number in the range $\pi/k \le |y| \le \pi$. This inequality may be deduced from the identity

$$\frac{1}{k} \sum_{j=1}^{k} (1 - \cos jy) = 1 + \frac{1}{2k} - \frac{\sin((2k + 1)y/2)}{2k \sin(y/2)},$$

by means of the inequality

$$|2k \sin(y/2)| \ge 2k|y|/\pi.$$

It follows from what we have proved so far that, for each positive integer k, the inequality

$$L\left(\frac{j}{k}\delta, \tau\left(\frac{j}{k}\delta\right)\right) \le 4K$$

holds uniformly for $j = 1, 2, \ldots, k$. We set $c = \tau(\delta)/\delta$, define $h(p) = f(p) - c \log p$, and deduce that

$$\sum_p{}' p^{-1}(1 - \cos(j\delta h(p)/k)) \leq 8K$$

where $'$ indicates that summation runs over those primes for which

$$\pi\delta^{-1} \leq |h(p)| \leq \pi k\delta^{-1}.$$

Summing over $j = 1, \ldots, k$, and utilising the previous remark, we deduce that

$$\sum{}' \frac{1}{p} \leq 16K,$$

and, since k may be chosen arbitrarily large,

$$\sum_{|h(p)| \geq \pi\delta^{-1}} \frac{1}{p} \leq 16K.$$

Moreover,

$$\frac{\delta^2}{\pi^2} \sum_{|h(p)| \leq \pi\delta^{-1}} p^{-1}h^2(p) \leq \sum_{|h(p)| \leq \pi\delta^{-1}} p^{-1} \sin^2(\delta h(p)/2)$$
$$\leq L(\delta, \tau(\delta)) \leq 4K.$$

This completes the proof of proposition (ii).

Proof of proposition (iii). Define the function

$$\lambda(x) = \sum_{p \leq x,\ |h(p)| \leq 1} \frac{h(p)}{p} \qquad (x \geq 1).$$

Then according to theorem (5.2) of Chapter five, there is a distribution function $G(z)$ so that as $x \to \infty$

$$\nu_x(n; h(n) - \lambda(x) \leq z) \Rightarrow G(z).$$

Let $\psi(t)$ be the characteristic function of this limiting distribution, and let

$$\psi(x, t) = x^{-1} \sum_{n \leq x} \exp(ith(n)).$$

Then we may write this last result in the form

$$\psi(x, t)\exp(-it\lambda(x)) \to \psi(t) \qquad (x \to \infty).$$

If we integrate by parts:

$$x^{-1}\sum_{n\leq x} n^{itc}\exp(ith(n)) = \psi(x, t)x^{itc} - itcx^{-1}\int_1^x y^{itc}\psi(y, t)dy.$$

However, uniformly for $x^{1/2} \leq y \leq x$,

$$|\lambda(x) - \lambda(y)|^2 \leq \sum_{x^{1/2}<p\leq x}\frac{1}{p}\cdot\sum_{\substack{p>x^{1/2}\\ |h(p)|\leq 1}}\frac{h^2(p)}{p} = o(1),$$

and, therefore,

$$\psi(y, t) = \psi(t)\exp(it\lambda(x)) + o(1),$$

as $x \to \infty$. Hence

$$\int_{x^{1/2}}^x y^{itc}\psi(y, t)dy = \psi(t)\exp(it\lambda(x))\int_{x^{1/2}}^x y^{itc}dy + o(x),$$

and

$$x^{-1}\sum_{n\leq x}\exp(itf(n)) = \frac{x^{itc}}{1 + itc}\psi(t)\exp(it\lambda(x)) + o(1) \qquad (x \to \infty).$$

It follows from this relation, by means of another appeal to lemma (1.11), that as $x \to \infty$ the frequencies

$$\nu_x(n; f(n) - c\log x - \lambda(x) \leq z)$$

converge weakly to a limiting distribution, $F(z)$ say, whose characteristic function has the form $(1 + ict)^{-1}\psi(t)$.

This completes the proof of proposition (iii).

We have now proved theorem (7.2), and theorem (7.1) save for the considerations concerning the nature of the limiting distribution $F(z)$. For the remainder of the proof of theorem (7.1) we shall adhere to the notation used in the proof of proposition (iii).

Suppose first that $c = 0$. Then $F(z) = G(z)$, and the desired results may be deduced directly from theorem (5.2).

Suppose next that $c > 0$. Then $F(z)$ is the convolution of $G(z)$ and the law $W(z)$ which is given by

$$W(z) = \begin{cases} e^{z/c} & \text{if } z < 0,\\ 1 & \text{if } z \geq 0.\end{cases}$$

It is clear that the law $W(z)$ is absolutely continuous, so that $F(z)$ must also be.

A similar argument may be given when $c < 0$, and it is straightforward to complete the proof of theorem (7.1) if we note that when $c \neq 0$ the inequality $|f(p)| \geq |c| \log p - 1$ holds for all primes p except for a set for which the series $\sum p^{-1}$ converges. Therefore, the series

$$\sum_{f(p) \neq 0} \frac{1}{p}$$

diverges.

This completes the proofs of theorems (7.1) and (7.2).

Remark. Assume that for suitable constants a_n and $b_n > 0$, the frequencies

$$\nu_n(m; f(m) - a_n \leq z b_n) \qquad (n = 1, 2, \ldots),$$

converge weakly. If there exists a sequence of integers so that the corresponding b_n are uniformly bounded, or, equivalently, lim inf $b_n < \infty$, then $f(n)$ is finitely distributed. Thus these frequencies will converge if we set $b_n \equiv 1$, and $a_n = \alpha(n)$, where $\alpha(x)$ is defined as in theorem (7.1).

Hence, if we seek constants a_n and b_n so that these frequencies converge, and if theorem (7.1) does not apply, then we may confine ourselves to the situation that $b_n \to \infty$ as $n \to \infty$.

This ends the remark.

It is convenient, at this point, to note that if a distribution function $F(z)$ is everywhere continuous, then it is uniformly continuous on the whole real line. For example, if $\varepsilon > 0$, then for a suitably chosen number $z_0(>0)$,

$$1 - F(z_0) + F(z_0) < \varepsilon.$$

$F(z)$ is uniformly continuous on the interval $-z_0 - 1 \leq z \leq z_0 + 1$. Suppose, now, that u and v are two real numbers which satisfy $|u - v| \leq \delta < 1/2$. Then either they both belong to the interval $[-z_0 - 1, z_0 + 1]$, or they satisfy $|u| > z_0$, $|v| > z_0$. In any case, provided that δ is sufficiently small, we shall have

$$|F(u) - F(v)| < \varepsilon.$$

Suppose that as $n \to \infty$ the frequencies

$$F_n(z) = \nu_n(m; f(m) \leq z)$$

converge weakly to the distribution function $F(z)$. Then we proved in theorem (5.1) that $F(z)$ is a continuous function of z if and only if the series

$$\sum_{f(p)\neq 0} \frac{1}{p}$$

diverges. Even when the functions $F_n(z)$ do not converge, a meaning can be given to this result if we replace the condition that $F(z)$ be (uniformly) continuous by

$$\limsup_{\delta\to 0+} \limsup_{n\to\infty} (F_n(z+\delta) - F_n(z-\delta)) = 0,$$

uniformly for all real values of z. We can then prove the following result:

Theorem 7.3 (Erdös; Elliott and Ryavec). *The following three conditions are equivalent:*

(a) $$\limsup_{\delta\to 0+} \sup_{z} \limsup_{n\to\infty} (F_n(z+\delta) - F_n(z-\delta)) = 0$$

(b) $$\limsup_{\delta\to 0+} \limsup_{n\to\infty} \sup_{z} (F_n(z+\delta) - F_n(z-\delta)) = 0$$

(c) *the series*

$$\sum_{f(p)\neq 0} \frac{1}{p}$$

diverges.

Remarks. This theorem has a number of useful applications to the theory of numbers, in particular to the study of the local behaviour of additive functions. (See, for example, Chapter 21.) Its content was succinctly summed up by Erdös in the following surrealistic manner ([10], p. 17):

"If

$$\sum_{f(p)\neq 0} \frac{1}{p} = \infty,$$

the distribution function tries to be continuous, whether it exists or not."

This ends the remarks.

We shall need the following simple result.

Lemma 7.4. *Let $G_n(z)$, $(n = 1, 2, \ldots)$, be a sequence of distribution functions which converges to a distribution function $G(z)$. Let δ be a positive number. Then*

$$\limsup_{n\to\infty} \sup_{z} (G_n(z+\delta) - G_n(z)) \le \sup_{z} (G(z+3\delta) - G(z)).$$

Proof. Let $\ldots < z_{-2} < z_{-1} < z_0 < z_1 < z_2 < \ldots$ be a doubly-infinite sequence of real numbers which are points of continuity of the distribution function $G(z)$, and which are chosen so that the intervals $(z_j, z_{j+1}]$ are each of length at least $\delta/2$ but no more than δ, and together cover the whole real line.

Let ε be a positive real number. Then, if the integer k is sufficiently large in absolute value,

$$\lim_{n\to\infty} (1 - G_n(z_k) + G_n(z_{-k+2})) = 1 - G(z_k) + G(z_{-k+2}) \leq \varepsilon.$$

Note that as $k \to \infty$, $z_{-k} \to -\infty$.

It is clear that

$$\sup_{z_{-k}<z<z_k} (G_n(z + \delta) - G_n(z)) \leq \max_{|j|\leq k} (G_n(z_{j+3}) - G_n(z_j))$$

so that

$$\limsup_{n\to\infty} \sup_z (G_n(z + \delta) - G_n(z)) \leq \max_{|j|\leq k} (G(z_{j+3}) - G(z_j)) + \varepsilon$$

$$\leq \sup_z (G(z + 3\delta) - G(z)) + \varepsilon.$$

Since ε may be chosen arbitrarily small, lemma (7.4) is proved.

Proof. We give a cyclic proof of this theorem.

It is clear that if condition (b) is satisfied, then so is (a).

Let us next assume that (a) is true, and deduce the validity of (c).

We assume to the contrary that

$$\sum_{f(p)\neq 0} \frac{1}{p} < \infty.$$

Let Δ denote the set of primes for which $f(p) \neq 0$. Then a simple application of the sieve of Eratosthenes (see also lemma (2.1)) shows that the number of integers in the interval $1 \leq m \leq n$ which are not divisible by any prime in Δ is equal to

$$(1 + o(1))n \prod_{p\in\Delta} \left(1 - \frac{1}{p}\right) \qquad (n \to \infty).$$

On each of these integers $f(m) = 0$. Taking $z = 0$ we see that

$$\limsup_{\delta \to 0+} \limsup_{n \to \infty} (F_n(\delta) - F_n(-\delta)) \geq \prod_{p \in \Delta} \left(1 - \frac{1}{p}\right) > 0,$$

which contradicts our temporary hypothesis (a).

Therefore (c) is true.

To complete the proof we shall assume (c) to be valid, and prove that (b) is true.

Suppose, instead, that proposition (b) fails. Then there exists a decreasing sequence $\delta_1 > \delta_2 > \cdots > 0$, and a constant γ, so that

$$\limsup_{n \to \infty} \sup_{z} (F_n(z + \delta_k) - F_n(z - \delta_k)) \geq \gamma > 0.$$

Thus we obtain a sequence of integers $n_1 < n_2 < \ldots$, and a sequence of numbers $z_1, z_2, \ldots$, so that

$$F_{n_k}(z_k + \delta_k) - F_{n_k}(z_k - \delta_k) \geq \gamma/2 \qquad (k = 1, 2, \ldots).$$

It follows that the intervals $1 \leq m \leq n_k$ contain at least $(\gamma/4)n_k$ integers a_i on which

$$|f(a_i) - f(a_j)| \leq 2\delta_k \leq 2\delta_1,$$

and so $f(m)$ is finitely distributed. Therefore $f(m)$ has the form

$$f(m) = c \log m + h(m),$$

where the series

$$\sum_{|h(p)| > 1} \frac{1}{p} \qquad \sum_{|h(p)| \leq 1} \frac{h^2(p)}{p}$$

converge.

We define

$$\alpha_n = c \log n + \sum_{p \leq n,\ |h(p)| \leq 1} \frac{h(p)}{p}$$

and deduce from theorem (7.1) that the frequencies

$$G_n(z) = \nu_n(m; f(m) - \alpha_n \leq z) \qquad (n = 1, 2, \ldots),$$

converge to a *continuous* limiting distribution, which we shall denote by $G(z)$.

It follows at once from lemma (7.4) that

$$\limsup_{n\to\infty} \sup_z (F_n(z+\delta) - F_n(z-\delta)) \leq \sup_z (G(z+6\delta) - G(z)),$$

and, since $G(z)$ is continuous, that condition (b) holds.

But this contradicts our temporary hypothesis that (b) fails.

Therefore (b) does, in fact, hold and the proof of theorem (7.3) is complete.

We state, without proof, another result of this type, and an analogue of theorem (7.2).

Theorem 7.5 (Elliott and Ryavec). *In order that*

$$\limsup_{z\to\infty} \limsup_{n\to\infty} \nu_n(m; |f(m)| < z) = 1$$

it is both necessary and sufficient that the series

$$\sum_p (f'(p))^2 p^{-1}$$

converges, and that

$$\liminf_{n\to\infty} \left| \sum_{p\leq n} f'(p)p^{-1} \right| < \infty,$$

where

$$f'(p) = \begin{cases} f(p) & \text{if } |f(p)| \leq 1, \\ 1 & \text{if } |f(p)| > 1. \end{cases}$$

A proof of this result may be found in the original paper [1].

A function $f(n)$ is said to be *finitely monotonic* if there is a constant c and an unbounded sequence of numbers $x_1 < x_2 < \dots$ so that for each member x of this sequence we can find positive integers $a_1 < a_2 < \cdots < a_k \leq x$, with $k > cx$, on which

$$f(a_1) \leq f(a_2) \leq \cdots \leq f(a_k).$$

Theorem 7.6 (Erdös and Ryavec). *A strongly additive function $f(n)$ is finitely monotonic if and only if it has the form $f(n) = A\log n + h(n)$, where A is a constant, and where the series*

$$\sum_{h(p)\neq 0} \frac{1}{p}$$

converges.

A proof of this theorem may be found in the original paper [1]. It makes use of the surrealistic continuity theorem (theorem (7.3)), together with a number of further arguments.

In theorem (5.5) of Chapter 5 it was shown that the additive function $f(n)$ has a limiting distribution with a finite mean and variance if and only if the series

$$\sum_{p,\, w} p^{-w} f(p^w) \qquad \sum_{p,\, w} p^{-w} f^2(p^w) \tag{1}$$

converge. We shall round out our consideration of the first and second moment behaviour of additive functions with the following result.

Theorem 7.7 (Elliott). *The additive function $f(n)$ possesses a limiting distribution with a finite mean and variance if and only if*

$$\lim_{n\to\infty} n^{-1} \sum_{m=1}^{n} f(m)$$

exists, and

$$\limsup_{n\to\infty} n^{-1} \sum_{m=1}^{n} f^2(m)$$

is finite.

Remark. Combining this result with theorem (5.5) we obtain a necessary and sufficient condition in order that an additive function possess limiting first and second moments.

Most of the proof of theorem (7.7) is contained in the following

Lemma 7.8. *Assume that, for positive constants c_1 and c_2, the additive function $f(n)$ satisfies*

$$\limsup_{x\to\infty} \nu_x(n; |f(n)| \le c_1) \ge c_2.$$

Then both the series

$$\sum_{|f(p)|>1} \frac{1}{p} \qquad \sum_{|f(p)|\le 1} \frac{f^2(p)}{p}$$

converge.

Proof of lemma (7.8). Since $f(n)$ is finitely distributed it has a representation of the form $f(n) = c \log n + h(n)$, where

$$\sum_{|h(p)|>1} \frac{1}{p} \qquad \sum_{|h(p)|\le 1} \frac{h^2(p)}{p}$$

are convergent. We shall prove that $c = 0$, and lemma (7.8) will follow at once.

Let Δ denote the set of primes p for which $|h(p)| > 1$. Let ε be a positive real number, $0 < \varepsilon < c_2/8$. Then for a suitably large (fixed) prime P,

$$\sum_{p>P,\, p\in\Delta} \frac{1}{p} < \varepsilon.$$

Moreover, if we choose P large enough to ensure that $P\varepsilon > 1$, then the frequency of those integers which are divisible by the square of a prime exceeding P is at most

$$\sum_{p>P} p^{-2} < \varepsilon.$$

The frequency of those integers which are divisible by a prime power p^m with p not exceeding P, and $m > P$, is at most

$$\sum_{p\le P} \sum_{m>P} p^{-m} \le 2 \sum_{p\le P} p^{-P} < P2^{1-P} < \varepsilon,$$

provided only that P is sufficiently large.

Consider now an unbounded sequence $x_1 < x_2 < \cdots < x_k < \ldots$ so that when $x = x_k$, $(k = 1, 2, \ldots)$,

$$\nu_x(n; |f(n)| \le c_1) \ge c_2 - \varepsilon.$$

Then we can find integers $a_1 < a_2 < \cdots < a_w \le x$, with $w > (c_2 - 4\varepsilon)x \ge (c_2/2)x$, on which both $|f(a_i)| \le c_1$, and $|h(a_i)| \le c_3\omega(a_i)$, where c_3 is a certain constant which depends upon $f(n)$ and P, and where $\omega(m)$ denotes the number of distinct prime divisors of the integer m.

Hence

$$|c| \sum_{m\le w} \log m \le |c| \sum_{i=1}^{w} \log a_i \le \sum_{i=1}^{w} (|f(a_i)| + |h(a_i)|)$$

$$\le c_1 x + c_3 \sum_{m\le x} \omega(m) = O(x \log\log x).$$

In view of the lower bound for w we obtain

$$c = O\left(\frac{\log \log x}{\log x}\right).$$

We set $x = x_k$ in turn, let $k \to \infty$, and deduce that $c = 0$.
Thus $f(n) = h(n)$, and lemma (7.8) is established.

Proof of theorem (7.7). In view of theorem (5.5) it will suffice to prove that the existence of the lim and lim sup in the hypotheses of theorem (7.7) guarantee the convergence of the series (1). Moreover, if we appeal to lemma (7.8) and then repeat part of the proof of theorem (5.5) we obtain easily the convergence of the series

$$\sum_{p,\, w} p^{-w} f^2(p^w).$$

Consider, now, the sum

$$\sum_{m=1}^{n} f(m) = \sum_{p^w \le n} f(p^w)\left(\left[\frac{n}{p^w}\right] - \left[\frac{n}{p^{w+1}}\right]\right).$$

An appeal to the Cauchy–Schwarz inequality shows that the error introduced into this last expression by removing the square brackets is at most

$$\sum_{p^w \le n} |f(p^w)| \le \left(\sum p^{-w} f^2(p^w) \cdot \sum_{p^w \le n} p^w\right)^{1/2} = O(n(\log n)^{-1/2}).$$

Then, as $n \to \infty$,

$$\sum_{p^w \le n} p^{-w} f(p^w) = n^{-1} \sum_{m=1}^{n} f(m) + O((\log n)^{-1/2}),$$

and from the hypothesis that

$$\lim_{n \to \infty} n^{-1} \sum_{m=1}^{n} f(m)$$

exists (and is finite) we deduce the convergence of the series

$$\sum_{p,\, w} p^{-w} f(p^w).$$

This completes our outline of the proof of theorem (7.7).

Distribution of Multiplicative Functions

For the duration of this section, which extends until the concluding remarks of the chapter, $g(n)$ will denote a multiplicative arithmetic function. We begin by determining a necessary and sufficient condition that such a function be essentially zero.

Theorem 7.9. *Let $g(n)$ be a complex-valued multiplicative function. Then the asymptotic density*

$$\lim_{x\to\infty} x^{-1} \sum_{\substack{n\le x\\ g(n)=0}} 1$$

always exists. It has the value zero if and only if $g(n) \neq 0$ for every integer n, and the value one if and only if the series

$$\sum_{g(p)=0} \frac{1}{p}$$

diverges.

Proof. Define the function $\delta(n)$ by

$$\delta(n) = \begin{cases} 1 & \text{if } g(n) \neq 0, \\ 0 & \text{if } g(n) = 0. \end{cases}$$

It is easy to check that this function is multiplicative, so that we may apply theorem (6.4) to deduce the existence of the limit

$$A = \lim_{x\to\infty} x^{-1} \sum_{n\le x} \delta(n).$$

It follows at once that the limit described in the statement of the theorem exists and has the value $1 - A$.

Since

$$A = \prod \left(1 - \frac{1}{p}\right)\left(1 + \sum_{k=1}^{\infty} \frac{g(p^k)}{p^k}\right),$$

the density $1 - A$ will have the respective values 0 and 1 if and only if $A = 1, 0$. The remaining assertions of theorem (7.9) are now readily justified.

Remark. An alternative proof that $g(n)$ is essentially zero if and only if the series

$$\sum_{g(p)=0} \frac{1}{p}$$

diverges may be constructed without appealing to the result of Wirsing.

By considering squarefree numbers one may show that the divergence of the infinite series is necessary for the 'almost sure' vanishing of $g(n)$. In order to show that it is sufficient, let $n_1 < n_2 < \cdots < n_k$ denote those integers, not exceeding x, which are not exactly divisible by any prime p for which $g(p) = 0$. If $p \| n$ and $g(p) = 0$, then

$$g(n) = g(p)g\left(\frac{n}{p}\right) = 0,$$

so that

$$v_x(n;\ g(n) \neq 0) \leq k[x]^{-1}.$$

We apply lemma (4.6) with

$$a_n = \begin{cases} 1 & \text{if } n \text{ is an } n_j, \\ 0 & \text{otherwise.} \end{cases}$$

When p is a prime for which $g(p) = 0$, then by definition of the n_j

$$p^{-1} \left| \sum_{\substack{n \leq x \\ p \| n}} a_n - p^{-1} \sum_{n \leq x} a_n \right|^2 = p^{-1} k^2,$$

so that

$$k^2 \sum_{\substack{p \leq x \\ g(p)=0}} \frac{1}{p} \leq 45kx.$$

It is clear from this inequality that as $x \to \infty$, $k[x]^{-1} \to 0$. This concludes our remark.

For the purposes of the theorems which conclude the present chapter we shall say that a sequence of distribution functions $F_n(z)$, $(n = 1, 2, \ldots)$, has a *modified-weak* limiting distribution if there is a distribution function $F(z)$ so that as $n \to \infty$ the $F_n(z)$ converge weakly to $F(z)$, together with the additional requirement that $F_n(0) \to F(0)$ and $F_n(0-) \to F(0-)$. If $F(z)$ is continuous at $z = 0$ these extra conditions are automatically satisfied. They are introduced in connection with the considerations of Zolotarev (lemma (1.45)) concerning the characteristic functions of products of random variables. The definition of *modified-weak* convergence may clearly be extended to distribution functions $F_x(z)$ which are indexed by a real parameter x $(\to \infty)$.

It is convenient to introduce the notation, defined for real numbers y,

$$\|y\| = \begin{cases} y & \text{if } |y| \le 1, \\ 1 & \text{if } |y| > 1. \end{cases}$$

Theorem 7.10 (Levin, Timofeev and Tuliaganov). *Let $g(n)$ be a real-valued multiplicative arithmetic function. In order that there exist functions $\alpha(x)$ and $\beta(x) \neq 0$, defined for all sufficiently large positive values of x, so that the frequencies*

$$\nu_x\left(n; \frac{g(n) - \alpha(x)}{\beta(x)} \le z\right)$$

possess a proper weak limiting distribution as $x \to \infty$, it is both necessary and sufficient that $g(n)$ not be identically one, that the series

$$\sum_{g(p)=0} \frac{1}{p} \tag{2}$$

converges, and that there is a constant c so that the series

$$\sum_{g(p)\neq 0} p^{-1} \|\log|g(p)|p^{-c}\|^2 \tag{3}$$

converges.

When these three conditions are satisfied one may take $\alpha(x) = 0$, and

$$\beta(x) = x^c \exp\left(\sum_{p \le x} p^{-1} \|\log|g(p)|p^{-c}\|\right).$$

The limit law will then be symmetric if and only if $g(2^k) = -2^{kc}$ for every positive integer k, or the series

$$\sum_{g(p)<0} \frac{1}{p}$$

diverges.

These results remain valid if "weak" is replaced by "modified-weak."

Remark. This theorem comprises theorems 1 and 2 of the joint paper [1] of Levin, Timofeev and Tuliagonov. The proof to be given here differs somewhat from theirs.

Proof of theorem (7.10) (Necessity). We assume that there is a proper distribution function $F(z)$ so that as $x \to \infty$

$$\nu_x\left(n; \frac{g(n) - \alpha(x)}{\beta(x)} \le z\right) \Rightarrow F(z). \tag{4}$$

It is convenient to define an additive function $f(n)$ by

$$f(p^k) = \begin{cases} \log|g(p^k)| & \text{if } g(p^k) \ne 0, \\ (\log p)^2 & \text{otherwise.} \end{cases}$$

We shall show that the function $f(n)$ is finitely distributed. There are two cases.

Case 1. Suppose that

$$\limsup_{x\to\infty} |\alpha(x)/\beta(x)| = \infty.$$

Let $x_1 < x_2 < \dots$ be an unbounded sequence of real numbers for which $|\alpha(x_l)/\beta(x_l)| \to \infty$, as $l \to \infty$.

If u is a suitably large real number which is a continuity point of $F(z)$, and $x = x_l$ is sufficiently large, then we have both $|\alpha(x)| \ge 2u|\beta(x)|$ and

$$\nu_x(n; |g(n) - \alpha(x)| \le u|\beta(x)|) \ge F(u) - F(-u) - 1/4 \ge 1/2.$$

If n is an integer which is counted in this frequency, then

$$||g(n)| - |\alpha(x)|| \le |g(n) - \alpha(x)| \le u|\beta(x)| \le |\alpha(x)|/2,$$

so that

$$|\alpha(x)|/2 \le |g(n)| \le 2|\alpha(x)|$$

and

$$-\log 2 \le f(n) - \log|\alpha(x)| \le \log 2.$$

Thus $f(n)$ is finitely distributed, and from theorem (7.2) we deduce the existence of a number c so that the series

$$\sum p^{-1} \| f(p) - c \log p \|^2$$

converges. In particular, if $g(p) = 0$ then

$$|f(p) - c \log p| = |(\log p)^2 - c \log p| > 1$$

for all sufficiently large primes p, so that the series (2) converges. We also obtain at once the convergence of the series (3).

Case 2. Suppose that

$$\limsup_{x\to\infty} |\alpha(x)/\beta(x)| < \infty.$$

Let $x_1 < x_2 < \dots$ be an unbounded sequence of real numbers for which $\alpha(x_l)/\beta(x_l) \to d$ as $l \to \infty$, where d is finite.

It follows from the hypothesis (4) that as $x = x_l \to \infty$

$$\nu_x(n; \beta(x)^{-1}g(n) \le z) \Rightarrow F(z - d).$$

Since $F(z - d)$ is a proper law it must have at least one point-of-increase, z_0 say, which is not the origin. In particular, therefore,

$$\limsup_{x\to\infty} \nu_x(n; |\beta(x)^{-1}g(n) - z_0| \le |z_0|/2) > 0.$$

For each integer n which is counted in this last frequency

$$\left| \left| \frac{g(n)}{\beta(x)} \right| - |z_0| \right| \le \left| \frac{g(n)}{\beta(x)} - z_0 \right| \le \frac{|z_0|}{2},$$

so that

$$|z_0|/2 \le \left| \frac{g(n)}{\beta(x)} \right| \le 3|z_0|/2$$

and

$$-\log 2 \le f(n) - \log|\beta(x)| - \log|z_0| \le \log 2.$$

Once again $f(n)$ is finitely distributed, and we deduce the validity of conditions (2) and (3).

It remains to show that $g(n)$ cannot be identically one in value. If in fact $g(n)$ is identically one, and z_1 is a point-of-increase of the limiting distribution $F(z)$, then for each $\varepsilon > 0$,

$$\liminf_{x\to\infty} \nu_x\left(n; \left| \frac{g(n) - \alpha(x)}{\beta(x)} - z_1 \right| \le \varepsilon \right) > 0,$$

so that since ε may be chosen arbitrarily small,

$$\frac{1-\alpha(x)}{\beta(x)} \to z_1 \qquad (x \to \infty).$$

However, the limit law must have at least one further point-of-increase z_2, $z_2 \neq z_1$, and the same argument shows

$$\frac{1-\alpha(x)}{\beta(x)} \to z_2 \qquad (x \to \infty).$$

This is impossible, therefore $g(n)$ is not identically zero.

Proof of theorem (7.10) (Sufficiency). We now assume that $g(n)$ is not identically one, that the series (2) converges, and that with a suitably chosen number c the series (3) converges. For the remainder of this section of the proof of theorem (7.10) we shall set $\alpha(x) = 0$ and

$$\beta(x) = x^c \exp\Bigg(\sum_{\substack{p \le x \\ g(p) \neq 0}} p^{-1} \| \log |g(p)|^{p^{-c}} \| \Bigg)$$

identically.

As before, we define the multiplicative function

$$\delta(n) = \begin{cases} 1 & \text{if } g(n) \neq 0, \\ 0 & \text{if } g(n) = 0. \end{cases}$$

For each real number t, and prime p for which $g(p) \neq 0$,

$$\begin{aligned} 1 - \operatorname{Re}(|g(p)|p^{-c})^{it} &= 2\sin^2(\tfrac{1}{2}t \log|g(p)|p^{-c}) \\ &\le \begin{cases} \tfrac{1}{2}t^2 |\log|g(p)|p^{-c}|^2 & \text{if } |\log|g(p)|p^{-c}| \le 1, \\ 2 & \text{if } |\log|g(p)|p^{-c}| > 1, \end{cases} \\ &\le \max(\tfrac{1}{2}t^2, 2) \| \log|g(p)|p^{-c}\|^2 \end{aligned}$$

so that the series

$$\sum p^{-1}(1 - \operatorname{Re} \delta(p)(|g(p)|p^{-c})^{it})$$

converges. It follows from theorem (6.2) that as $x \to \infty$

$$\sum_{n \le x} \delta(n)|g(n)|^{it} = \frac{x^{1+ic}}{1+ict} \prod_{p \le x} \left(1 - \frac{1}{p}\right)\left(1 + \sum_{k=1}^{\infty} \frac{\delta(p^k)|g(p^k)|^{it}}{p^{k(1+ict)}}\right) + o(x).$$

It is now straightforward to check that

$$[x]^{-1}\sum_{n\le x}\delta(n)\left(\frac{|g(n)|}{\beta(x)}\right)^{it}\to w_0(t)\qquad(x\to\infty),$$

where

$$w_0(t)=\frac{1}{1+ict}\prod\nolimits_1 r(p)\prod\nolimits_2 r(p)\exp\left(-\frac{it}{p}\|\log|g(p)|p^{-c}\|\right)$$

$$r(p)=\left(1-\frac{1}{p}\right)\left(1+\sum_{k=1}^{\infty}\frac{\delta(p^k)|g(p^k)|^{it}}{p^{k(1+ict)}}\right),$$

the product $\prod_1$ is taken over those primes for which

$$e^{-1}\le|g(p)|p^{-c}\le e$$

and the product $\prod_2$ is taken over those primes for which

$$|g(p)|p^{-c}<e^{-1}\quad\text{or}\quad|g(p)|p^{-c}>e.$$

These products converge uniformly on any bounded set of t-values, so that $w_0(t)$ is a continuous function of t.

In a similar way we may establish the existence of the limit

$$w_1(t)=\lim_{x\to\infty}[x]^{-1}\sum_{n\le x}\operatorname{sign} g(n).\left(\frac{|g(n)|}{\beta(x)}\right)^{it}.$$

If the series

$$\sum_{g(p)<0}\frac{1}{p}\tag{5}$$

converges then one may adapt the argument given above to prove that

$$\sum p^{-1}(1-\operatorname{Re}\operatorname{sign} g(p).(|g(p)|p^{-c})^{it})\tag{6}$$

converges uniformly on every bounded interval of t-values, so that

$$w_1(t)=\frac{1}{1+ict}\prod\nolimits_1 y(p)\prod\nolimits_2 y(p)\exp\left(-\frac{it}{p}\|\log|g(p)|p^{-c}\|\right)$$

where

$$y(p)=\left(1-\frac{1}{p}\right)\left(1+\sum_{k=1}^{\infty}\frac{\operatorname{sign} g(p^k).|g(p^k)|^{it}}{p^{k(1+ict)}}\right)$$

with the same summation conventions as in the product representation of $w_0(t)$.

If the series (5) diverges, then it follows from our hypothesis that (3) converges that the series

$$L_\varepsilon = \sum_{\substack{|\log|g(p)|p^{-c}| \le \varepsilon \\ g(p) < 0}} \frac{1}{p}$$

diverges for each fixed positive number ε. For any particular t we may then choose a value of ε so that $\cos(t\varepsilon/2) \ge 1/2$, and (with an obvious interpretation)

$$\sum p^{-1}(1 - \operatorname{sign} g(p)(|g(p)|p^{-c})^{it}) \\ \ge \sum_{g(p)<0} 2p^{-1} \cos^2(\tfrac{1}{2}t \log|g(p)|p^{-c}) \ge \tfrac{1}{2}L_\varepsilon = \infty.$$

In this case we may apply theorem (6.2) to deduce that $w_1(t)$ is identically zero.

In either case $w_1(t)$ is a continuous function of t.

It follows from lemma (1.45) (Zolotarev) that as $x \to \infty$ the frequencies

$$\nu_x(n; \beta(x)^{-1}g(n) \le z)$$

possess a modified-weak limit law whose Mellin characteristic function is

$$\begin{pmatrix} w_0(t) & 0 \\ 0 & w_1(t) \end{pmatrix}.$$

It will be symmetric if and only if $w_1(t)$ is identically zero, a condition which is satisfied if and only if *either* the series (5) diverges, *or* the series (5) converges but one of the individual terms in the product representation of $w_1(t)$ vanishes. This last situation can occur only when

$$\frac{\operatorname{sign} g(2^k)|g(2^k)|^{it}}{2^{kict}} = -1$$

for every positive integer k and real number t, a condition which is equivalent to

$$g(2^k) = -2^{kc} \qquad (k = 1, 2, \ldots).$$

In order to complete the proof of theorem (7.10) we must show that in the above circumstances the limit law will be proper unless $g(n)$ is identically one in value. Indeed if the limit law is improper then $w_0(t)$ will have the form $|a|^{it}$

for some real number a. Since $w_0(0) = 1$, a cannot be zero, therefore $|w_0(t)| = 1$. But $w_0(t)$ is the product of characteristic functions, each of which does not exceed one in absolute value. Hence $c = 0$, and $|r(p)| = 1$ for each prime p and all real numbers t. It is easy to deduce from this fact that $|g(p^k)| = 1$ holds for every prime-power.

With $c = 0$ and $|g(p^k)| = 1$, either $w_1(t)$ has the value zero, or

$$w_1(t) = \prod_p \left(1 - \frac{1}{p}\right)\left(1 + \sum_{k=1}^{\infty} \frac{\operatorname{sign} g(p^k)}{p^k}\right).$$

However, since the jump in our improper limit law occurs at a point $a \neq 0$, the limit law cannot be symmetric, so that for some value of t we must have $w_1(t) \neq 0$, in fact $|w_1(t)| = 1$. Therefore

$$\operatorname{sign} g(p^k) = 1 \qquad (k = 1, 2, \ldots),$$

and $g(n)$ is identically one in value.

This completes the proof of theorem (7.10).

Theorem 7.11 (Bakstys; Galambos; Levin, Timofeev and Tuliaganov). *In order that the real-valued arithmetic function $g(n)$ possess a weak limiting distribution it is both necessary and sufficient that the three series*

$$(7) \qquad \sum_{g(p)=0} \frac{1}{p} \qquad \sum_{g(p)\neq 0} \frac{1}{p} \|\log|g(p)|\| \qquad \sum_{g(p)\neq 0} \frac{1}{p} \|\log|g(p)|\|^2$$

converge.

When these conditions are satisfied, the limit law is symmetric if and only if $g(2^k) = -1$ for every integer k, or the series

$$\sum_{g(p)<0} \frac{1}{p}$$

diverges.

The limit law will be continuous if and only if $g(n)$ is never zero, and the series

$$\sum_{|g(p)|\neq 1} \frac{1}{p}$$

diverges.

These results remain valid if "weak" is replaced by "modified-weak."

Remarks. For unsymmetric limit laws theorem (7.11) was proved by Bakstys [1]. This case corresponds to the Erdös–Wintner theorem (theorem (5.1)).

For symmetric limit laws, subject to the additional hypotheses that $g(n)$ be strongly multiplicative and that $g(2) \neq -1$, it was proved by Galambos [3]. In the generality which we give here it was first established by Levin, Timofeev and Tuliaganov, as theorem 3 of their joint paper [1]. At the present time, a satisfactory treatment of the symmetric case is only possible by means of the application of the results of Halász, of Chapter six.

We have added the criterion for the limit law to be continuous.

This ends our remarks.

Proof of theorem (7.11), (Necessity). We assume that as $x \to \infty$ the frequencies

$$\nu_x(n;\, g(n) \leq z)$$

converge weakly to a proper law. It follows at once from theorem (7.10) that the series

$$\sum_{g(p)=0} \frac{1}{p} \tag{8}$$

converges, and that for a suitable constant c so does the series

$$\sum_{g(p)\neq 0} p^{-1} \|\log|g(p)|p^{-c}\|^2. \tag{9}$$

Moreover, if we define

$$\beta(x) = x^c \exp\left(\sum_{\substack{p \leq x \\ g(p) \neq 0}} p^{-1} \|\log|g(p)|p^{-c}\| \right),$$

then the frequencies

$$\nu_x(n;\, \beta(x)^{-1} g(n) \leq z)$$

also converge to a proper law. In accordance with lemma (1.9) this fact, in combination with the initial assumption, guarantees the existence of a positive number b so that $\beta(x) \to b$ as $x \to \infty$. Hence, as $x \to \infty$,

$$c \log x = - \sum_{\substack{p \leq x \\ g(p) \neq 0}} p^{-1} \|\log|g(p)|p^{-c}\| + \log b + o(1) \tag{10}$$

and

$$|c| \leq \frac{1}{\log x} \sum_{p \leq x} \frac{1}{p} + o(1) = O\left(\frac{\log \log x}{\log x} \right).$$

Clearly $c = 0$, and from (8), (9) and (10) we obtain the convergence of the three series (7).

Proof of theorem (7.11), (Sufficiency). Assuming the convergence of the series (7) we may apply theorem (7.10) to deduce the existence of a (modified-) weak limit for the frequencies

$$\nu_x(n; \beta(x)^{-1}g(n) \leq z).$$

If this limit law is $F(z)$, and if $\lim \beta(x) = b$, $(x \to \infty)$, then $g(n)$ has the (modified-) weak limiting distribution $F(z/b)$.

To complete the proof of theorem (7.11) we note that in these circumstances the law $F(z/b)$ will be symmetric if and only if $F(z)$ is. The criterion for $F(z/b)$ to be symmetric is therefore obtained directly from theorem (7.10).

If there is an integer m for which $g(m) = 0$, then those integers nm where n is prime to m have the positive density $\varphi(m)/m^2$, and satisfy $g(nm) = g(n)g(m) = 0$. In order that $g(n)$ possess a limit law $G(z)$ which is continuous at the origin it is therefore necessary that $g(n)$ not vanish for any positive integer. Assuming this to be the case it is clear that whenever $\pm z$ are continuity points of $G(z)$ we have

$$\nu_x(n; \log|g(n)| \leq z) \to G(z) - G(-z) \qquad (x \to \infty).$$

Since the function $\log|g(n)|$ is additive the remaining criterion for the continuity of $G(z)$ follows from that of theorem (5.1).

This completes the proof of theorem (7.11).

We continue this chapter with a number of examples.

EXAMPLES

(11) Any example of an additive function $f(n)$ which satisfies the conditions of theorem (5.2) automatically satisfies the conditions of theorem (7.1) if we add to it the function $c \log n$, where c is a constant.

(12) The function $\varphi(n)n^{-1}$ satisfies the conditions of theorem (7.11), and in fact has a continuous non-symmetric limiting distribution.

(13) If $\mu(n)$ denotes Möbius' function then it is easy to apply theorem (7.11) to show that the function $\mu(n)\varphi(n)n^{-1}$ has a continuous symmetric limiting distribution.

According to this same theorem the function $\mu(n)$ has itself a symmetric but non-continuous limiting distribution, since the series

$$\sum_{\mu(p)=-1} \frac{1}{p}$$

diverges. This is as it should be, indeed

$$\lim_{x\to\infty} x^{-1} \sum_{\substack{n\le x\\ \mu(n)=r}} 1 = \begin{cases} \dfrac{3}{\pi^2} & \text{if } r = 1, \\[2ex] 1 - \dfrac{6}{\pi^2} & \text{if } r = 0, \\[2ex] \dfrac{3}{\pi^2} & \text{if } r = -1. \end{cases}$$

Here the existence of a symmetric limit law amounts to the assertion that

$$\sum_{n\le x} \mu(n) = o(x) \qquad (x \to \infty).$$

Concluding Remarks

Finitely distributed additive functions were introduced and characterized in the manner of theorem (7.2) by Erdös [10]. His proof was elementary, but very complicated. It was his privately expressed opinion that since its publication no one had read it. The proof of the essential step which is embodied in proposition (ii) is due to Ryavec [1], and is along quite different lines. We have incorporated a further simplifying device of Delange [9]. Ryavec's method is interesting in that it reduces a problem of additive functions to a problem in measure theory. We mention that he rediscovered the theorem of Steinhaus (see lemma (1.1) of Chapter one) along the way.

Note that the function $\tau(t)$ which is defined during the course of the proof of proposition (ii) clearly satisfies Cauchy's functional equation, and is eventually shown to have the form $\tau(t) = ct$ for some constant c. It appears that there is not only a close analogue, but also a close link between the behaviour of additive functions defined on the semigroup of positive integers, and additive functions defined on the additive group of real numbers. We shall encounter this phenomenon again in later chapters.

The aforementioned paper of Erdös [10] contained a large number of results and conjectures concerning additive functions, and, in particular, various possible characterisations of the logarithmic function. Some of these are as yet unsolved. In particular, we mention the poor knowledge concerning the behaviour of differences of additive functions, such as $f(n + 1) - f(n)$. For specific questions we refer to the chapter, at the end of volume two, in which we discuss unsolved problems.

A suitably general form of theorem (7.3) appears as theorem IV of Erdös [10], where it is proved under the additional assumption that the $f(p)$ are uniformly bounded. The present unconditional form was given by Elliott and Ryavec [1].

Theorem (7.7) is contained in the author's paper [13].

Chapter 8

Distribution of Additive Functions (mod 1)

In this chapter, largely under the impulse of Weyl's classic paper on uniform distribution, we make an excursion into the study of additive functions (mod 1).

For the duration of the chapter $\{y\}$ will denote the fractional part of the real number y, and

$$\|y\| = \min(\{y\}, 1 - \{y\}),$$

the distance from y to its nearest integer. *This notation differs from that used in theorems* (7.10) *and* (7.11).

Theorem 8.1. *In order that the (real-valued) additive function $f(n)$ should be uniformly distributed* (mod 1), *it is both necessary and sufficient that for each positive integer k the series*

$$\sum p^{-1}\|kf(p) - \tau \log p\|^2 \tag{1}$$

diverges for every real number τ.

Theorem 8.2. *In order that the additive function $f(n)$ should possess a non-uniform limiting distribution* (mod 1), *it is both necessary and sufficient that for some positive integer k the series*

$$\sum p^{-1}\|kf(p)\|^2 \qquad \sum p^{-1}\|kf(p)\|\operatorname{sign}(\tfrac{1}{2} - \{kf(p)\}) \tag{2}$$

converge. When this condition is satisfied the limit law will be continuous if and only if the series

$$\sum_{\|mf(p)\| \neq 0} \frac{1}{p} \tag{3}$$

diverges for every positive integer m.

Remarks. We shall show that the simultaneous convergence of the two series in (2) is equivalent to the convergence of the series

$$\sum p^{-1}(1 - e^{2\pi i k f(p)}).$$

Define the function $F(m) = f(m) - [f(m)]$. In his paper, [10], of 1946, Erdös states and later proves the following result (we make a small change in notation):

"*Let $f(p) \to 0$ as $p \to \infty$ and assume that $\sum f^2(p)^{-1} = \infty$. Then the distribution function of $F(m)$ is x. In other words the density of integers m for which*

$$F(m) \leq c$$

equals c."

He goes on to say:

"If we do not assume that $f(p) \to 0$, the situation becomes rather complicated. First it is clear that the distribution function of $F(m)$ does not have to be x. Put $f(p) = 1/2$, $f(p^\alpha) = 0$. Then it follows from the prime number theorem that $\psi(x) = 1/2$, $0 \leq x \leq 1/2$, $\psi(x) = 1$, $1/2 \leq x \leq 1$. If $f(p) = \alpha$, α irrational, it can be shown that the distribution function of $F(m)$ is again x. The proof is not easy and we do not discuss it here.

It can be conjectured that $F(m)$ always has a distribution function. This if true must be very deep, since it contains the prime number theorem."

There is then a footnote, added in proof, to the effect that this conjecture is false, as may be seen from an example of Wintner, (p. 48 "II, bis" of [4]) which shows that not even

$$\lim_{n \to \infty} n^{-1} \sum_{m=1}^{n} F(m)$$

exists.

It is not clear what Erdös means here by the distribution function of $F(m)$; for example two values are given for $\psi(1/2)$. Moreover, the example of Wintner to which he refers concerns itself with complex-valued *multiplicative* functions which do not possess mean-values.

We shall return to consider what happens to this conjecture of Erdös in that part of the present chapter which follows theorem (8.8).

In the years following Erdös' paper interest was confined to proving that various classes of additive functions were uniformly distributed (mod 1). We mention here the papers [2] and [1] of Delange, in which he showed, in particular, that the function $\alpha\omega(n)$ is uniformly distributed (mod 1) when α is irrational, thus justifying one of Erdös' remarks; with a similar, appropriate, result when α is a non-zero rational number. Further sufficient conditions for an additive function to be uniformly distributed (mod 1) were given by Kubilius [5], and Levin and Faĭnleĭb [4], [5], some of the results of the latter being quantitative.

The possibility of additive functions possessing non-uniform limit laws (mod 1) was considered, independently, by Delange [8], and Elliott [8]. Whilst the reduction of the problem to the consideration of the mean-value of multiplicative functions was given in a satisfactory manner in the author's paper [8], owing to a hasty reference to the paper [1] of Halász, the final statement of the theorem was only valid if, for example, $f(n)$ were assumed to be strongly (completely) additive. This mistake was rectified, and the result generalised to cover translations of additive functions, by Manstavičius [1].

We shall largely follow the elegant treatment of Delange. In his paper he did not consider the nature of the limit law, and we shall consider that part of theorem (8.2) a little later.

For $0 \le z < 1$ define frequencies (mod 1) by

$$F_n(z) = \nu_n(m; f(m) \le z \pmod 1) \qquad (n = 1, 2, \ldots),$$

so that

$$\int_{0-}^{1+} e^{2\pi i k z}\, dF_n(z) = n^{-1} \sum_{m=1}^{n} e^{2\pi i k f(m)}.$$

According to lemma (1.46), $f(m)$ will possess a limiting distribution (mod 1) if and only if

$$\beta_k = \lim_{n \to \infty} n^{-1} \sum_{m=1}^{n} e^{2\pi i k f(m)}.$$

exists for each non-zero integer k. It will clearly suffice to consider the cases when k is positive, and we shall do this by applying theorem (6.3).

This ends the remarks.

Before we give the proofs of these theorems it is convenient to gather together a number of elementary inequalities.

Lemma 8.3. *If z is a complex number, $|z| \le 1$, then*

$$1 - \operatorname{Re} z^2 \le 4(1 - \operatorname{Re} z).$$

For any real number y,

$$\tfrac{1}{2}\|y\| \le |\sin \pi y| \le \pi\|y\|.$$

For any pair of real numbers u, v, and integer m,

$$|\sin mu| \le |m||\sin u|,$$

$$\sin(u + v) - \sin u - \sin v \le 2\left(\sin^2 \frac{u}{2} + \sin^2 \frac{v}{2}\right),$$

$$\sin^2(u + v) \le 2(\sin^2 u + \sin^2 v).$$

Proof. Of these inequalities perhaps only the first and fourth call for comment.
If $z = x + iy$, x, y real, then

$$\begin{aligned} 1 - \operatorname{Re} z^2 &= 1 + x^2 + y^2 - 2x^2 \le 2(1 - x^2) \\ &= 2(1 - x)(1 + x) \le 4(1 - x), \end{aligned}$$

which yields the first inequality. The fourth inequality follows from the identities

$$\begin{aligned} \sin(u + v) - \sin u - \sin v &= \sin u(\cos v - 1) + \sin v(\cos u - 1) \\ &= -2 \sin u \sin^2 \frac{v}{2} - 2 \sin v \sin^2 \frac{u}{2}. \end{aligned}$$

Proof of theorem (8.1), (Sufficiency). Assume first that for some positive integer k the series

$$\sum \frac{1}{p} \|kf(p) - \tau \log p\|^2 \tag{1}$$

diverges for every real number τ. Then, according to the second inequality of lemma (8.3), so does the series

$$\sum p^{-1}(1 - \operatorname{Re} e^{2\pi i k f(p)} p^{-i\tau}). \tag{4}$$

We may now apply part (iv) of theorem (6.3) to deduce that β_k exists, and has the value zero.

If we can do this for every $k > 0$, then $\beta_k = 0$ for all non-zero integers k, and it follows from the classical criterion of Weyl (see lemma (1.46)) that $f(n)$ is uniformly distributed (mod 1).

Proof of theorem (8.1), (Necessity). Suppose that $f(n)$ is uniformly distributed (mod 1), but that for some positive integer k, and real number τ, the series (1), and so (4), converges. For the particular k in question there can only be one real value of τ which will meet this requirement. Since $\beta_k = 0$, it follows from part (iii) of theorem (6.3) that the relations

$$2^{-im\tau}e^{2\pi ikf(2^m)} = -1 \qquad (m = 1, 2, \ldots), \tag{5}$$

must be satisfied.

Moreover, from the third inequality of lemma (8.3) applied to our temporary hypothesis, we obtain

$$\begin{aligned}\sum p^{-1}(1 - \operatorname{Re} e^{2\pi i(2k)f(p)}p^{-i2\tau}) &= 2\sum p^{-1}\sin^2(2\pi kf(p) - \tau\log p)\\ &\leq 8\sum p^{-1}\sin^2\left(\pi kf(p) - \frac{\tau}{2}\log p\right)\\ &= 4\sum p^{-1}(1 - \operatorname{Re} e^{2\pi ikf(p)}p^{-i\tau}) < \infty.\end{aligned}$$

However, $\beta_{2k} = 0$, and we may argue as before to deduce that

$$2^{-2im\tau}e^{4\pi ikf(2^m)} = -1 \qquad (m = 1, 2, \ldots). \tag{6}$$

The relations (5) and (6) are incompatible, and we deduce that the series (1) must diverge for all positive integers k and real numbers τ.

Proof of theorem (8.2), (Necessity). If the function $f(n)$ has a non-uniform limiting distribution (mod 1), then according to lemma (1.46) every limit β_k exists, and for at least one positive integer k, $\beta_k \neq 0$. It follows from theorem (6.3) part (ii) that the series

$$\sum_p p^{-1}(1 - e^{2\pi ikf(p)})$$

converges.

By taking real parts, and applying the second inequality of lemma (8.3), we obtain the convergence of the series

$$\sum_p p^{-1}\|kf(p)\|^2.$$

The convergence of the second of the two series which appear in (2) is obtained by means of the simple results

$$\operatorname{Im}(e^{2\pi iw}) = i\sin 2\pi w,$$

$$|\sin 2\pi w - 2\pi\|w\|\operatorname{sign}(\tfrac{1}{2} - \{w\})| \leq \pi^3\|w\|^2,$$

which are valid for all real numbers w.

Proof of theorem (8.2), (Sufficiency). To begin with, consider the set K of those integers k with the property that the series

$$\sum p^{-1}(1 - \text{Re}\ e^{2\pi i k f(p)} p^{-i\tau}) \tag{7}$$

converges for some (necessarily unique) real number τ. We shall prove that K is, in the algebraic sense, a $\mathbb{Z}$-module. The argument which we use to prove this fact is an analogue of that used in the consideration of the set E which occurs in our present, and Ryavec's original, proof of proposition (ii) of theorem (7.2). In each case a form of Cauchy's functional equation is implicitly discussed.

Let the integers k_1 and k_2 belong to K, the corresponding values of τ being τ_1 and τ_2. By making use of the last of the inequalities in lemma (8.3) we deduce that

$$\sum p^{-1} \sin^2(\pi(k_1 + k_2)f(p) - \tfrac{1}{2}(\tau_1 + \tau_2) \log p)$$
$$\leq 2 \sum_{j=1}^{2} \sum p^{-1} \sin^2 (\pi k_j f(p) - \tfrac{1}{2}\tau_j \log p) < \infty,$$

so that $k_1 + k_2$ belongs to K. Likewise $k_1 - k_2$ also belongs to K. Thus K is a $\mathbb{Z}$-module, and so consists of all integral multiples of its least positive member q, say. We notice also from this proof that if $k = mq$ for some integer m, then $\tau_k = m\tau_q$.

We have therefore shown that if $k = mq$, and $\lambda = \tau_q$, then the series

$$\sum p^{-1}(1 - \text{Re}\ e^{2\pi i m q f(p)} p^{-im\lambda})$$

converges for $m = 1, 2, \ldots$. On the other hand, if k is not of the form mq then the series (5) diverges for all real values of τ, and so by theorem (6.3)

$$n^{-1} \sum_{m=1}^{n} e^{2\pi i k f(m)} \to 0 \qquad (n \to \infty).$$

We shall now prove that if the series

$$\sum p^{-1}(1 - e^{2\pi i k f(p)}) \tag{8}$$

converges for some $k = k_0$, this being the same as the convergence of the series (2) for $k = k_0$, then it converges for each of the integers k in K. This will then establish the sufficiency of the condition stated in theorem (8.2). For, by theorem (6.3), the limit β_k will exist for every k in K; and the argument concerning the incompatibility of conditions (5) and (6), with τ now having the value zero, shows that not both of β_q and β_{2q} can be zero. Since $\beta_k = 0$

when k does not belong to K, the existence of a non-uniform limiting distribution (mod 1) will follow from lemma (1.46).

We note straightaway that if the series (8) converges, then so does the series (7), and with $\tau = 0$. Since k_0 must belong to K, and τ is uniquely determined, $0 = \tau_{k_0} = (k_0 q^{-1})\tau_q$, so that $\lambda = \tau_q = 0$. The series

$$\sum p^{-1}(1 - \operatorname{Re} e^{2\pi i m q f(p)})$$

therefore converges for $m = 0, \pm 1, \pm 2, \ldots$.

For each integer m define the functions

$$H(m, x) = \sum_{p \le x} p^{-1} \sin 2\pi m q f(p)$$

and

$$M(m, y, x) = \sum_{y < p \le x} p^{-1} \sin 2\pi m q f(p) \qquad 0 \le y \le x < \infty.$$

If m_1 and m_2 are two positive integers, applying the fourth inequality of lemma (8.3) yields

$$|M(m_1 + m_2, y, x) - M(m_1, y, x) - M(m_2, y, x)|$$

$$\le 2 \sum_{j=1}^{2} \sum_{y < p \le x} p^{-1} \sin^2 \pi m_j q f(p)$$

$$= \sum_{j=1}^{2} \sum_{y < p \le x} p^{-1}(1 - \operatorname{Re} e^{2\pi i m_j q f(p)}) \to 0 \qquad (y \to \infty).$$

Therefore, by Cauchy's criterion of convergence,

$$\lim_{x \to \infty} (H(m_1 + m_2, x) - H(m_1, x) - H(m_2, x))$$

exists. If we set $m_1 = m_2 = 1$ then we deduce that

$$\lim_{x \to \infty} (H(2, x) - 2H(1, x))$$

exists. Setting $m_1 = 2$, $m_2 = 1$ we see that

$$\lim_{x \to \infty} (H(3, x) - 3H(1, x))$$

exists. Proceeding by induction we prove that for each positive integer m

$$\lim_{x\to\infty} (H(m, x) - mH(1, x))$$

exists. From the hypothesis that the series (8) converges with $k = k_0$ we know, in particular, that $\lim H(k_0 q^{-1}, x)$, $(x \to \infty)$, exists. Therefore $\lim H(1, x)$, $(x \to \infty)$, exists, and so does $\lim H(m, x)$, $(x \to \infty)$, for any integer m.

We have now established the convergence of the series

$$\sum p^{-1}(1 - e^{2\pi imqf(p)}) = \sum p^{-1}(1 - \operatorname{Re} e^{2\pi imqf(p)}) - i \sum p^{-1} \sin 2\pi mqf(p)$$

for $m = 0, \pm 1, \pm 2, \ldots$, and with it guaranteed the existence of a non-uniform limiting distribution (mod 1).

We round out our proof of theorem (8.2) by considering when the limit law can be continuous.

The Nature of the Limit Law

For each integer k define

$$\eta_k = 1 + \sum_{m=1}^{\infty} 2^{-m} e^{2\pi ikf(2^m)}.$$

Assume that $f(n)$ has the limiting distribution $F(z)$ (mod 1), so that all of the characteristic coefficients β_k exist. It is easy to see that the β_k have a representation

$$\beta_k = \gamma_k \cdot \eta_k \cdot \exp\left(\sum_{p \geq 3} p^{-1}(1 - \operatorname{Re} e^{2\pi ikf(p)})\right),$$

where the γ_k are bounded by $0 < c_1 \leq |\gamma_k| \leq c_2$, for certain positive numbers c_1 and c_2, uniformly for all integers k. Here one interprets β_k to be zero if the series

$$\sum_{p \geq 3} p^{-1}(1 - \operatorname{Re} e^{2\pi ikf(p)})$$

diverges. For example, one may employ the representation which follows part (ii) of theorem (6.3), together with the method of proof of lemma (6.6).

It follows at once from lemma (1.46) that $F(z)$ is absolutely continuous with a derivative which belongs to the class $L^2(0, 1)$ if and only if the series

$$\sum_{k=-\infty}^{\infty} |\eta_k|^2 \exp\left(-4 \sum_{p \geq 3} p^{-1} \sin^2 \pi kf(p)\right)$$

converges. Moreover $F(z)$ is continuous if and only if

$$\lim_{N\to\infty} (2N+1)^{-1} \sum_{k=-N}^{N} |\eta_k|^2 \exp\left(-4\sum_{p\geq 3} p^{-1}\sin^2 \pi k f(p)\right) = 0.$$

These conditions, which were first stated in the author's paper [8], are not elegant. Fortunately, in the case of continuity it is possible to give a simpler (equivalent) condition. In the form which we give it in theorem (8.2), the criterion for the limit law to be continuous was first correctly stated and proved in a paper by P. Hartman [1], as a motivational example in connection with some work on locally compact groups. He also considered the purity of type of the limit law. We note here that as reference number 5 (p. 231) of his account Hartman refers to a paper of the present author entitled "The continuity of the limiting distribution of additive functions (mod 1)." This paper will not appear.

We shall, instead, give a number-theoretical treatment which depends upon the use of Schnirelmann density, and is analagous to the considerations which were made at the close of Chapter 5.

It would be very interesting to have a simple condition concerning absolute continuity.

Condition necessary for continuity.

Assume that for some positive integer m the series

$$\sum_{\|mf(p)\|=0} \frac{1}{p}$$

converges. Then, according to lemma (5.10), those squarefree integers n which are not divisible by any prime p for which $\|mf(p)\| \neq 0$ have the positive asymptotic density

$$\delta = \frac{6}{\pi^2} \prod_{\|mf(p)\|\neq 0} \left(1+\frac{1}{p}\right).$$

For each such integer n we have $mf(n) \equiv 0 (\text{mod } 1)$. It follows that

$$\sum_{l=0}^{m-1} \limsup_{x\to\infty} \nu_x\left(n; f(n) \equiv \frac{l}{m} (\text{mod } 1)\right) \geq \delta,$$

so that the limiting distribution of $f(n)$ must be discontinuous at at least one of the points l/m, $l = 0, 1, \ldots, m-1$.

The necessity of the given condition in order that $F(z)$ be continuous is now clear.

Condition sufficient for continuity: Beginning of proof. Most of the argument will be devoted to establishing an appropriate analogue of lemma (1.6).

It is convenient to recall the notion of *Schnirelmann density.*

If $a_1 < a_2 < \ldots$ is a sequence of positive integers, let $A(n)$ denote the number of its members which do not exceed the integer n. The Schnirelmann density of this sequence is defined to be

$$\alpha = \inf_{n \geq 1} n^{-1} A(n),$$

the infimum being taken over all positive integers n. We shall also use $\sigma(A)$ to denote the Schnirelmann density of the sequence A. A similar notation will be used for other sequences. Notice that this particular density will be zero unless the sequence contains the integer 1.

The *(Schnirelmann) sum,* $A + B$, of two sequences of positive integers A and B, is defined to be that sequence, each of whose members are of the form a_i, b_j, or $a_i + b_j$, for some i, j.

We shall need the following well-known result of Schnirelmann.

Lemma 8.4. *For any two integer sequences A and B we have*

$$\sigma(A + B) \geq \sigma(A) + \sigma(B) - \sigma(A)\sigma(B).$$

Remark. As was pointed out in the remarks which were made at the end of Chapter one, concerning lemma (1.6), there is an improved form of this result, due to H. B. Mann, which asserts that

$$\sigma(A + B) \geq \min(1, \sigma(A) + \sigma(B)).$$

For our present purposes this stronger result would bring no advantages.

Proof of lemma (8.4). Let $a_1 < a_2 < \cdots < a_k$ be the members of the sequence A which do not exceed a given positive integer n. If $1 \leq i \leq k - 1$, and $1 \leq b_j \leq (a_{i+1} - a_i - 1)$, then no $a_i + b_j$ belongs to A. Similarly, if $1 \leq b_r \leq (n - a_k - 1)$, then no $a_k + b_r$ belongs to A.

Let $C = A + B$. Then from what we have shown so far

$$\begin{aligned} C(n) &\geq k + \sum_{i=1}^{k-1} B(a_{i+1} - a_i - 1) + B(n - a_k - 1) \\ &\geq k + \beta \sum_{i=1}^{k-1} (a_{i+1} - a_i - 1) + \beta(n - a_k - 1) \\ &\geq k(1 - \beta) + \beta n \geq \alpha(1 - \beta)n + \beta n. \end{aligned}$$

Hence, the inequality

$$n^{-1}C(n) \geq \alpha + \beta - \alpha\beta$$

is satisfied for every $n \geq 1$, and lemma (8.4) is established.

We say that a sequence A is a *basis* if there is a (fixed) number l so that every positive integer may be expressed as the sum of not more than l members of A. As a corollary of lemma (8.4) we obtain, (a result also due to Schnirelmann),

Lemma 8.5. *Any sequence of integers with a positive Schnirelmann density is a basis.*

Proof of lemma (8.5). Let A be a sequence of integers, with $\alpha = \sigma(A) > 0$. We may apply lemma (8.4) with $A = B$, and write it in the form

$$1 - \sigma(A + A) \leq (1 - \alpha)^2.$$

Indeed, a simple inductive argument shows that for each positive integer r

$$1 - \sigma(\underbrace{A + A + \cdots + A}_{r \text{ terms}}) \leq (1 - \alpha)^r.$$

Choose r so large that $(1 - \alpha)^r < 1/3$, and let C denote the sequence $A + A + \cdots + A$ where r summands are taken. Thus $\sigma(C) > 2/3$.

If n is any positive integer, we note that the two rows of non-negative integers

$$n, n - c_1, n - c_2, n - c_3, \ldots$$

$$0, c_1, c_2, c_3, \ldots$$

where the c_i run through the members of C which do not exceed n, are distinct in each row. Since they contain between them at least $2C(n) + 2 > (4n/3) + 2 > n + 1$ integers, each of which lies in the interval $[0, n]$, at least one member of the first row coincides with a member of the second row.

Thus n belongs to $C + C$, and A is a basis.

Remark. The proof given here shows that not more than

$$2r < 2 - \frac{2 \log 3}{\log(1 - \alpha)}$$

summands taken from A are needed to represent each positive integer.

We need two further definitions.

We define the *lower asymptotic density*, or more shortly the *lower density*, $d(A)$, of the integer sequence A, to be

$$\liminf_{n \to \infty} n^{-1} A(n).$$

Thus, in every case $d(A) \geq \sigma(A)$. However, $d(A) > 0$ may hold even when A has Schnirelmann density zero.

We say that d is the *highest common factor of the sequence* A if d is the largest positive integer which divides every member of the sequence A.

Lemma 8.6. *Let A be an integer sequence with positive lower asymptotic density. Let d be the highest common factor of A. Then there is a (fixed) number l so that all sufficiently large integral multiples of d may be written as the sum of not more than l integers taken from A.*

Remarks. Lemma (8.6) asserts that a sequence A which has positive lower density is as near to being a basis as may be. For example, the sequence $A = \{2, 4, 6, \ldots\}$ has lower density $1/2$, but cannot represent any odd integer, no matter how many summands are taken. This result is a special case of proposition (3.1) of the author's first paper [1].

Proof of lemma (8.6). Dividing the members of the sequence A by d, if necessary, it is clear that we may assume that $d = 1$; and we shall.

Since the highest common factor A is (now) 1, there are members b_i, $(i = 1, \ldots, r)$, of A so that the highest common factor of these b_i is 1. Hence, there are integers s_i, $(i = 1, \ldots, r)$, not all zero, so that

$$1 = \sum_{i=1}^{r} s_i b_i'.$$

Let

$$m = \sum_{i=1}^{r} 2|s_i| b_i,$$

so that

$$m + 1 = \sum_{i=1}^{r} (2|s_i| + s_i) b_i > 0,$$

and $(m, m + 1) = 1$.

It follows from the hypothesis of the present lemma that for some real number $c > 0$, and all integers $n \geq n_0$, the inequality $A(n) \geq cn$ is satisfied.

The sequence D which is obtained by adjoining 1 to A (if necessary), therefore has a positive Schnirelmann density. Indeed, $D(n) \geq cn$ if $n > n_0$, and $D(n) \geq 1$ if $n \leq n_0$, so that $\sigma(D) \geq \min(c, n_0^{-1})$.

We deduce from lemma (8.5) that D is a basis.

Hence there is an integer t so that the equation

$$n - 3tm(m + 1) = a_{i_1} + a_{i_2} + \cdots + a_{i_w} + u$$

where w and u are non-negative integers, and $w + u \leq t$, is certainly soluble for all sufficiently large integers n. We shall complete the proof by showing that $3tm(m + 1) + u$ may be written as the sum of a bounded (in terms of t) number of integers taken from A.

Since m and $(m + 1)$ are coprime, there is an integer h, in the range $0 \leq h \leq m$, for which the congruence $u \equiv hm(\bmod(m + 1))$ is satisfied. Therefore

$$u = hm + k(m + 1),$$

where

$$|k| = \frac{|u - hm|}{m + 1} \leq \frac{t + m^2}{m + 1} < t + m.$$

We can now write

$$3tm(m + 1) + u = hm + (3tm + k)(m + 1), \tag{9}$$

where our construction of k has ensured that

$$3tm + k > 3tm - t - m \geq tm > 0.$$

The expression appearing on the left-hand side of equation (9) has been represented as the sum of positive integral multiples of m and $(m + 1)$, each of which is the sum of finitely many members of A.

This completes our proof of lemma (8.6).

We can now quickly complete the proof of theorem (8.2).

Condition sufficient for continuity: Completion of proof. Suppose that

$$\liminf_{N \to \infty} (2N + 1)^{-1} \sum_{k=-N}^{N} |\beta_k|^2 = \delta > 0.$$

Since $|\beta_k| = |\beta_{-k}|$ we may write this hypothesis in the form

$$\liminf_{N\to\infty} N^{-1}\sum_{k=1}^{N}|\beta_k|^2 = \delta.$$

According to a remark made at the beginning of the section on the nature of the limit law, these β_k are bounded by the inequality, suitably interpreted,

$$|\beta_k| \le 2c_2 \exp\left(-\sum_{p\ge 3} p^{-1}(1 - \operatorname{Re} e^{2\pi i k f(p)})\right).$$

Let m be a positive real number, chosen so that $2c_2 e^{-m} \le (\delta/2)^{1/2}$, and let $A;\, a_1 < a_2 < \dots$ denote the sequence of positive (integers) k for which the inequality

$$\sum_{p\ge 3} p^{-1}(1 - \operatorname{Re} e^{2\pi i k f(p)}) \le m$$

is satisfied. Then, bearing in mind that $|\beta_k| \le 1$ for every integer k, we have

$$\liminf_{N\to\infty} N^{-1}A(N) \ge \liminf N^{-1}\sum_{a_j\le N}|\beta_{a_j}|^2$$

$$\ge \liminf_{N\to\infty} N^{-1}\sum_{k=1}^{N}|\beta_k|^2 - (2c_2e^{-m})^2 \ge \delta/2.$$

Since A has positive lower density, we may apply lemma (8.6). This guarantees the existence of positive integers d and l so that all sufficiently large integral multiples of d may be expressed as the sum of not more than l members of the sequence A. By replacing d with a larger integer, if necessary, we may without loss of generality assume that all positive integral multiples of d are so representable. This we shall do.

Suppose that, typically,

$$md = b_1 + \cdots + b_w,$$

where the b_i are taken from A, and $w \le l$. Then from each prime p

$$\sin^2(\pi m d f(p)) = \sin^2(\pi b_1 f(p) + \cdots + \pi b_w f(p))$$

$$\le \left(\sum_{j=1}^{w}|\sin \pi b_j f(p)|\right)^2 \le w\sum_{j=1}^{w}\sin^2 \pi b_j f(p),$$

so that

$$\sum_{p\geq 3} p^{-1}(1 - \operatorname{Re} e^{2\pi imdf(p)}) = 2\sum p^{-1}\sin^2 \pi mdf(p)$$

$$\leq 2w\sum_{j=1}^{w}\sum p^{-1}\sin^2 \pi b_j f(p)$$

$$= w\sum_{j=1}^{w}\sum_{p\geq 3} p^{-1}(1 - \operatorname{Re} e^{2\pi ib_j f(p)})$$

$$\leq w^2 m \leq l^2 m.$$

We now make use of the sum-formula

$$\sum_{m=-M}^{M} e^{2\pi imy} = \begin{cases} \dfrac{\sin(2M+1)\pi y}{\sin \pi y} & y \neq 0, \pm 1, \pm 2, \ldots, \\ 2M+1 & \text{otherwise,} \end{cases}$$

which is valid for all real numbers y.

For any positive real number t we have

$$\sum_{\substack{3\leq p\leq t \\ \|df(p)\|\neq 0}} \frac{1}{p} = \lim_{M\to\infty} \frac{1}{2M+1}\sum_{m=-M}^{M}\sum_{3\leq p\leq t} p^{-1}(1 - \operatorname{Re} e^{2\pi imdf(p)})$$

$$\leq l^2 m,$$

and, since t may be chosen arbitrarily large, obtain the convergence of the series

$$\sum_{\|df(p)\|\neq 0} \frac{1}{p}.$$

To complete the proof of theorem (8.2) we note that if the series (3) diverge for every positive integer m, then this last situation cannot occur. Therefore,

$$\liminf_{N\to\infty} (2N+1)^{-1}\sum_{k=-N}^{N} |\beta_k|^2 = 0,$$

and it follows at once from lemma (1.46) that $F(z)$, the limiting distribution (mod 1) of $f(n)$, is continuous.

The application of theorem (8.1) is aided by the following result.

Lemma 8.7. *Let $\{c_p\}$ be a sequence of real numbers, one for each prime p. Let τ be a non-zero real number, and suppose that*

$$\lim_{x\to 0} \frac{1}{\log\log x} \sum_{p\le x} p^{-1}\|c_p - \tau \log p\|^2 = 0.$$

Then the c_p are well distributed (mod 1) *in the sense that if*

$$L(x) = \sum_{p\le x} \frac{1}{p},$$

then

$$L(x)^{-1} \sum_{\substack{p\le x\\ c_p\le w (\mathrm{mod}\ 1)}} \frac{1}{p} \to w \qquad (x\to\infty),$$

holds for every real number w in the range $0 \le w \le 1$.

Remark. In a weighted sense the conclusion of lemma (8.7) is that the c_p must be uniformly distributed (mod 1).

As an example in the use of lemma (8.7) consider the additive function $\alpha\omega(n)$, where α is an irrational number. For each (fixed) integer k the sequence $c_p = k\alpha\omega(p) = k\alpha$ is clearly not well distributed (mod 1). Moreover, since α is irrational, $\|k\alpha\| > 0$. The series (1) in theorem (8.1) therefore diverges for every pair (k, τ) not both of which are zero. Hence, the function $\alpha\omega(n)$ is uniformly distributed (mod 1).

If α is rational then there is an integer k so that $\|k\alpha\| = 0$. It follows from theorem (8.2) that $\alpha\omega(n)$ then has a non-uniform limiting distribution (mod 1).

Proof of lemma (8.7). We first prove that the sequence $\tau \log p$ is well distributed (mod 1). According to lemma (1.46) it will suffice to prove that for each positive integer k

$$L(x)^{-1} \sum_{p\le x} p^{-1} e^{2\pi i k\tau \log p} \to 0 \qquad (x\to\infty).$$

For this we shall need the estimate

$$L(x) = \log\log x + A + O((\log x)^{-2}),$$

which may be deduced from lemma (2.6) by partial integration. By means of another partial integration we see that if $\beta \neq 0$,

$$\sum_{p \le x} p^{-1-i\beta} = \int_{2-}^{x} y^{-i\beta}\, dL(y) = x^{-i\beta}L(x) + \int_{2}^{x} i\beta y^{-i\beta-1}L(y)dy$$

$$= x^{-i\beta} \log\log x + i\beta \int_2^x y^{-i\beta-1}(\log\log y + A)dy + O(1)$$

$$= \int_2^x y^{-i\beta-1}(A - (\log y)^{-1})dy + O(1),$$

since

$$\frac{d}{dy}(y^{-i\beta}\log\log y) = i\beta y^{-i\beta-1}\log\log y + y^{-i\beta-1}(\log y)^{-1}.$$

An integration by parts also shows that the integral

$$J = \int_2^x \frac{y^{-i\beta}}{\log y}\cdot y^{-1}\, dy$$

satisfies

$$J = [y^{-i\beta}]_2^x + \int_2^x i\beta y^{-i\beta-1}\, dy - J,$$

so that

$$\sum_{p \le x} p^{-1-i\beta} = \left(A + \frac{i\beta}{2}\right)\int_2^x y^{-i\beta-1}\, dy + O(1) = O(1),$$

and

$$L(x)^{-1}\sum_{p \le x} p^{-1-i\beta} \to 0 \qquad (x \to \infty).$$

With $\beta = 2\pi k\tau$ this is our required result.

Applying the Cauchy–Schwarz inequality gives

$$\left|\sum_{p \le x} p^{-1}e^{2\pi i k c_p} - \sum_{p \le x} p^{-1}e^{2\pi i k\tau \log p}\right|^2$$

$$\le L(x)\sum_{p \le x} p^{-1}|\exp(2\pi i k(c_p - \tau\log p)) - 1|^2$$

$$\le L(x)\sum_{p \le x} p^{-1}4\pi^2\|c_p - \tau\log p\|^2 = o(L^2(x)),$$

so that

$$L(x)^{-1}\sum_{p\le x} p^{-1}e^{2\pi ikc_p} \to 0 \qquad (x\to\infty).$$

It follows from lemma (1.46) that the c_p are well distributed (mod 1).

It is now easy to deduce the validity of

Theorem 8.8. *Let* $\|f(p)\| \to 0$ *as* $p\to\infty$. *Then the additive function* $f(n)$ *is uniformly distributed* (mod 1) *if and only if the series*

$$\sum p^{-1}\|f(p)\|^2$$

diverges; and has a non-uniform limiting distribution (mod 1) *if and only if both the series*

$$\sum p^{-1}\|f(p)\| \qquad \sum p^{-1}\|f(p)\|^2$$

converge.

In the latter case the limiting distribution will be continuous if and only if the series $\sum p^{-1}$, *taken over those primes for which* $\|f(p)\| \neq 0$, *diverges.*

Let us now return to the early result of Erdös, which we discussed in the remarks which followed the statement of theorem (8.2). If $f(p)\to 0$ as $p\to\infty$, then $f(n)$ will be uniformly distributed (mod 1) if and only if the series

$$\sum p^{-1}f^2(p)$$

diverges. This result follows from lemma (8.7) and was first pointed out by Delange [8].

Concerning Erdös' example with $f(p) = 1/2$, we note that every positive integer n may be expressed uniquely in the form $n = n_1n_2$, where $p|n_1$ implies $p^2\nmid n$, and $p|n_2$ implies $p^2|n$. If $\mu(m)$ denotes the usual Möbius function, then $F(n) = 1/2$ if $\mu(n_1) = -1$, and $F(n) = 0$ if $\mu(n_1) = +1$. It is straightforward to establish the estimate

$$\sum_{n\le x}\mu(n_1) = o(x) \qquad (x\to\infty).$$

For example, one can make use of the fact that the function $n\mapsto\mu(n_1)$ is multiplicative, so that there is a Dirichlet series representation

$$\sum_{n=1}^{\infty}\mu(n_1)n^{-s} = \prod_p\left(1 - \frac{1}{p^s} + \frac{1}{p^s(p^s-1)}\right) \qquad (\sigma > 1).$$

Alternatively, one may apply theorem (6.4) directly. Hence the function $\omega(n_1)/2$ has the limiting distribution $G(z)$ (mod 1), where

$$G(z) = \begin{cases} 1 & \text{if } 1/2 \le z \le 1, \\ 1/2 & \text{if } 0 \le z < 1/2. \end{cases}$$

Concerning Erdös' conjecture that every additive function possess a limiting distribution (mod 1), define an additive function $h(n)$ by

$$h(p) = \begin{cases} \dfrac{1}{\log\log p} & \text{if } p > 3^9, \\ 0 & \text{if } p \le 3^9, \end{cases}$$

where 3^9 has been chosen since $\log\log 3^9 > 2$; and set $h(p^l) = h(p)$ if $l \ge 2$.

Integration by parts, using the above estimate for $L(x)$, shows that the series

$$\sum p^{-1}\|h(p)\|^2$$

converges, whilst

$$\sum_{p \le x} p^{-1}\|h(p)\| = \log\log\log x + \text{constant} + o(1) \qquad (x \to \infty).$$

Neither of the conditions of theorem (8.8) are satisfied. The function

$$h(n) = \sum_{\substack{p|n \\ p > 3^9}} \frac{1}{\log\log p}$$

therefore has no limiting distribution (mod 1), so that Erdös' conjecture was, indeed, false.—However, it was nearly true, as the following result shows!

Theorem 8.9. *Whatever the additive function $f(n)$, there is always a real function $\alpha(x)$ so that the function $f(n) - \alpha(x)$ has a limiting distribution* (mod 1). *That is to say, the frequencies*

$$\nu_x(n; f(n) - \alpha(x) \le z (\text{mod } 1)) \qquad (x \to \infty),$$

converge weakly to a limiting distribution (mod 1).

In each case there is a real number c, and a non-negative integer q, so that a suitable function $\alpha(x)$ is given by

$$\alpha(x) = c \log x + \eta(x),$$

where

$$\eta(x) = \sum_{p \le x} p^{-1} \| qf(p) - c \log p \| \operatorname{sign}(\tfrac{1}{2} - \{qf(p) - c \log p\})$$

and satisfies

$$\eta(x^y) - \eta(x) \to 0 \qquad (x \to \infty),$$

for each fixed $y > 0$.

The limit law is continuous if and only if the series

$$\sum_{\|mf(p)\| \neq 0} \frac{1}{p} \tag{3}$$

diverges for every positive integer m.

Proof. In the notation of the proof of theorem (8.2), if the set K is empty then $f(m)$ is uniformly distributed (mod 1), and we may set $c = 0$, $q = 0$, so that $\eta(x) = 0$ identically.

Otherwise, let K be non-empty, and generated by the integer q, and define c by $2\pi qc = \tau_q$. Then the series

$$\sum p^{-1}(1 - \operatorname{Re} e^{2\pi imq(f(p) - c \log p)})$$

converges for each integer m.

Define the additive function

$$h(n) = f(n) - c \log n.$$

Then, according to theorem (6.2),

$$x^{-1} \sum_{n \le x} e^{2\pi imqh(n)} = \prod_{p \le x} \left(1 - \frac{1}{p}\right)\left(1 + \sum_{r=1}^{\infty} p^{-r} e^{2\pi imqh(p^r)}\right) + o(1)$$

as $x \to \infty$. It is easy to see that

$$\delta(mq) = \lim_{x \to \infty} x^{-1} \sum_{n \le x} \exp(2\pi imq(h(n) - \eta(x)))$$

exists (cf. step one of the proof of theorem (5.2)).

An integration by parts now shows that if k belongs to K then

$$\sum_{n \le x} e^{2\pi ikf(n)} = \int_{1-}^{x} w^{2\pi ikc} d\left(\sum_{n \le w} e^{2\pi ikh(n)}\right)$$

$$= x^{1 + 2\pi ikc} \sum_{n \le x} e^{2\pi ikh(n)} - \int_{1}^{x} \sum_{n \le w} e^{2\pi ikh(n)} \cdot 2\pi ikc \cdot w^{2\pi ikc - 1}\, dw.$$

We next prove that uniformly over the range $x^{1/2} \leq w \leq x$

$$\eta(w) = \eta(x) + o(1) \qquad (x \to \infty).$$

Indeed, by an application of the Cauchy–Schwarz inequality

$$|\eta(x) - \eta(w)|^2 \leq \sum_{w<p\leq x} \frac{1}{p} \cdot \sum_{w<p\leq x} \frac{1}{p} \|qf(p) - c \log p\|^2.$$

The first of the sums which appear on the right-hand side of this inequality does not exceed

$$\sum_{x^{1/2}<p\leq x} \frac{1}{p} = \log 2 + o(1),$$

whilst the second is $o(1)$ since k belongs to K. Therefore, as $x \to \infty$

$$\int_{x^{1/2}}^{x} \sum_{n\leq w} e^{2\pi ikh(n)} \cdot 2\pi ikc \cdot w^{2\pi ikc-1}\, dw$$
$$= \delta(k)e^{2\pi ik\eta(x)} \int_{x^{1/2}}^{x} 2\pi ikc \cdot w^{2\pi ikc}\, dw + o(1),$$

so that

$$(10) \quad x^{-1} \sum_{n\leq x} \exp(2\pi ik(f(n) - \eta(x) - c \log x)) \to \delta(k)(1 + 2\pi ikc)^{-1}.$$

Moreover, if k is an integer which does not belong to K, then for each real number τ the series

$$\sum p^{-1}(1 - \operatorname{Re} e^{2\pi ikf(p)}p^{-i\tau})$$

diverges. Hence, by part (iv) of theorem (6.3), the sum which appears in (10) approaches zero as $x \to \infty$.

It follows from lemma (1.46) that the frequencies

$$\nu_x(n; f(n) - c \log x - \eta(x) \leq z (\text{mod } 1))$$

possess a limiting distribution (mod 1) as $x \to \infty$; and that the characteristic function β_k of this distribution is given by

$$\beta_k = \begin{cases} \delta(k)(1 + 2\pi ikc)^{-1} & \text{if } k \text{ belongs to } K, \\ 0 & \text{otherwise.} \end{cases}$$

As in the proof of theorem (8.2), an inequality of the form

$$|\beta_k| \leq \text{constant} \,.\, \exp\left(\sum_{p\geq 3}(1 - \text{Re } e^{2\pi i k f(p)}\right),$$

which is uniform in all integers k, still obtains. As before, we may show that the condition stated in theorem (8.9) is indeed sufficient for the limiting distribution to be continuous.

Concerning its necessity, let numbers $l(r)$ be chosen for $r = 1, 2, \ldots$, so that $l(r) \equiv \eta(r) + c \log r \pmod 1$, and $0 \leq l(r) < 1$. Considered in the usual topology on the real line, the numbers $l(r)$ will possess at least one convergent subsequence. Let $l(r_j) \to l$ as $j \to \infty$, say. Proceeding as in the corresponding part of theorem (8.2) we see that if for some positive integer m the series

$$\sum_{\|mf(p)\|=0} \frac{1}{p}$$

converges, then

$$\sum_{w=0}^{m-1} \limsup_{j\to\infty} \nu_{r_j}\left(n; f(n) \equiv \frac{w-l}{m} \pmod 1\right) \geq \delta > 0.$$

Thus the limit law (mod 1) of $f(n)$ will have a discontinuity at at least one of the points $(w - l)/m \pmod 1$, $(w = 0, \ldots, m - 1)$.

This shows that the divergence of the series (3) is also necessary for the continuity of the limit law.

The proof of theorem (8.9) is now complete.

Concluding Remarks

The existence of a limiting distribution (mod 1) for the (not-necessarily additive) function $f(n) - c \log n - \eta(n)$ was indicated by Delange, in his aforementioned paper [8]. It turns out that the conditions given by Manstavičius [1] which are necessary and sufficient in order that the function $f(n) - \alpha(x)$ have a limiting distribution (mod 1), for a suitably chosen translation $\alpha(x)$, are in fact always satisfied, although this is not immediately apparent from their form.

In practice the integer q in theorem (8.9) may sometimes be determined by finding the least positive integer k for which the Fourier–Stieltjes coefficient

$$\beta_k = \int_0^1 e^{2\pi i k z}\, dF(z)$$

of the limit law does not vanish. Then $q = k$ will hold. Theorem (8.9) does not predict an algorithm for the determination of q and/or c.

Concerning our example $h(n)$, the convergence of the series

$$\sum p^{-1}\|h(p)\|^2$$

shows that $\beta_1 \neq 0$, so that $q = 1$. Moreover, we have $c = 0$, and, as $x \to \infty$,

$$\alpha(x) = \log\log\log x + A + o(1),$$

for a certain constant A. Thus, the frequencies

$$\nu_x\left(n;\ \sum_{\substack{p|n\\ p>3^9}} \frac{1}{\log\log p} - \log\log\log x \le z (\mathrm{mod}\ 1)\right) \qquad (x \to \infty),$$

possess a continuous non-uniform limiting distribution (mod 1).

The satisfactory form of theorem (8.9) is due to the simple topological state of affairs which underlies the study of distributions (mod 1). Let $f(n)$ be an additive function, and consider the characteristic functions (of distributions on the whole real line) given by

$$\phi_n(t) = n^{-1} \sum_{m=1}^{n} e^{itf(m)} \qquad (n = 1, 2, \ldots).$$

It was proved in the paper of Elliott and Ryavec [1] that *either* $\phi_n(t) \to 0$ as $n \to \infty$, almost surely in the real variable t; *or* $f(n)$ is finitely distributed. In the latter case numbers α_n exist so that the function

$$\lim_{n\to\infty} \phi_n(t)e^{-it\alpha_n}$$

is the characteristic function of a distribution function on the whole real line. In the former case the (almost surely defined) function $\lim \phi_n(t)$, $n \to \infty$, is certainly not continuous at the point $t = 0$, so that no such sequence α_n can be found.

Considered (mod 1), the Fourier–coefficients

$$\beta_k(n) = n^{-1} \sum_{m=1}^{n} e^{2\pi i k f(m)}$$

have a similar behaviour to the extent that *either* numbers δ_n exist so that

$$\lim_{n\to\infty} \beta_k(n)e^{-2\pi i k \delta_n}$$

is the characteristic function of a distribution function (mod 1), *or*

$$\lim_{n\to\infty} \beta_k(n) = 0.$$

In this case, however, the character group of the additive group of real numbers (mod 1) is discrete, so that any requirement of continuity of a characteristic function β_k, $(k = 0, \pm 1, \pm 2, \ldots)$, is trivially met. In particular the case $\beta_0 = 1$, $\beta_k = 0$ when $k \neq 0$, corresponds to a genuine law, the uniform law (mod 1). Either of the two alternatives thus leads to a satisfactory conclusion. An analogue of theorem (8.9) is therefore to be expected whenever the character group is discrete, or a similar situation prevails. For example, one might study the limiting distribution of additive functions (mod D), for a fixed integer modulus $D > 1$; or the distribution of $f(n)$ (mod λ), for a fixed real number $\lambda > 0$.

It does not seem possible to study the distribution of multiplicative functions (mod 1) by means of a straightforward analogue of Zolotarev's use of the Mellin transform. For example, the (multiplicative) functions $|z|^{2\pi ik}$ are not periodic in z (considered additively). One may apply the Stone–Weierstrass theorem to show that the functions $e^{2\pi ikz}$ can be uniformly approximated by sums of the form

$$\sum_{k=-N}^{N} c_k z^{2\pi ik}$$

on intervals of the form $[\alpha, \beta]$ where $0 < \alpha < \beta < \infty$, or $-\infty < \alpha < \beta < 0$. In this manner satisfactory analogues of the theorems in Chapter 8 may be obtained for that class of multiplicative functions $g(n)$ which satisfy the auxiliary condition

$$\lim_{A\to\infty} \limsup_{n\to\infty} v_n(m; |g(m)| > A) = 0.$$

This concludes Chapter eight.

Chapter 9

Mean Values of Multiplicative Functions, Halász' Method

In this and the following chapter we shall continue with the study of multiplicative arithmetic functions. We shall weaken the assumption on the size of $f(n)$ to requiring only that in some average sense $f(p)$ or $f(n)$ is bounded. This will allow the consideration of a class of functions much wider than that of Chapter 6.

The main aim in the present chapter will be to establish the following theorem (9.1) of Halász [1]. We follow the method of his paper. Apart from the intrinsic interest of this theorem, on the way we establish two inequalities, lemma (9.4) and lemma (9.5), of which we shall make considerable use in certain of the later chapters.

As in Chapter 6, $g(n)$ will denote a complex-valued multiplicative arithmetic function. We define

$$G(s) = \sum_{n=1}^{\infty} g(n)n^{-s} \qquad s = \sigma + i\tau.$$

Each of the Dirichlet series to appear in this chapter will be convergent in the half-plane $\sigma > 1$.

Theorem 9.1 (Halász). *Let $g(n)$ be a multiplicative arithmetic function. Let δ and c_1 be positive constants so that the inequality*

(i) $$\sum_{p} \frac{|g(p)|^{1+\delta} \log p}{p^{\sigma}} \leq \frac{c_1}{\sigma - 1}$$

holds uniformly for $1 < \sigma \leq 2$. Suppose further that

(ii) $$\sum_{p, m \geq 2} \sum |g(p^m)| p^{-m} < \infty,$$

(iii) *that none of the series*

$$1 + \sum_{m=1}^{\infty} g(p^m)p^{-ms}$$

has the sum zero on the line $\sigma = 1$,

(iv) *and that there is a constant* A *so that as* $\sigma \to 1+$

$$G(s) = \frac{A}{s-1} + o\left(\frac{|s|}{\sigma - 1}\right),$$

uniformly for $\operatorname{Re}(s) = \sigma$.
Then, as $x \to \infty$,

$$x^{-1} \sum_{n \le x} g(n) \to A.$$

Remarks. Assuming the validity of the conclusion of the theorem, integration by parts shows that the condition (iv) is necessary. In the next chapter we shall consider a modified situation in which we can investigate the necessity of conditions of the type (i), (ii) and (iii).

We shall, without loss of generality, assume that $0 < \delta \le 1$.

This ends our remarks.

We begin the proof of theorem (9.1) with a useful estimate for the mean-value of $|g(n)|$, which we shall deduce from assumptions (i) and (ii) only. This will give us a feeling for the (average) size of $g(n)$. We obtain such an estimate in two steps.

Lemma 9.2. *Let* $g(n)$ *satisfy the hypothesis* (i) *of theorem* (9.1). *Then, for a certain constant* c_2 *which depends upon* c_1,

$$\sum_p \frac{|g(p)|}{p^\sigma} \le c_2 \log\left(\frac{e}{\sigma - 1}\right) \qquad (1 < \sigma \le 2).$$

Proof. Define the function

$$h(\sigma) = \sum_p p^{-\sigma} |g(p)|^{1+\delta} \qquad (1 < \sigma \le 2).$$

Then by hypothesis (i)

$$0 \le -h'(\sigma) = \sigma \sum_p p^{-\sigma} |g(p)|^{1+\delta} \log p \le 2c_1(\sigma - 1)^{-1},$$

so that

$$h(\sigma) - h(2) = -\int_\sigma^2 h'(w)dw \le 2c_1 \int_\sigma^2 \frac{dw}{w-1} = 2c_1 \log\left(\frac{1}{\sigma - 1}\right).$$

Moreover,

$$h(2) \leq \frac{1}{\log 2} \sum p^{-2} |g(p)|^{1+\delta} \log p \leq \frac{c_1}{\log 2}.$$

Thus, on the one hand

$$\sum_{|g(p)| \leq 1} |g(p)| p^{-\sigma} \leq \sum p^{-\sigma} \leq c_3 \log\left(\frac{1}{\sigma - 1}\right),$$

and on the other hand

$$\sum_{|g(p)| > 1} |g(p)| p^{-\sigma} \leq \sum |g(p)|^{1+\delta} p^{-\sigma} \leq 2c_1 \log\left(\frac{1}{\sigma - 1}\right) + h(2).$$

We set

$$c_2 = c_1\left(2 + \frac{1}{\log 2}\right) + c_3,$$

and lemma (9.2) is proved.

Lemma 9.3. *Let $g(n)$ satisfy the hypotheses* (i) *and* (ii) *of theorem* (9.1). *Then, for a certain constant c_4,*

$$\sum_{n=1}^{\infty} |g(n)| n^{-\sigma} \leq c_4(\sigma - 1)^{-c_2} \qquad (1 < \sigma \leq 2),$$

and

$$\sum_{n \leq x} |g(n)| \leq c_4 x (\log x)^{c_2} \qquad (x \geq 2).$$

Proof. For each $\sigma > 1$

$$\begin{aligned}\sum_{n=1}^{\infty} |g(n)| n^{-\sigma} &= \prod_p (1 + |g(p)| p^{-\sigma} + \cdots) \\ &\leq \exp\left(\sum_p |g(p)| p^{-\sigma} + \sum_{p,\, m \geq 2}\sum |g(p^m)| p^{-m}\right).\end{aligned}$$

By lemma (9.2) and hypothesis (ii) of theorem (9.1) this expression is

$$O\left(\exp\left(c_2 \log \frac{1}{\sigma - 1}\right)\right),$$

which justifies the first assertion.

As the the second,

$$\sum_{n \le x} |g(n)| \le \sum_{n=1}^{\infty} |g(n)| \left(\frac{x}{n}\right)^{\sigma} = O\left(x^{\sigma} \frac{1}{(\sigma - 1)^{c_2}}\right),$$

and we set $\sigma = 1 + (\log x)^{-1}$, which is permissible if $x \ge 2$.

This proves lemma (9.3).

Remark. Apart from an improvement in the value of the constant c_2 in this lemma, little more may be expected on the strength of only the hypotheses (i) and (ii) of theorem (9.1). For example, one might set $g(p^m) = m + 1$, for $p \ge 2$, $m \ge 1$. This clearly satisfies the hypotheses (i) and (ii), whilst it is a classical result of Dirichlet that

$$\sum_{n \le x} g(n) = \sum_{n \le x} d(n) \sim x \log\ x \qquad (x \to \infty).$$

This ends our remark.

To continue, let $\Lambda(n)$ denote von Mangoldt's function

$$\Lambda(n) = \begin{cases} \log n & \text{if } n = p^m, \\ 0 & \text{otherwise.} \end{cases}$$

Lemma 9.4. *Let $b_1, b_2, \ldots$ be a sequence of complex numbers, and define the function*

$$B(s) = \sum_{n=1}^{\infty} b_n \Lambda(n) n^{-s}.$$

Assume that for a positive constant c_1

$$\text{(1)} \qquad \sum_{n=1}^{\infty} |b_n|^{1+\delta} \Lambda(n) n^{-\sigma} \le \frac{c_1}{\sigma - 1} \qquad (1 < \sigma \le 2).$$

Let D be a positive number, which we shall regard as fixed in what follows, and set

$$T = \frac{1}{(\sigma - 1)^D}.$$

Then there is a positive constant α_0, *so that for each real number* $\alpha > \alpha_0$, *there is a further real number* c_2, *depending upon* c_1, α, D *and* δ, *so that the inequality*

$$\int_{\sigma - iT}^{\sigma + iT} |B(s)|^\alpha d\tau \le \frac{c_2}{(\sigma - 1)^{\alpha - 1}}$$

holds uniformly for $1 < \sigma \le 2$.

Remarks. Important aspects of this result are the value of the exponent $-\alpha + 1$, and the fact that the constant c_2 does not depend upon the value of σ, so that the inequality is uniform for $1 < \sigma \le 2$. This is the case even if the numbers b_n are defined in terms of σ. We have, of course, renumbered the constants c_1, c_2.

This ends our remarks.

Proof. We shall first give a self-contained proof of lemma (9.4) on the lines of Halász' proof of the analogous result in his paper [1]. Then, in the remarks following this proof, we sketch how this fits in, mathematically and historically, with the duality principle which was considered in Chapter 4.

Let σ be a real number in the range $1 < \sigma \le 2$, temporarily regarded as fixed. Let w be a further positive real number. We shall begin by obtaining an upper bound for the measure of the set Ω of τ-values which lie in the interval $-T \le \tau \le T$, and for which

$$|B(\sigma + i\tau)| \ge \frac{w}{\sigma - 1}.$$

Let η be a positive real number, and suppose that we can find N points $\tau_1, \ldots, \tau_N$, all of which lie in the set Ω, and which satisfy $|\tau_k - \tau_l| \ge \eta(\sigma - 1)$ when $k \ne l$. We shall obtain an upper bound for N in terms of w.

From our (temporary) hypothesis

$$\left| \sum_{n=1}^{\infty} \frac{b_n \Lambda(n)}{n^{\sigma + i\tau_k}} \right| \ge \frac{w}{\sigma - 1},$$

so that for suitably chosen complex numbers ρ_k, $(k = 1, \ldots, N)$, $|\rho_k| = 1$, we have

$$\sum_{k=1}^{N} \rho_k \sum_{n=1}^{\infty} \frac{b_n \Lambda(n)}{n^{\sigma + i\tau_k}} \ge \frac{Nw}{\sigma - 1}.$$

Therefore, inverting the order of summation,

$$\sum_{n=1}^{\infty} \frac{|b_n|\Lambda(n)}{n^\sigma}\left|\sum_{k=1}^{N} \rho_k n^{-i\tau_k}\right| \geq \frac{Nw}{\sigma - 1}.$$

To estimate the left-hand side from above, we apply Hölder's inequality with the exponent pair $p = 1 + \delta$ and $q = (1 + \delta)/\delta$, so that

$$(2) \qquad \sum_{n=1} \frac{|b_n|\Lambda(n)^{1/p}}{n^{\sigma/p}} \cdot \frac{\Lambda(n)^{1/q}}{n^{\sigma/q}}\left|\sum_{k=1}^{N} \rho_k n^{-i\tau_k}\right|$$

$$\leq \sum_{n=1}^{\infty}\left(\frac{|b_n|^{1+\delta}\Lambda(n)}{n^\sigma}\right)^{1/p} \cdot \left(\sum_{n=1}^{\infty} \frac{\Lambda(n)}{n^\sigma}\left|\sum_{k=1}^{N} \rho_k n^{-i\tau_k}\right|^q\right)^{1/q}.$$

The first of these two last sums is by hypothesis at most $c_1(\sigma - 1)^{-1}$. As for the second (final) sum, involving the exponent q, we note that $q \geq 2$, and trivially

$$\left|\sum_{k=1}^{N} \rho_k n^{-i\tau_k}\right| \leq N,$$

so that

$$\left|\sum_{k=1}^{N} \rho_k n^{-i\tau_k}\right|^q \leq N^{q-2}\left|\sum_{k=1}^{N} \rho_k n^{-i\tau_k}\right|^2$$

$$= N^{q-2}\sum_{k=1}^{N}\sum_{l=1}^{N} \rho_k \bar{\rho}_l n^{-i(\tau_k - \tau_l)}.$$

The final sum in (2) therefore does not exceed

$$N^{q-2}\sum_{k=1}^{N}\sum_{l=1}^{N} \rho_k \bar{\rho}_l \sum_{n=1}^{\infty} \Lambda(n) n^{-\sigma - i(\tau_k - \tau_l)} = -N^{q-2}\sum_{k=1}^{N}\sum_{l=1}^{N} \rho_k \bar{\rho}_l \frac{\zeta'}{\zeta}(\sigma + i(\tau_k - \tau_l)).$$

Consider first those terms for which $k = l$. Since each ρ_k has absolute value 1 these terms contribute at most

$$-N^{q-2}N\frac{\zeta'}{\zeta}(\sigma) \leq c_3 N^{q-1}(\sigma - 1)^{-1}.$$

If, however, $k \neq l$, then $\eta(\sigma - 1) \leq |\tau_k - \tau_l| \leq 2T$, and according to lemma (2.14),

$$-\frac{\zeta'}{\zeta}(\sigma + i\tau) \leq \frac{c_4}{|s - 1|} + c_4 \log(2 + |\tau|),$$

so that the terms corresponding to such pairs (k, l) contribute not more than

$$N^{q-2} \cdot N^2 c_5 \left(\frac{1}{\eta(\sigma - 1)} + \log\left(\frac{e}{\sigma - 1} \right) \right) \leq \frac{c_6 N^q}{\eta(\sigma - 1)},$$

provided that $\eta(\sigma - 1)^{1/2} \leq 1$, say. Assuming for the moment that this is indeed the case, we have established that

$$\frac{Nw}{\sigma - 1} \leq \left(\frac{c_1}{\sigma - 1} \right)^{1/p} \cdot \left(\frac{c_3 N^{q-1}}{\sigma - 1} + \frac{c_6 N^q}{\eta(\sigma - 1)} \right)^{1/q}.$$

Hence for a certain constant c_7

$$(wN)^q \leq c_7(N^{q-1} + \eta^{-1} N^q).$$

This constant c_7 depends upon the constants c_1, δ and D only.

We choose η so that $2c_7 = \eta w^q$, and deduce that

$$N \leq 2c_7 w^{-q},$$

this choice of η being permissible if $w \geq c_8(\sigma - 1)^{1/(2q)}$ for a suitably chosen constant c_8; in fact $c_8 = (2c_7)^{1/q}$ will do.

We now construct such a set of values τ_k. Since $B(s)$ is a continuous function of τ, the set Ω (of real numbers) is clearly closed in the usual topology of the real numbers. Let τ_1 denote the least member of Ω, so that $\tau_1 \geq -T$. Then define $\tau_2, \tau_3, \ldots,$ inductively. Assuming $\tau_1, \tau_2, \ldots, \tau_{k-1}$, $k \geq 2$, to be already defined we choose τ_k to be the least member of Ω for which the inequality $\tau_k - \tau_{k-1} \geq \eta(\sigma - 1)$ is satisfied.

The measure of that part of the set Ω which lies in a typical interval $[\tau_{k-1}, \tau_k]$ is at most $\eta(\sigma - 1)$. Since the whole of Ω lies in the union of N such intervals, the total measure $|\Omega|$ of that set is bounded above by

$$|\Omega| \leq N\eta(\sigma - 1) \leq 4c_7^2 w^{-2q}(\sigma - 1),$$

provided only that w is not too small, as indicated earlier. This will suffice for our purposes.

From the hypotheses of lemma (9.4)

$$\begin{aligned} |B(s)| &\leq \sum_{\substack{n=1 \\ |b_n| \leq 1}}^{\infty} |b_n| \Lambda(n) n^{-\sigma} + \sum_{\substack{n=1 \\ |b_n| > 1}}^{\infty} |b_n| \Lambda(n) n^{-\sigma} \\ &\leq \sum_{n=1}^{\infty} \Lambda(n) n^{-\sigma} + \sum_{n=1}^{\infty} |b_n|^{1+\delta} \Lambda(n) n^{-\sigma} \leq \frac{c}{\sigma - 1} \end{aligned}$$

say, uniformly for $1 < \sigma \leq 2$.

For each positive integer m let Ω_m denote that set of τ-values, $|\tau| \leq T$, for which the inequalities

$$2^{-m}c < (\sigma - 1)|B(\sigma + i\tau)| \leq 2^{-m+1}c$$

hold. Let $\Omega_0 = \varnothing$, the empty set.

According to our previous argument, (replacing Ω by Ω_m), if $2^{-m}c > c_8(\sigma - 1)^{1/(2q)}$ then

$$|\Omega_m - \Omega_{m-1}| \leq |\Omega_m| \leq 4c_7^2\left(\frac{2^m}{c}\right)^{2q}(\sigma - 1).$$

Define the integer r by

$$2^{-r}c > c_8(\sigma - 1)^{1/(2q)} \geq 2^{-r-1}c.$$

Then

$$\sum_{m=1}^{r}\int_{\Omega_m - \Omega_{m-1}} |B(s)|^\alpha d\tau \leq \sum_{m=1}^{r}\left(\frac{2c}{2^m(\sigma - 1)}\right)^\alpha 4c_7^2\left(\frac{2^m}{c}\right)^{2q}(\sigma - 1)$$
$$\leq c_9(\sigma - 1)^{-\alpha+1}\sum_{m=1}^{\infty} 2^{(2q-\alpha)m} \leq c_{10}(\sigma - 1)^{-\alpha+1}$$

provided only that α is ultimately chosen to have a fixed value which satisfies $\alpha > 2q$, so as to ensure the convergence of the geometric series.

On that part Γ of the interval $[-T, T]$ which is not covered by the union of the Ω_m with $1 \leq m \leq r$, the estimate

$$|B(s)| \leq 2^{-r}c(\sigma - 1)^{-1} \leq 2c_8(\sigma - 1)^{-1+1/(2q)}$$

is valid. Therefore

$$\int_\Gamma |B(s)|^\alpha d\tau \leq \left(\frac{2c_8(\sigma - 1)^{1/(2q)}}{\sigma - 1}\right)^\alpha \cdot \frac{2}{(\sigma - 1)^D} \leq c_{11}(\sigma - 1)^{-\alpha+1}$$

if α is ultimately chosen to satisfy $\alpha/(2q) \geq D + 1$.

We define $\alpha_0 = 2q(2 + D) = (4 + 2D)\delta^{-1}(1 + \delta)$ and $c_2 = c_{10} + c_{11}$.

This completes the proof of lemma (9.4).

Remarks. If we choose for α a (fixed) value $\alpha > \max(\alpha_0, 2)$, then we deduce that the measure of that subset of $[-T, T]$ on which the inequality

$$|B(\sigma + i\tau)| \geq \frac{w}{\sigma - 1}$$

is satisfied, does not exceed $c_2 w^{-2}(\sigma - 1)$. This bound is valid uniformly for all positive numbers w. We shall make use of it in the next lemma.

If necessary an explicit value could be given for the constant c_2.

In order to view the proof of lemma (9.4) from the standpoint of the duality principle of Chapter 4 let us replace the hypothesis of lemma (9.4) by

$$\sum_{n=1}^{\infty} \Lambda(n)|b_n|^2 n^{-\sigma} \le c_{12}(\sigma - 1)^{-1} \qquad (1 < \sigma \le 2).$$

This amounts to the ability to set $\delta = 1$. Then the sum

$$\sum_{k=1}^{N} \left| \sum_{n=1}^{\infty} b_n \Lambda(n) n^{-\sigma - i\tau_k} \right|^2$$

can be viewed as an example of a sum of the form

$$\sum_{k=1}^{N} \left| \sum_{n=1}^{\infty} a_n \Lambda(n)^{1/2} n^{-(\sigma/2) - i\tau_k} \right|^2.$$

Here we understand $0^{1/2} = 0$. The choice $a_n = \Lambda(n)^{1/2} b_n n^{-\sigma/2}$ reconciles these two expressions.

We consider the (dual) inequality

$$\sum_{n=1}^{\infty} \left| \sum_{k=1}^{N} \Lambda(n)_n^{1/2 - (\sigma/2) - i\tau_k} c_k \right|^2 \le \lambda \sum_{k=1}^{N} |c_k|^2,$$

and look for a value of λ so that this inequality will be valid for all complex numbers c_k, $(k = 1, \ldots, N)$. Expanding the expression to be majorised we are led to the consideration of the bilinear form

$$\sum_{k=1}^{N} \sum_{l=1}^{N} c_k \bar{c}_l \sum_{n=1}^{\infty} \Lambda(n)_n^{1/2} n^{-(\sigma/2) - i\tau} \left. c_k \right|^2 \le \lambda \sum_{k=1}^{N} c_k|^2,$$

According to the remarks made in Chapter 4, concerning Gershgorin discs, we may choose for λ the value

$$-\frac{\zeta'}{\zeta}(\sigma) + \max_k \sum_{\substack{l=1 \\ l \ne k}}^{N} \left| -\frac{\zeta'}{\zeta}(\sigma + i(\tau_k - \tau_l)) \right|$$

$$\le c_{13}\left(\frac{1}{\sigma - 1} + \frac{N}{\eta(\sigma - 1)}\right).$$

By the duality principle the inequality

$$\sum_{k=1}^{N}\left|\sum_{n=1}^{\infty} a_n \Lambda(n)^{1/2} n^{-(\sigma/2)-i\tau_k}\right|^2 \le \frac{c_{13}}{\sigma-1}\left(1+\frac{N}{\eta}\right)\sum_{n=1}^{\infty}|a_n|^2$$

is valid for all sequences $(a_1, a_2, \ldots)$ for which the right-hand series is convergent. We set $a_n = b_n \Lambda(n)^{1/2} n^{-\sigma/2}$ and apply our hypothesis (that $\delta = 1$) to obtain

$$\sum_{k=1}^{N}\left|\sum_{n=1}^{\infty} b_n \Lambda(n) n^{-\sigma-i\tau_k}\right|^2 \le \frac{c_{13}}{\sigma-1}\left(1+\frac{N}{\eta}\right)\sum_{n=1}^{\infty}|b_n|^2\Lambda(n)n^{-\sigma}$$

$$\le \frac{c_{14}}{(\sigma-1)^2}\left(1+\frac{N}{\eta}\right).$$

From the Cauchy–Schwarz inequality

$$\left(\frac{Nw}{\sigma-1}\right)^2 \le \left(\sum_{k=1}^{N}|B(\sigma+i\tau_k)|\right)^2 \le N\sum_{k=1}^{N}|B(\sigma+i\tau_k)|^2 \le \frac{c_{14}N}{(\sigma-1)^2}\left(1+\frac{N}{\eta}\right)$$

so that if η is chosen by $2\eta w^2 = c_{14}$ then

$$N^2w^2 \le c_{14}(N+\eta^{-1}N^2) \qquad N \le 2c_{14}w^{-2},$$

this inequality being valid for $w \ge c_{15}(\sigma-1)^{1/4}$, for a certain positive constant c_{15}. If $\delta = 1$ then $q = \delta^{-1}(1+\delta) = 2$, so that the direct argument given earlier yields $N \le 2c_7 w^{-2}$. The power of w^{-1} is not of importance except in its effect upon the value of α_0, and this is of no consequence in the applications which we have in mind.

As was pointed out in the concluding remarks of Chapter 4, the large sieve was first understood and applied with regard to the (discrete) characters of the finite groups of residue classes (mod D), for integers $D > 1$. We mention, in particular, the work of Linnik's pupil Rényi [2]. The argument used in the proof of lemma (9.4) was introduced by Halász in his paper on multiplicative functions [1]. It gave a new impetus to the study of Dirichlet series, especially of their zeros, since it applied to the (continuous) characters n^{-it}, t real. Subsequently the methods of Halász and the (Linnik-) large sieve were combined by Montgomery [2]. All of these ideas have now been subsumed under the notion of the large sieve as an exercise in duality, and it is standard to consider at the outset mixed characters of the form $\chi(n)n^{-it}$, where $\chi(n)$ is a Dirichlet character. See, for example, Bombieri; Forti and Viola [1].

Lemma 9.5. *Let σ^* be a positive real number, $1 < \sigma^* \le 2$.*

Let $E(x)$ be analytic in a domain which contains the strip $1 < \sigma \le \sigma^*$, and satisfy there the inequalities

$$|E(s)| \le \frac{c_3|s|}{\sigma - 1} \qquad |E(s)| \le \frac{c_3}{(\sigma - 1)^B},$$

where c_3 and B are positive constants.

Assume further that in this strip there is a representation

$$\frac{E'(s)}{E(s)} = \sum_{n=1}^{\infty} b_n \Lambda(n) n^{-s}$$

where the coefficients b_n satisfy the condition (1) of lemma (9.4).

Let $\beta(>1)$ and D be (fixed) numbers, and set $T = (\sigma - 1)^{-D}$.

Then there is a constant c_4 so that

$$\int_{\sigma - iT}^{\sigma + iT} \left|\frac{E(s)}{s}\right|^{\beta} d\tau \le \frac{c_4}{(\sigma - 1)^{\beta - 1}} \tag{3}$$

holds uniformly for $1 < \sigma \le \sigma^*$.

Remarks. We retain the meaning of the constants c_1 and c_2 in lemma (9.4), but re-label otherwise.

In the present lemma the b_n are assumed *not* to vary with σ over the range $1 < \sigma \le \sigma^*$; we shall need to make use of the uniform nature of the hypotheses. The constant c_4 in the statement of lemma (9.5) then depends upon the constants c_1, c_2 and δ of lemma (9.4), and c_3, β and D.

It is understood (implicitly) that $E(s)$ does not vanish in the strip $1 < \sigma \le \sigma^*$.

It will be important in the applications of lemma (9.5) that β may be chosen arbitrarily close to the value $\beta = 1$. The information which this lemma gives is more valuable the nearer β is to this value. For $\beta = 1$ the inequality can be false, as may be seen by considering the case $E(s) = \zeta(s)$. By means of the estimate $2|s - 1||\zeta(s)| \ge 1$, which is valid in a neighbourhood of the point $s = 1$, it is straightforward to prove that

$$\int_{-1}^{1} \left|\frac{\zeta(s)}{s}\right| d\tau \ge \frac{1}{3}\int_{-1}^{1} |\zeta(s)|\, d\tau > c_5 \log\left(\frac{e}{\sigma - 1}\right) \qquad (1 < \sigma \le 2).$$

This concludes our remarks.

Proof. We must be careful in computing the dependence of various functions upon σ. It is convenient to single out a value σ_0, $1 < \sigma_0 \le \sigma^*$, and to define the function

$$U(\sigma) = \int_{|\tau| \le (\sigma_0 - 1)^{-D}} \left| \frac{E(\sigma + i\tau)}{\sigma + i\tau} \right|^{\beta} d\tau.$$

Thus $U(\sigma)$ is a continuous and, as presently show, differentiable function of σ. We shall regard σ_0 as (temporarily) fixed.

We shall estimate the size of $U'(\sigma)$, where $'$ denotes differentiation with respect to the variable σ. We make use of the easily proved fact that for any function $g(\sigma)$, ($\mathbb{R} \to \mathbb{C}$), the inequality $||g(\sigma)|'| \le |g'(\sigma)|$ is satisfied whenever the derivatives exist.

In the present situation we can integrate the expression $E'(s)/E(s)$ along the line from the point $\sigma^* + i\tau$ to $\sigma + i\tau$, to deduce the representation

$$|E(s)| = |E(\sigma^* + i\tau)| \exp\left(\operatorname{Re} \int_{\sigma^*}^{\sigma} \frac{E'}{E}(w + i\tau) dw \right).$$

It is clear from this that $|E(\sigma + i\tau)|$ is a differentiable function of σ. Moreover

$$\begin{aligned} \left| \frac{\partial}{\partial \sigma} \left| \frac{E(s)}{s} \right|^{\beta} \right| &\le \left| \frac{\partial}{\partial \sigma} \left(\frac{E(s)}{s} \right)^{\beta} \right| \\ &= \left| \beta \left(\frac{E(s)}{s} \right)^{\beta - 1} \left\{ \frac{E'(s)}{s} - \frac{E(s)}{s^2} \right\} \right| \\ &\le \beta |E(s)|^{\beta - 1} |E'(s)| |s|^{-\beta} + \beta |E(s) s^{-1}|^{\beta}. \end{aligned}$$

Integrating over the interval $[-T, T]$ yields

$$|U'(\sigma)| \le \beta \int_{-T}^{T} \left| \frac{E'(s)}{E(s)} \right| \cdot \left| \frac{E(s)}{s} \right|^{\beta} d\tau + \beta U(\sigma).$$

To estimate the size of the integral which appears here we shall make use of lemma (9.4).

We shall need an estimate for the size of $E'(s)$ in the strip $1 < \sigma \le \sigma^*$. In fact

$$\sum_{|b_n| \ge 1} |b_n| \Lambda(n) n^{-\sigma} \le \sum_{n=1}^{\infty} |b_n|^{1+\delta} \Lambda(n) n^{-\sigma} \le \frac{c_1}{\sigma - 1}$$

and

$$\sum_{|b_n| < 1} |b_n| \Lambda(n) n^{-\sigma} \le \sum_{n=1}^{\infty} \Lambda(n) n^{-\sigma} = -\frac{\zeta'}{\zeta}(\sigma) = O((\sigma - 1)^{-1}),$$

so that

$$|E'(s)| = \left|\frac{E'(s)}{E(s)} \cdot E(s)\right| \leq \sum_{n=1}^{\infty} |b_n|\Lambda(n)n^{-\sigma} \cdot |E(s)| = O\left(\frac{|s|}{(\sigma - 1)^2}\right).$$

It is convenient to introduce a further real number σ_1 in the range $\sigma_0 \leq \sigma_1 \leq \sigma^*$. *We assume for the moment* that there is a constant $C > 0$ so that $(\sigma_0 - 1)^{-D} = T_0 \leq (\sigma - 1)^{-C}$ holds for every value of σ in the interval $\sigma_0 \leq \sigma \leq \sigma_1$.

Let $I(\sigma)$ denote the interval $[-(\sigma - 1)^{-C}, (\sigma - 1)^{-C}]$, and let Ω denote its subset of τ-values on which the inequality

$$\left|\frac{E'}{E}(\sigma + i\tau)\right| > \frac{\beta - 1}{3\beta(\sigma - 1)}$$

is satisfied. Then according to lemma (9.4) with $w = (\beta - 1)/3\beta$, there is a constant c_6 so that $|\Omega| \leq c_6(\sigma - 1)$. The value of this constant may depend upon C, β, c_1 and δ, but not upon σ, σ_0 or σ_1. Using this estimate together with the bound for $E'(s)$ we obtain

$$\int_{\tau \in \Omega} |E'(s)||E(s)|^{\beta - 1}|s|^{-\beta} d\tau \leq c_7 \int_{\Omega} \frac{|s|}{(\sigma - 1)^2}\left(\frac{|s|}{\sigma - 1}\right)^{\beta - 1} |s|^{-\beta} d\tau$$

$$\leq \frac{c_7}{(\sigma - 1)^{\beta + 1}} |\Omega| \leq \frac{c_8}{(\sigma - 1)^{\beta}}.$$

However,

$$\int_{[-T_0, T_0] - \Omega} \left|\frac{E'(s)}{E(s)}\right| \cdot \left|\frac{E(s)}{s}\right|^{\beta} d\tau \leq \frac{\beta - 1}{3\beta(\sigma - 1)} \int_{-T_0}^{T_0} \left|\frac{E(s)}{s}\right|^{\beta} d\tau$$

$$= \frac{\beta - 1}{3\beta(\sigma - 1)} U(\sigma).$$

These results together show that

$$|U'(\sigma)| \leq \beta \cdot \frac{\beta - 1}{3\beta(\sigma - 1)} U(\sigma) + \frac{\beta c_8}{(\sigma - 1)^{\beta}} + \beta U(\sigma),$$

uniformly for $\sigma_0 \leq \sigma \leq \sigma_1$.

Assume, further, that $\sigma_1 \leq 1 + (\beta - 1)/(6\beta)$.

Set $\theta = (\beta - 1)/2$. Then since $-U'(\sigma) \leq |U'(\sigma)|$ we have

$$-U'(\sigma) \leq \frac{\theta}{\sigma - 1} U(\sigma) + \frac{c_9}{(\sigma - 1)^{\beta}}.$$

Multiplying by $(\sigma - 1)^\theta$, and collecting together the terms involving $U'(\sigma)$ and $U(\sigma)$, we can express this inequality in the form

$$-[U(\sigma)(\sigma - 1)\theta]' \leq \frac{C_9}{(\sigma - 1)^{\beta - \theta}}.$$

Integrating over the range $\sigma_0 \leq \sigma \leq \sigma_1$ we obtain

$$-U(\sigma_1)(\sigma_1 - 1)^\theta + U(\sigma_0)(\sigma_0 - 1)^\theta \leq c_9 \int_{\sigma_0}^{\infty} \frac{d\sigma}{(\sigma - 1)^{\beta - \theta}}$$

$$= \frac{c_9}{\beta - \theta - 1} \cdot \frac{1}{(\sigma_0 - 1)^{\beta - \theta - 1}},$$

this last step being permissible since $\beta - \theta = (\beta + 1)/2 > 1$.

We have therefore proved that

$$U(\sigma_0) \leq \frac{c_{10}}{(\sigma_0 - 1)^{\beta - 1}} + U(\sigma_1)\left(\frac{\sigma_1 - 1}{\sigma_0 - 1}\right)^\theta.$$

The first of the two majorants given here is of the desired type. The number σ_1 is at our disposal (subject to the early condition involving T_0), and we shall choose it so as to make the second term on the right-hand side also small.

In fact

$$U(\sigma_1) = \int_{-T_0}^{T_0} \left|\frac{E(s)}{s}\right|^\beta d\tau \leq \max_{\sigma = \sigma_1} |E(s)|^\beta \int_{-T}^{T} \frac{d\tau}{|s|^\beta}$$

which by an hypothesis of lemma (9.5) does not exceed

$$c_{11}(\sigma_1 - 1)^{-B\beta}.$$

We shall assume, without loss of generality, that $B \geq 1$. We choose σ_1 so that

$$\frac{1}{(\sigma_1 - 1)^{B\beta}} = \frac{1}{(\sigma_0 - 1)^{\theta/2}}.$$

Note that $\theta/2 < \beta \leq B\beta$, so that $\sigma_1 > \sigma_0$. Moreover, if $y = \theta(2B\beta)^{-1}(<1)$, and

$$\sigma_0 \leq 1 + (\sigma^* - 1)^{1/y},$$

then $\sigma_1 - 1 = (\sigma_0 - 1)^y \leq \sigma^* - 1$, so that $\sigma_1 \leq \sigma^*$, *We shall assume* that this upper bound on σ_0 is satisfied. This is our third (temporary) assumption.

With our choice of σ_1 we have

$$U(\sigma_1)\left(\frac{\sigma_1 - 1}{\sigma_0 - 1}\right)^{\theta} \leq \frac{c_{11}}{(\sigma_0 - 1)^{3\theta/2}} \leq \frac{c_{11}}{(\sigma_0 - 1)^{\beta - 1}},$$

since $3\theta/2 = 3(\beta - 1)/4 < \beta - 1$, and $0 < \sigma_0 - 1 \leq 1$. Hence

$$U(\sigma_0) \leq (c_{10} + c_{11})(\sigma_0 - 1)^{-\beta + 1}. \tag{3}$$

Concerning our three (temporary) assumptions, set $C = 2B\beta\theta D^{-1}$. If $\sigma_0 \leq \sigma \leq \sigma_1$, then owing to the monotone nature of the functions involved

$$(\sigma_0 - 1)^{-D} = (\sigma_1 - 1)^{-C} \leq (\sigma - 1)^{-C}$$

and the first hypothesis is satisfied. The remaining two hypotheses will certainly be valid if σ_0 lies in the range $1 < \sigma_0 \leq c_{12}$ for a suitably chosen number $c_{12} > 1$.

We have thus established an inequality of the desired type (namely (3)) which is valid uniformly in the strip $1 < \sigma \leq c_{12}$. Such an inequality is trivially valid over the range $c_{12} < \sigma \leq \sigma^*$.

This completes the proof of lemma (9.5).

Remark. We can regard the argument used in the above proof as a form of that given for lemma (9.2), but "integrated with respect to τ."

We shall next adapt a formula of Perron, concerning Mellin inversion, that if $\sigma > 0$ then

$$\frac{1}{2\pi i}\int_{\sigma - i\infty}^{\sigma + i\infty} \frac{x^s}{s}\,ds = \begin{cases} 1 & \text{if } x > 1, \\ 0 & \text{if } 0 < x < 1. \end{cases}$$

The traditional quantitative formulation of this result, as embodied in lemma (2.12) for example, assumes sharper estimates for the individual coefficients of the Dirichlet series under study than are available to us under the conditions of theorem (9.1). We therefore argue from first principles.

Lemma 9.6. *Let l be a (fixed) positive real number. Let the function $G(s)$ satisfy the condition* (i) *of theorem* (9.1), *together with the condition*

$$\sum_{p,\, m \geq 2} |g(p^m)|^{1+\delta} p^{-m} < \infty.$$

Then there are further positive numbers D and r, independent of the function g(n), so that the estimate

$$\sum_{n \le x} g(n) \log n = -\frac{1}{2\pi i} \int_{\sigma_0 - iT}^{\sigma_0 + iT} \frac{y^s}{s} G'(s) ds + O(x(\log x)^{-l})$$

with $\sigma_0 = 1 + (\log x)^{-1}$, $T = (\sigma_0 - 1)^{-D}$, *and some value of y in the interval* $x - x(\log x)^{-r} < y \le x$, *holds uniformly for all* $x \ge 3$.

Remark. The numbers D and r depend upon δ, C_1, and the sum of the series involving the $|g(p^m)|^{1+\delta}$ with p, $m \ge 2$, but not otherwise upon the individual values of the function $g(n)$.

Proof. Let $\sigma_0 = 1 + (\log x)^{-1}$, so that $1 < \sigma_0 < 2$ when $x \ge 3$. Suppose, for the moment, that T is a positive real number. We shall presently fix its value in terms of σ_0 and r.

We begin with the representation

$$-\frac{1}{2\pi i} \int_{\sigma_0 - iT}^{\sigma_0 + iT} \frac{y^s}{s} G'(s) ds = \sum_{n=1}^{\infty} g(n) \log n \frac{1}{2\pi i} \int_{\sigma_0 - iT}^{\sigma_0 + iT} \left(\frac{y}{n}\right)^s \frac{ds}{s}.$$

Consider first an integer n for which $n > y$. We deform the path of integration to the union of line-segments $[\sigma_0 - iT, w - iT]$, $[w - iT, w + iT]$, $[w + iT, \sigma_0 + iT]$, and let $w \to \infty$. The corresponding integral on the right-hand side of the above equation is seen, in this manner, not to exceed

$$\frac{1}{\pi} \left(\frac{y}{n}\right)^{\sigma_0} \frac{1}{T|\log y/n|}$$

in absolute value. If $n < y$ we carry out a similar procedure, but with $w < 0$, $w \to -\infty$, and so pass over the simple pole at the point $s = 0$, where there is a residue 1. We define

$$S(y) = \sum_{n \le y} g(n) \log n \qquad y > 0,$$

and have proved that if y is not an integer, then

$$\left| S(y) + \frac{1}{2\pi i} \int_{\sigma_0 - iT}^{\sigma_0 + iT} \frac{y^s}{s} G'(s) ds \right| \le \frac{y^{\sigma_0}}{T} \sum_{n=1}^{\infty} \frac{|g(n)| \log n}{n^{\sigma_0} |\log y/n|}.$$

We denote the upper bound in this inequality by $R(y)$, and consider the sum

$$\sum_{x - x(\log x)^{-r} < y \le x} R(y),$$

where y is restricted to run through numbers of the form $\frac{1}{2}$ (odd integer). We shall obtain an upper bound for this sum, and so prove that for one of the values of y, $R(y)$ is $O(x(\log x)^{-r})$.

Consider the expression

$$\lambda(\theta) = \log(1+\theta) - \frac{\theta}{1+\theta} \qquad \theta > 0.$$

Differentiating with respect to θ yields the positive function $\theta(1+\theta)^{-2}$. Since $\lambda(0) = 0$ we must have $\lambda(\theta) > 0$ for all $\theta > 0$.

If now $1 < n/y \le 2$, then with $\theta = (n/y) - 1$,

$$\left|\log \frac{n}{y}\right| = \log \frac{n}{y} \ge \frac{\dfrac{n}{y} - 1}{1 + \dfrac{n}{y} - 1} \ge \frac{|n-y|}{2y} \ge \frac{|n-y|}{2x}.$$

If, however, $1/2 \le n/y < 1$, then with $\theta = (y/n) - 1$,

$$\left|\log \frac{n}{y}\right| \ge \frac{|y-n|}{2n} \ge \frac{|y-n|}{4y} \ge \frac{|y-n|}{4x}.$$

For values of n, y which satisfy $n/y < 1/2$ or $n/y > 2$ it is immediately clear that

$$\left|\log \frac{n}{y}\right| \ge \log 2.$$

Inverting the order of summation, we have

$$\sum_{x(\log x)^{-r} < y \le x} R(y) \le \frac{x^{\sigma_0}}{T} \sum_{n=1}^{\infty} \frac{|g(n)|\log n}{n^{\sigma_0}} \left(\sum_{n/2 \le y \le 2n} \frac{4x}{|y-n|} + \sum_{y} \frac{1}{\log 2} \right).$$

The expression inside the brackets is (typically)

$$O\left(x \sum_{0 \le m \le 2n} \frac{1}{\dfrac{1}{2} + m} + x(\log x)^{-r} \right) = O(x \log 2n).$$

Therefore, with y restricted to the interval $x - x(\log x)^{-r} < y \le x$,

$$\min_{y} R(y) \le c_3 \frac{x(\log x)^r}{T} \left(1 + \sum_{n=2}^{\infty} \frac{|g(n)|}{n^{\sigma_0}} (\log n)^2 \right).$$

Set $\sigma_1 = \sigma_0 - (\sigma_0 - 1)/2 = 1 + (\sigma_0 - 1)/2$. Then for $n \geq 2$

$$n^{(\sigma_0 - 1)/2} \geq \frac{1}{2!}\left(\frac{\sigma_0 - 1}{2}\log n\right)^2,$$

and we may estimate the sum involving $|g(n)|(\log n)^2$ by means of lemma (9.3); it does not exceed

$$8(\log x)^2 \sum_{n=2}^{\infty} \frac{|g(n)|}{n^{\sigma_1}} = O((\log x)^{c_2 + 2}).$$

There is therefore a value of y in the interval $x - x(\log x)^{-r} < y \leq x$ for which

$$R(y) = O(T^{-1}x(\log x)^{r + c_2 + 2}). \tag{4}$$

We next show that for any such y the values of $S(y)$ and $S(x)$ are close. We apply Hölder's inequality with exponent pair $p = 1 + \delta$, and $q = (1 + \delta)/\delta$, to the sum

$$\sum_{y < n \leq x} |g(n)| n^{-\sigma/(1+\delta)}.$$

In this way we obtain

$$\begin{aligned}\sum_{y<n\leq x} |g(n)| &\leq x^{\sigma_0/(1+\delta)} \sum_{y<n\leq x} |g(n)| n^{-\sigma_0/(1+\delta)} \\ &\leq x^{\sigma_0/(1+\delta)} \left(\sum_{n=1}^{\infty} \frac{|g(n)|^{1+\delta}}{n^{\sigma_0}}\right)^{1/(1+\delta)} (x - y + 1)^{\delta/(1+\delta)} \\ &\leq c_4 x(\log x)^{-r\delta/(1+\delta)} \left(\prod_p \left(1 + \frac{|g(p)|^{1+\delta}}{p^{\sigma_0}} + \cdots\right)\right)^{1/(1+\delta)}.\end{aligned}$$

According to the hypotheses of lemma (9.6),

$$\sum_{p,\, m \geq 2} |g(p^m)|^{1+\delta} p^{-m\sigma_0} \leq c_5 < \infty,$$

whilst the estimate for the function $h(\sigma)$ which was obtained during the proof of lemma (9.2) shows that

$$\prod_p (1 + p^{-\sigma_0}|g(p)|^{1+\delta}) \leq \exp\left(h(2) + 2c_1 \log\left(\frac{1}{\sigma_0 - 1}\right)\right)$$

so that altogether

$$S(x) - S(y) = O(x(\log x)^{(2c_1)/(\delta+1) - (r\delta)/(\delta+1)}).$$

We choose a value of r so that the exponent of the logarithm which appears in this last expression is $-l$, and with this value of r set $T = (\sigma_0 - 1)^{-D}$, where $D = 2r + c_2 + 2$.

According to this last result together with (4), lemma (9.6) is proved.

Proof of theorem (9.1) Let the function $g(n)$ satisfy the hypotheses of theorem (9.1). According to hypotheses (i) and (ii) there is a prime number q so that the inequality

$$\sum_{m=1}^{\infty} |g(p^m)| p^{-m} < 1/2$$

holds uniformly for all primes $p > q$.

It is convenient to introduce a multiplicative function $h(n)$, defined by

$$h(p) = \begin{cases} g(p) & \text{if } p > q, \\ 0 & \text{if } p \le q, \end{cases}$$

and $h(p^m) = 0$ for all prime powers with $m \ge 2$. Corresponding to this function we introduce the Dirichlet series

$$H(s) = \sum_{n=1}^{\infty} h(n) n^{-s}.$$

It is technically convenient to consider the function $h(n)$ rather than $g(n)$.

It is clear that $h(n)$ satisfies the first three conditions of theorem (9.1), and (trivially) the requirement that

$$\sum_{p,\, m \ge 2} |h(p^m)|^{1+\delta} p^{-m} < \infty.$$

We shall now show that as $\sigma \to 1+$, $H(s)$ satisfies a condition of the type (iv).

By considering their associated Euler products we obtain a representation

$$W(s)H(s) = G(s),$$

where

$$W(s) = \prod_{p \le q} \left(1 + \sum_{m=1}^{\infty} g(p^m) p^{-ms}\right) \prod_{p > q} \left(1 + \sum_{m=1}^{\infty} g(p^m) p^{-ms}\right)(1 + g(p)p^{-s})^{-1}.$$

If p is a prime exceeding q, a typical term of this last product has the form $1 + \eta(p)$, where

$$\eta(p) = \sum_{m=2}^{\infty} p^{-ms}(g(p^m) - g(p^{m-1})g(p) + g(p^{m-2})g^2(p) \cdots).$$

In absolute value this sum does not exceed

$$\sum_{m=2}^{\infty} p^{-m\sigma}(|g(p^m)| + |g(p^{m-1})||g(p)| + \cdots)$$

$$\leq \sum_{m=2}^{\infty} p^{-m\sigma}|g(p^m)| \sum_{k=0}^{\infty} (|g(p)|p^{-\sigma})^k$$

$$\leq (1 - |g(p)|p^{-1})^{-1} \sum_{m=2}^{\infty} p^{-m}|g(p^m)|,$$

so that, in view of our choice of q,

$$\sum_{p>q} |\eta(p)| \leq 2 \sum_{p,\, m\geq 2} p^{-m}|g(p^m)| < \infty.$$

Indeed, it follows readily from an application of lemma (2.13) that there is a representation

$$W(s) = \sum_{n=1}^{\infty} w(n)n^{-s} \qquad (\sigma \geq 1),$$

in terms of a multiplicative function $w(n)$, where the series converges absolutely, and to a non-zero sum, at the point $s = 1$.

For each fixed prime p the sum

$$1 + \sum_{m=1}^{\infty} g(p^m)p^{-m(1+i\tau)}$$

is a continuous function of the real variable τ, and by condition (iii), uniformly bounded away from zero on the interval $|\tau| \leq \pi/\log p$. Since it is periodic in τ, of period $2\pi/\log p$, it is therefore uniformly bounded away from zero on the whole real line $-\infty < \tau < \infty$. Taken together with the results established above, this shows that there are positive constants c and σ^* (>1), so that

$$|H(s)| \geq c \qquad (1 < \sigma \leq \sigma^*).$$

Since $G(s)$ satisfies condition (iv) of theorem (9.1), if $s = \sigma + i\tau$, $|\tau| \leq (\sigma - 1)^{1/2}$, $1 < \sigma \leq \sigma^*$, then as $\sigma \to 1+$ we have

$$H(s) = (1 + o(1))W(1)^{-1} \cdot G(s) = \frac{F}{s-1} + o\left(\frac{|s|}{\sigma - 1}\right).$$

If, however, $|\tau| > (\sigma - 1)^{1/2}$, and $\sigma \to 1+$, then

$$\left|H(s) - \frac{F}{s-1}\right| \leq |W(s)|^{-1}|G(s)| + \frac{F}{|s-1|}$$

$$\leq c^{-1}\left(\frac{A}{|s-1|} + o\left(\frac{|s|}{\sigma - 1}\right)\right) + \frac{F}{(\sigma - 1)^{1/2}} = o\left(\frac{|s|}{\sigma - 1}\right).$$

In either case we have shown that $H(s)$ satisfies (iv) with the constant $F = W(1)^{-1}A$ in place of A.

We are now ready to complete the proof of theorem (9.1). This will be easy enough, since the hard work was already done in lemmas (9.4) and (9.5).

We apply lemma (9.6) to the function $H(s)$, choosing $l = 1$. Let values of D and r in that lemma be suitably fixed. Then, in each interval of the type $x - x(\log x)^{-r} < y \leq x$, there is a number y for which

$$\sum_{n \leq x} h(n)\log n = -\frac{1}{2\pi i}\int_{\sigma_0 - iT}^{\sigma_0 + iT} \frac{y^s}{s} H'(s)ds + O(x(\log x)^{-1}).$$

Let K be a (temporarily fixed) positive number. We divide the range of integration into the two parts $|\tau| \leq K(\sigma_0 - 1)$, and $K(\sigma_0 - 1) < |\tau| \leq T$. Let us denote these sets by R_1 and R_2, respectively.

Consider that part of the integral over the range R_1. It follows from condition (iv) of theorem (9.1), as satisfied by $H(s)$, that if $|z - s| = (\sigma - 1)/2$, then

$$H(z) = \frac{F}{z-1} + o\left(\frac{|s|}{\sigma - 1}\right)$$

as $\sigma \to 1+$, uniformly in z, τ. Hence, by Cauchy's integral representation theorem,

$$H'(s) = -\frac{1}{2\pi i}\int_{|z-s|=(\sigma-1)/2} \frac{H(s)}{(z-s)^2}\,dz = -\frac{F}{(s-1)^2} + o\left(\frac{|s|}{(\sigma-1)^2}\right).$$

By means of this estimate we obtain, in an obvious notation,

$$-\frac{1}{2\pi i}\int_{R_1} \frac{y^s}{s} H'(s)ds = \frac{1}{2\pi i}\int_{-K(\sigma_0-1)}^{K(\sigma_0-1)} \frac{Fy^s}{s(s-1)^2}\,ds + o(x\log x).$$

We deform the path of integration into a semi-circle on the line-segment of R_1 as a diameter, passing over the double pole at the point $s = 1$. This yields for the integral over R_1 the estimate

$$Fy\log y + O(K^{-1}x\log x) = Fx\log x + O(K^{-1}x\log x).$$

As for the integral over R_2, we apply Hölder's integral inequality with exponent pair $p = \alpha$, $q = \alpha(\alpha-1)^{-1}$. Then

$$\begin{aligned} \left|-\frac{1}{2\pi i}\int_{R_2}\frac{y^s}{s}H'(s)ds\right| &\le \frac{x^{\sigma_0}}{2\pi}\int_{R_2}\left|\frac{H'(s)}{H(s)}\right|\left|\frac{H(s)}{s}\right|d\tau \\ &\le x\left(\int_{-T}^{T}\left|\frac{H'(s)}{H(s)}\right|^{\alpha}d\tau\right)^{1/\alpha}\left(\int_{R_2}\left|\frac{H(s)}{s}\right|^{q}d\tau\right)^{1/q}. \end{aligned} \tag{5}$$

It is readily checked that the function $H(s)$ satisfies the hypotheses of both lemma (9.4) and lemma (9.5). We note here only that if $\sigma > 1$, then

$$\begin{aligned} \frac{H'(s)}{H(s)} &= \frac{d}{ds}\sum_{p>q}\log(1+g(p)p^{-s}) \\ &= -\sum_{p>q} g(p)p^{-s}\log p\cdot(1-g(p)p^{-s}+(g(p)p^{-s})^2-\cdots), \end{aligned}$$

and

$$\begin{aligned} &\sum_{p>q}|g(p)p^{-s}\log p\cdot(1-g(p)p^{-s}+\cdots)| \\ &\quad\le \sum_{p>q}|g(p)|p^{-\sigma}\log p\cdot(1-|g(p)|p^{-\sigma})^{-1} \\ &\quad\le 2\sum_{p>q}(1+|g(p)|^{1+\delta})p^{-\sigma}\log p = O((\sigma-1)^{-1}). \end{aligned}$$

Accordingly, if α is fixed at a sufficiently large value, the first of the two integrals in (5) is $O((\sigma_0 - 1)^{-\alpha+1})$. As for the second of them,

$$\int_{R_2} \left|\frac{H(s)}{s}\right|^q d\tau \le \max_{s\in R_2} \left|\frac{H(s)}{s}\right|^{(q-1)/2} \int_{-T}^{T} \left|\frac{H(s)}{s}\right|^{(q+1)/2} d\tau$$

$$\le \max_{s\in R_2} \left|\frac{F}{|s-1|} + o\left(\frac{1}{\sigma_0 - 1}\right)\right|^{(q-1)/2} \cdot O((\sigma_0 - 1)^{-((q+1)/2)+1})$$

$$= O(K^{-((q-1)/2)}(\sigma_0 - 1)^{-q+1}).$$

Altogether the integral over R_2 contributes

$$O(K^{-(q-1)/(2q)} x \log x)$$

so that

$$\sum_{n\le x} h(n)\log n = Fx\log x + O(K^{-(q-1)/(2q)} x \log x).$$

Since K may be chosen arbitrarily large, we have proved that as $x \to \infty$

$$\sum_{n\le x} h(n)\log n = (F + o(1))x\log x.$$

Integrating by parts:

$$\sum_{n\le x} h(n) = (F + o(1))x + \int_2^x \frac{1}{w(\log w)^2} \sum_{n\le w} h(n)\log n \, dw = (F + o(1))x.$$

Recalling that $G(x) = W(s)H(s)$, and that $W(1)$ is defined by an absolutely convergent series, we apply lemma (2.19) to deduce that

$$x^{-1}\sum_{n\le x} g(n) \to F \qquad (x \to \infty).$$

This completes the proof of theorem (9.1).

Remarks. In his paper [1], Halász furnishes an example of a non-negative function $g(n)$ which satisfies the hypotheses (ii)–(iv) of theorem (9.1), with $A = 0$, but for which $g(p) \to \infty$ (as $p \to \infty$) in as slow a manner as desired, and so that

$$\sum \frac{g(p)\log p}{p^\sigma} = O\left(\frac{1}{\sigma - 1}\right).$$

Nevertheless, for his particular function the result

$$x^{-1}\sum_{n\le x} g(n) \to 0 \qquad (x \to \infty),$$

fails to be true. This shows that the hypotheses in theorem (9.1) cannot be appreciably weakened.

If $g(p)$ is uniformly bounded, and the hypothesis (iv) is replaced by

$$G(s) = O\left(\frac{|s|}{\sigma - 1}\right),$$

to hold uniformly in a strip $1 < \sigma \le \sigma^*$, then, in a subsequent paper [2], Halász showed that one may conclude that

$$\sum_{n\le x} g(n) = O(x).$$

It is interesting to see what modifications could be made to the arguments of the present chapter in order to establish the theorems of Wirsing, say Satz (1.1), as mentioned in the remarks at the end of Chapter 6. The main change would involve the result of lemma (9.5). In the case of Wirsing's theorem we should be investigating the Dirichlet series

$$E(s) = \sum_{n=1}^{\infty} \lambda(n)n^{-s},$$

where $\lambda(n)$ is a non-negative multiplicative function. In place of the inequality which is obtained in the present lemma (9.5), the inequality

$$\int_{\sigma-iT}^{\sigma+iT} \left|\frac{E(s)}{s}\right|^{\beta} d\tau \le c_4(\sigma - 1)E(\sigma)^{\beta}$$

would be appropriate.

One may set about obtaining an estimate of this type by following the proof of lemma (9.5), but making use of the inequality $|E(s)| \le E(\sigma)$ in place of $|E(s)| \le c_3|s|/(\sigma - 1)$. We replace the condition for τ to belong to Ω by

$$\left|\frac{E'}{E}(\sigma + i\tau)\right| > \frac{\varepsilon}{\sigma - 1},$$

where ε is to be given a (small) fixed value presently. Then, for example,

$$\int_{\tau\in\Omega} |E'(s)||E(s)|^{\beta-1}|s|^{-\beta}d\tau \le c_7 \int_{\Omega} \frac{E'(\sigma)}{E(\sigma)} \cdot E(\sigma)^{\beta}|s|^{-\beta}d\tau$$

$$\le \frac{c_8}{\sigma - 1} E(\sigma)^{\beta}|\Omega| = O(E(\sigma)^{\beta}),$$

and we obtain the differential inequality

$$-U'(\sigma) \le \frac{2\varepsilon}{\sigma - 1} U(\sigma) + c_9 E(\sigma)^\beta.$$

We multiply by $(\sigma - 1)^{2\varepsilon}$ and integrate over the range $\sigma_0 \le \sigma \le \sigma_1$ to obtain

$$(6) \quad U(\sigma_0) \le U(\sigma_1)\left(\frac{\sigma_1 - 1}{\sigma_0 - 1}\right)^{2\varepsilon} + c_9(\sigma_0 - 1)^{-2\varepsilon} \int_{\sigma_0}^{\sigma_1} (\sigma - 1)^{2\varepsilon} E(\sigma)^\beta d\sigma.$$

It is readily checked that

$$(7) \quad \frac{d}{d\sigma}((\sigma - 1)^{1+3\varepsilon} E(\sigma)^\beta) = \beta(\sigma - 1)^{1+3\varepsilon} E(\sigma)^\beta \left(\frac{1 + 3\varepsilon}{\beta(\sigma - 1)} + \frac{E'(\sigma)}{E(\sigma)}\right).$$

Suppose now that

$$(8) \qquad \sum_{p \le x} \frac{\lambda(p)\log p}{p} \sim \eta \log p \qquad (x \to \infty),$$

where the constant η satisfies $\eta \ge 1$. Then an integration by parts shows that as $\sigma \to 1+$

$$\frac{E'}{E}(\sigma) \sim \frac{-\eta}{\sigma - 1}.$$

Since $\beta > 1$, if we choose for ε a sufficiently small value the derivative in (7) will be negative. The integral in (6) will therefore not exceed

$$c_9(\sigma_0 - 1)^{-2\varepsilon} \cdot (\sigma_0 - 1)^{1+3\varepsilon} E(\sigma_0)^\beta \int_{\sigma_0}^{\sigma_1} (\sigma - 1)^{-1-\varepsilon} d\sigma$$
$$\le c_9 \varepsilon^{-1}(\sigma_0 - 1) E(\sigma_0)^\beta.$$

This leads to an inequality for $U(\sigma_0)$ which is of the desired type.

The hypothesis (8) ensures that as $\sigma \to 1+$

$$\frac{E'}{E}(s) = \eta \frac{\zeta'}{\zeta}(s) + o\left(\frac{1}{\sigma - 1}\right),$$

uniformly over each interval $|\tau| \le K(\sigma - 1)$, and we are left to obtain an appropriate upper bound for $E(s)$ over the range $K(\sigma - 1) < |\tau| \le T$. We do not pursue this topic here, as it would take us too far afield. A related result under slightly different hypotheses may be found as lemma (19.6) in Chapter 19. See also the concluding remarks to Chapter 10.

Chapter 10

Multiplicative Functions with First and Second Means

In Chapter six we proved, amongst other things, the following elegant result of Delange: Let $g(n)$ be a complex-valued multiplicative arithmetic function for which $|g(n)| < 1$, $(n = 1, 2, \ldots)$. Then there is a non-zero mean-value

$$A = \lim_{n\to\infty} n^{-1} \sum_{m=1}^{n} g(m)$$

if and only if the series

$$\sum p^{-1}(g(p) - 1)$$

taken over all the primes p is convergent, *and* for at least one positive integer k, $g(2^k) \neq -1$. It is important that the limit A is assumed non-zero.

In this chapter we establish a more general result in which the explicit condition $|g(n)| \leq 1$ is removed. We allow $g(n)$ to be sometimes large, provided that it is in a certain average sense bounded. The proof depends upon some of the considerations of Chapters 4, 6 and 8, and differs from that of Delange.

The presentation is a modified version of that given in the author's paper [21].

Theorem 10.1 (Elliott). *Let $g(n)$ be a complex-valued multiplicative function. Let there be a non-zero mean-value*

(1)
$$A = \lim_{n\to\infty} n^{-1} \sum_{m=1}^{n} g(m).$$

Furthermore, let

(2)
$$\limsup_{n\to\infty} n^{-1} \sum_{m=1}^{n} |g(m)|^2$$

be finite. Then the following conditions are satisfied: The series

$$\sum \frac{g(p)-1}{p} \qquad \sum \frac{|g(p)-1|^2}{p} \qquad \sum_{p,\,k\geq 2}\sum \frac{|g(p^k)|^2}{p^k} \tag{3}$$

are convergent, and for each prime p,

$$\sum_{k=1}^{\infty} g(p^k)p^{-k} \neq -1. \tag{4}$$

Conversely, if conditions (3) *and* (4) *are satisfied, then so is* (2), *and* $g(n)$ *has a non-zero mean-value* (1).

Whenever conditions (1) *and* (2) *are satisfied then the mean-value*

$$\lim_{n\to\infty} n^{-1} \sum_{m=1}^{n} |g(m)|^2 \tag{5}$$

actually exists.

Remarks. The last part of the condition (3) ensures that the series

$$\sum_{k=1}^{\infty} g(p^k)p^{-ks}$$

converges in the half-plane $\sigma > 1/2$.

An important feature of theorem (10.1) *is that the conditions* (3) *and* (4) *are both necessary and sufficient.*

It would seem that the strengthening of "lim sup" in (2) to "lim" in (5) is possible because of the hypothesis that $A \neq 0$. It would be interesting to establish (5) directly from the hypotheses (1) and (2).

Theorems (5.5) and (7.7) together give necessary and sufficient conditions in order that

$$A = \lim_{n\to\infty} n^{-1} \sum_{m=1}^{n} f(m) \qquad \limsup_{n\to\infty} n^{-1} \sum_{m=1}^{n} |f(m)|^2 < \infty$$

exist for additive functions $f(m)$. Theorem (10.1) may be viewed as the analogue of this result for multiplicative functions, when $A \neq 0$. It is a classical problem to give necessary and sufficient conditions in order that a finite mean-value

$$\lim_{n\to\infty} n^{-1} \sum_{m=1}^{n} g(m)$$

exist for multiplicative functions. A similar question may be asked concerning additive functions. We discuss these questions a little in the later chapter on problems.

Theorem (10.1) clarifies to a certain extent the need for the conditions (i), (ii), and (iii) if theorem (9.1).

This ends our remarks.

Most of the proof will be devoted to showing that the conditions (3) and (4) are necessary. That they are sufficient will be obtained from relatively straightforward applications of theorem (9.1).

Outline of the argument:

Assuming the validity of (1) *and* (2):

§§2. Proof of the second and third parts ($p > p_0$) of condition (3).
3. Study of Dirichlet series
4. Removal of the condition $p > p_0$ of §2.
5. Proof that condition (4) and the first part of condition (3) are satisfied.

Assuming the validity of (3) *and* (4):

§§6. Proof that (1) is satisfied.
7. Proof that (5), and so (2), is satisfied.

Application of the Dual of the Turán–Kubilius Inequality [§2]

In this and the following sections we assume that the conditions (1) and (2) hold, and establish the validity of (3) and (4).

It is convenient to recall the notation $p^k \| n$ which denotes that the prime-power p^k divides n, but p^{k+1} does not.

We begin by applying lemma (4.6) with $a_n = g(n)$ for $n = 1, 2, \ldots, [x]$, where x is a positive real number, appealing to the hypotheses (1) and (2). Let $'$ denote that a side condition, to be specified presently, is in force upon the prime-powers p^k. Then, for a certain constant c_1,

$$\sum_{p^k \le x^{1/2}}{}' p^k \left| g(p^k) \sum_{\substack{m \le xp^{-k} \\ p \nmid m}} g(m) - (1 + o(1))Axp^{-k} \right|^2 \le c_1 x^2 \qquad (x \ge 1). \tag{6}$$

We next prove that as $x \to \infty$ the estimate

$$\sum_{\substack{m \le xp^{-k} \\ p \nmid m}} g(m) = (1 + o(1))Axp^{-k} + O(xp^{-k-1/2}) \tag{7}$$

holds uniformly for all prime-powers p^k not exceeding $x^{1/2}$. Indeed, the sum to be estimated may be represented in the form

$$\sum_{m \le xp^{-k}} g(m) - \sum_{\substack{m \le xp^{-k} \\ m \equiv 0 (\mathrm{mod}\ p)}} g(m).$$

Applying the Cauchy–Schwarz inequality to the second of these latter sums, and appealing to hypothesis (2):

$$\left| \sum_{\substack{m \le xp^{-k} \\ m \equiv 0 (\mathrm{mod}\ p)}} g(m) \right| \le \left(\sum_{\substack{m \le xp^{-k} \\ m \equiv 0 (\mathrm{mod}\ p)}} 1 \cdot \sum_{m \le xp^{-k}} |g(m)|^2 \right)^{1/2} = O((xp^{-k-1}xp^{-k})^{1/2}).$$

It is now clear that the estimate (7) holds. Using it in (6), dividing by x, and letting $x \to \infty$, we see that for each positive real number y:

$$\sum_{p^k \le y}{}' p^{-k} | A(g(p^k) - 1) + O(|g(p^k)| p^{-1/2})|^2 \le c_1.$$

Here the constant c_1 depends at most upon the function $g(n)$. We make use of the fact that $A \neq 0$, and express this inequality in the different manner

$$\sum_{p^k \le y}{}' p^{-k} |g(p^k) - 1|^2 \le c_2 \left(1 + \sum_{p^k \le y}{}' |g(p^k)|^2 p^{-k-1} \right). \tag{8}$$

Let p_0 be a prime number, chosen so that $4c_2 p_0^{-1} < 1$. Define

$$S(y) = \sum_{p^k \le y}{}' p^{-k} |g(p^k)|^2,$$

where $'$ denotes that $k \ge 2$ and $p > p_0$. Then from this last inequality:

$$\begin{aligned} S(y) &\le 2 \sum_{p^k \le y}{}' p^{-k}(|g(p^k) - 1|^2 + 1) \le c_3 + 2c_2 \left(1 + \sum_{p^k \le y}{}' |g(p^k)|^2 p^{-k-1} \right) \\ &\le c_4 + S(y)/2, \end{aligned}$$

so that $S(y)$ is uniformly bounded for all values of y. Letting $y \to \infty$ we deduce that the series

$$\sum_{p > p_0} \sum_{k \ge 2} p^{-k} |g(p^k)|^2 \tag{9}$$

is convergent. We shall defer the removal of the restriction $p > p_0$ until §4.

Returning to the inequality (8) let $'$ now denote that the condition $k = 1$, $p_0 < p \leq y$ is in force. Arguing as in the estimation of $S(y)$ we see that

$$D(y) = \sum_{p_0 < p \leq y} p^{-1}|g(p)|^2 \leq 2 \sum_{p_0 < p \leq y} p^{-1}|g(p) - 1|^2 + 2\sum_{p \leq y} p^{-1}$$
$$\leq \tfrac{1}{2} \sum_{p_0 < p \leq y} p^{-1}|g(p)|^2 + O(\log\log y),$$

so that

$$D(y) = O(\log\log y).$$

An integration by parts now shows that

$$\sum_{p_0 < p \leq y} p^{-2}|g(p)|^2 = y^{-1}D(y) + \int_{p_0}^{y} w^{-2}D(w)dw = O(1).$$

Using this estimate in inequality (8) we obtain

$$\sum_{p_0 < p \leq y} p^{-1}|g(p) - 1|^2 \leq c_5 < \infty,$$

where the constant c_5 is independent of y. Letting $y \to \infty$ we conclude that the series

$$\sum p^{-1}|g(p) - 1|^2$$

is convergent; which is the second part of condition (3).

This appears to be as far as elementary arguments will take us. In particular, in order to remove the restriction $p > p_0$ in (9) we shall apply the theory of Dirichlet series. The study of these series will occupy us for most of the next two sections.

Study of Dirichlet Series [§3]

Let $h(n)$ be a multiplicative function which satisfies the conditions $h(p^m) = 0$ if $m \geq 2$, and $2|h(p)p^{-1}| \leq 1$ for all primes p. Define the Dirichlet series

$$H(s) = \sum_{n=1}^{\infty} h(n)n^{-s},$$

which is certainly absolutely convergent in the half-plane $\sigma > 2$.

Lemma 10.2. *Let the series*

$$\sum_{p} p^{-1}|h(p) - 1|^2$$

be convergent. Define the function

$$w(\sigma) = \sum_p p^{-\sigma}(h(p) - 1) \qquad (\sigma > 1).$$

Then $H(s)$ is analytic in the half-plane $\sigma > 1$. Moreover, there is a non-zero constant B, so that for each fixed number $K > 0$, the estimate

$$H(s) = (1 + o(1))\frac{B}{s - 1}\exp(w(\sigma)) \qquad (\sigma \to 1+),$$

holds uniformly in the Stolz angle $|\tau| \le K(\sigma - 1)$.

Remark. It follows from an hypothesis of the present lemma, by means of an application of the Cauchy–Schwarz inequality, that

$$\sum p^{-\sigma}|h(p) - 1| \le \left(\sum p^{-1}|h(p) - 1|^2 \sum p^{-\sigma}\right)^{1/2}$$

so that $w(\sigma)$ is defined by a series which is absolutely convergent for $\sigma > 1$.

Proof. According to this same remark

$$\sum p^{-\sigma}|h(p)| \le \sum p^{-\sigma}(1 + |h(p) - 1|).$$

We deduce from lemma (2.13) that $H(s)$ is analytic in the half-plane $\sigma > 1$, and has an absolutely convergent Euler product there. Moreover, in terms of the principal value of the logarithms,

$$\sum_p |\log(1 + h(p)p^{-s}) - h(p)p^{-s}| \le \sum_p |h(p)p^{-\sigma}|^2,$$

this last series being convergent if $\sigma > 1/2$. There is thus a representation

$$H(s) = H_1(s)\exp\left(\sum p^{-s}h(p)\right),$$

where the function $H_1(s)$ is analytic in the half-plane $\sigma > 1/2$, and uniformly bounded, both above, and below away from zero, in the half-plane $\sigma \ge 1$. In particular $H_1(1) \ne 0$.

Lemma (6.6) guarantees a similar representation for the Riemann zeta function:

$$\zeta(s) = H_2(s)\exp\left(\sum p^{-s}\right).$$

Bearing in mind that $(s - 1)\zeta(s) \to 1$ as $s \to 1$, we see that if $\sigma \to 1+$ with $|\tau| \le K(\sigma - 1)$, we have

$$H(s) = \zeta(s) \cdot H(s)\zeta(s)^{-1} = (1 + o(1))\frac{B}{s - 1}\exp\left(\sum p^{-s}(h(p) - 1)\right),$$

where $B = H_1(1)H_2(1)^{-1} \neq 0$. In order to complete the proof of the present lemma it will therefore suffice to show that with the same uniformity

$$\sum_p |h(p) - 1||p^{-s} - p^{-\sigma}| \to 0 \qquad (\sigma \to 1+). \tag{10}$$

For each fixed prime p it is clear that $|p^{-s} - p^{-\sigma}| \to 0$ as σ (and so s) $\to 1$. If y is any fixed prime, then by the Cauchy–Schwarz inequality

$$\sum_{p>y} |h(p) - 1||p^{-s} - p^{-\sigma}| \le \left(\sum_{p>y} p^{-\sigma}|h(p) - 1|^2 \sum_p p^{\sigma}|p^{-s} - p^{-\sigma}|^2\right)^{1/2}.$$

Here

$$\sum_p p^{\sigma}|p^{-s} - p^{-\sigma}|^2 = \sum_p p^{-\sigma}|\exp((\sigma - s)\log p) - 1|^2$$

$$\le \sum p^{-\sigma}|(\sigma - s)\log p|^2 = O\left((\sigma - 1)^2 \frac{d}{d\sigma}\left(-\frac{\zeta'}{\zeta}(\sigma)\right)\right) = O(1),$$

so that

$$\limsup_{\sigma \to 1+} \sum_{p>y} |h(p) - 1||p^{-s} - p^{-\sigma}| \le c_6\left(\sum_{p>y} p^{-1}|h(p) - 1|^2\right)^{1/2}.$$

Letting $y \to \infty$, and appealing to the hypothesis of the lemma, we establish (10), and complete the proof of lemma (10.2).

Lemma 10.3. *Let the conditions of lemma* (10.1) *be satisfied. Assume further that*

$$\sup_{\sigma>1} \sum_p p^{-\sigma}(|h(p)|^2 - 1) < \infty.$$

Then the inequality

$$|H(s)| \le c_7|\zeta(s)\zeta(\sigma)|^{1/2}$$

holds in the half-plane $\sigma > 1$.

Proof. We begin with the following remark. If r and θ are real numbers, then

$$|1 - re^{i\theta}|^2 = 1 + r^2 - 2r\cos\theta = (r^2 - 1) + 2(1 - r\cos\theta).$$

If, now, τ is a real number, then

$$|1 - p^{-i\tau}|^2 = |1 - p^{i\tau}|^2 \leq 2|1 - h(p)|^2 + 2|h(p) - p^{i\tau}|^2$$
$$\leq 2|1 - h(p)|^2 + 2|1 - h(p)p^{-i\tau}|^2,$$

which by our remark does not exceed

$$2|1 - h(p)|^2 + 2(|h(p)|^2 - 1) + 4(1 - \operatorname{Re} h(p)p^{-i\tau}).$$

We multiply by $p^{-\sigma}$, sum over all primes p, and exponentiate. Thus

$$\exp(\textstyle\sum p^{-\sigma}|1 - p^{-i\tau}|^2) \leq H_3(\sigma)H_4(\sigma)\exp(\sum 4p^{-\sigma}(1 - \operatorname{Re} h(p)p^{-i\tau})),$$

where

$$H_3(\sigma) = \exp(2\textstyle\sum p^{-\sigma}|h(p) - 1|^2) \leq c_8$$

by the hypothesis of lemma (10.2), and

$$H_4(\sigma) = \exp(2\textstyle\sum p^{-\sigma}(|h(p)|^2 - 1)) \leq c_9$$

from the hypotheses of the present lemma. Altogether

$$\exp(\textstyle\sum 2p^{-\sigma}(1 - \operatorname{Re} p^{-i\tau})) \leq c_{10}\exp(\sum 4p^{-\sigma}(1 - \operatorname{Re} h(p)p^{-i\tau})).$$

By means of the representations for $\zeta(s)$ and $H(s)$ which were introduced during the proof of lemma (10.2), we deduce that

$$\zeta(\sigma)^2|\zeta(\sigma + i\tau)|^{-2} \leq c_{11}\zeta(\sigma)^4|H(\sigma + i\tau)|^{-4}.$$

Lemma (10.3) is proved.

Removal of the Condition $p > p_0$ [§4]

In this section we remove the condition $p > p_0$ in (9), by proving that for each prime p the series

$$\sum_{k=2}^{\infty} |g(p^k)|^2 p^{-k}$$

is convergent. We shall apply the results of §3, but need a number of additional results.

Lemma 10.4.

$$\sup_{\sigma > 1} \sum_{|g(p)| \leq 2} p^{-\sigma}(|g(p)|^2 - 1) < \infty.$$

Proof. Define the multiplicative functions $r(n)$ by $r(p^m) = 0$ if $m \geq 2$, and

$$r(p) = \begin{cases} g(p) & \text{if } p > 8p_0 \text{ and } |g(p)| \leq 2, \\ 0 & \text{otherwise.} \end{cases}$$

Consider the Dirichlet series

$$R(\sigma) = \sum_{n=1}^{\infty} |r(n)|^2 n^{-\sigma},$$

and its Euler product

$$\prod_p (1 + |r(p)|^2 p^{-\sigma}) \qquad (\sigma > 1).$$

Clearly

$$R(\sigma) \leq \exp\left(\sum_{|g(p)| \leq 2} |g(p)|^2 p^{-\sigma} \right).$$

Introducing the Riemann zeta function in the form

$$\zeta(\sigma) = H_2(\sigma) \exp\left(\sum p^{-\sigma}\right),$$

and noting that from what we have so far established

$$\sum_{|g(p)| > 2} p^{-\sigma} \leq \sum_p p^{-1} |g(p) - 1|^2 < \infty,$$

we deduce that

$$\exp\left(\sum_{|g(p)| \leq 2} (|g(p)|^2 - 1) p^{-\sigma} \right) \leq c_{12} R(\sigma) \zeta(\sigma)^{-1}.$$

However, integrating by parts and making use of hypothesis (2), we see that

$$\begin{aligned} R(\sigma) &= \sigma \int_1^{\infty} y^{-\sigma-1} \sum_{n \leq y} |r(n)|^2 dy \leq \sigma \int_1^{\infty} y^{-\sigma-1} \sum_{n \leq y} |g(n)|^2 \mu^2(n) dy \\ &\leq c_{13} \sigma \int_1^{\infty} y^{-\sigma} dy = O\left(\frac{\sigma}{\sigma - 1} \right). \end{aligned}$$

If δ is a suitably chosen positive number, then $R(\sigma)\zeta(\sigma)^{-1} \leq c_{14}$ holds in the range $1 < \sigma \leq 1 + \delta$. We now deduce the inequality stated in lemma (10.4), first in the range $1 < \sigma \leq 1 + \delta$, and then (trivially) in the range $\sigma > 1 + \delta$.

We shall continue to study the function $r(n)$ which was defined during the proof of lemma (10.4).

Define the Dirichlet series

$$E(s) = \sum_{n=1}^{\infty} |r(n)| n^{-s}.$$

From the bound

$$\sum_{n \le x} |r(n)| = O(x)$$

we see that $E(s)$ is well defined in the half-plane $\sigma > 1$, where it represents an analytic function. Indeed, it follows from the assumption of (2), by means of an integration by parts, that

$$|E(s)| \le E(\sigma) = O\left(\frac{1}{\sigma - 1}\right),$$

whilst, since $r(p^m) = O(1)$ for $m \ge 1$,

$$\frac{E'(s)}{E(s)} = O\left(\frac{1}{\sigma - 1}\right),$$

these estimates holding uniformly in the strip $1 < \sigma \le 2$. Since

$$\sum p^{-1} |r(p) - 1|^2 \le \sum p^{-1} |g(p) - 1|^2 < \infty$$

we may apply lemmas (10.2) amd (10.3) with $h(p) = r(p)$, and so deduce

Lemma 10.5. *Define the function*

$$l(\sigma) = \sum_{p} (|r(p)| - 1) p^{-\sigma} \qquad (\sigma > 1).$$

Then there is a non-zero constant F so that

$$E(s) = (1 + o(1)) \frac{F}{s - 1} \exp(l(\sigma)) \qquad (\sigma \to 1+),$$

holds uniformly in any fixed angle $|\tau| \le K(\sigma - 1)$, K a fixed positive number. Moreover, the inequality

$$|E(s)| \le c_1 |\zeta(s) \zeta(\sigma)|^{1/2}$$

holds in the strip $1 < \sigma \le 2$.

We are now in a position to apply the method of Halász as exposed in Chapter 9.

Set $\sigma_0 = 1 + (\log x)^{-1}$, $(x \geq 3)$.

Lemma 10.6. *As* $x \to \infty$,

$$\sum_{n \leq x} |r(n)| \log n = (1 + o(1))Fx \log x \cdot \exp(l(\sigma_0)) + o(x \log x).$$

Proof. From lemma (9.6) we obtain a representation

$$\sum_{n \leq x} |r(n)| \log n = -\frac{1}{2\pi i} \int_{\sigma_0 - iT}^{\sigma_0 + iT} \frac{y^s}{s} E'(s) ds + O(x), \tag{11}$$

where $T = (\sigma_0 - 1)^{-D}$, and y is a number lying in the interval $x - x(\log x)^{-r_0} < y \leq x$; r_0 and D being fixed positive numbers. We shall estimate the behaviour of this integral as x (and so y) $\to \infty$.

Let K be a (temporarily fixed) positive real number.

Let $\alpha > 1$ be a positive constant. Define q so that $\alpha^{-1} + q^{-1} = 1$, and therefore $q > 1$. Applying Hölder's inequality, making use of the identity

$$E(s) = \frac{E'(s)}{E(s)} \cdot E(s),$$

yields

$$\int_{K(\sigma_0 - 1) \leq |\tau| \leq T} \left| \frac{E'(s)}{s} \right| d\tau \leq \left(\int_{\sigma_0 - iT}^{\sigma_0 + iT} \left| \frac{E'(s)}{E(s)} \right|^{\alpha} d\tau \right)^{1/\alpha} \left(\int_{K(\sigma_0 - 1) \leq |\tau| \leq T} \left| \frac{E(s)}{s} \right|^{q} d\tau \right)^{1/q}.$$

If α is fixed at a sufficiently large value we may apply lemma (9.4), and obtain for the integral involving α the upper bound $O((\sigma_0 - 1)^{-1 + 1/\alpha})$. As for the final integral, involving q, it does not exceed

$$\max_{K(\sigma_0 - 1) \leq |\tau| \leq T} \left| \frac{E(s)}{s} \right|^{(q-1)/2} \int_{\sigma_0 - iT}^{\sigma_0 + iT} \left| \frac{E(s)}{s} \right|^{(q+1)/2} d\tau.$$

From lemma (10.5), over the range $K(\sigma_0 - 1) \leq |\tau| \leq T$ we have

$$\left| \frac{E(s)}{s} \right| \leq \frac{c_1 |\zeta(s)\zeta(\sigma_0)|^{1/2}}{|s|} = O\left(\frac{1}{K^{1/2}(\sigma_0 - 1)} \right).$$

Therefore, applying lemma (9.5), the contribution towards the integral in (11) which arises from the range $K(\sigma_0 - 1) \leq |\tau| \leq T$ is at most

$$O(K^{-\delta} y(\sigma_0 - 1)^{-1}) = O(K^{-\delta} x \log x),$$

where $\delta = (q - 1)/(2q) > 0$.

We estimate the behaviour of $E'(s)$ on the line-segment $\operatorname{Re} s = \sigma_0, |\tau| \le K(\sigma_0 - 1)$, by means of Cauchy's integral formula

$$G'(s) = -\frac{1}{2\pi i}\int_{|z-s|=\rho} \frac{G(z)}{(z-s)^2}\,dz,$$

where we choose $\rho = (\sigma_0 - 1)/2$. According to the first part of lemma (10.5), on the circle $|z - s| = \rho$ we have the estimate

$$G(z) = (1 + o(1))\frac{F}{z-1}\exp(l(\sigma_1)),$$

where $\sigma_1 = \operatorname{Re} z$. We note that, as $x \to \infty$,

$$|l(\sigma_1) - l(\sigma_0)| \le \sum_p ||g(p)| - 1||p^{-\sigma_1} - p^{-\sigma_0}| \to 0,$$

just as was the case in (10). Thus we may replace $l(\sigma_1)$ by $l(\sigma_0)$, and deduce that on the line-segment in which we are interested

$$E'(s) = \frac{F}{(s-1)^2}\exp(l(\sigma_0)) + o\left(\frac{\exp(l(\sigma_0))}{(\sigma_0 - 1)^2}\right).$$

Hence

$$\int_{|\tau| \le K(\sigma_0 - 1)} \frac{y^s}{s} E'(s)ds = F\exp(l(\sigma_0))\int_{|\tau| \le K(\sigma_0 - 1)} \frac{y^s}{s(s-1)^2}\,ds$$
$$+ o(x\exp(l(\sigma_0))\log x).$$

We replace the segment in the range of summation in the integral which appears on the right-hand side of this approximate equation by a semi-circle centre $s = \sigma_0$, radius $K(\sigma_0 - 1)$, to the left-hand side of the line $\operatorname{Re} s = \sigma_0$. If K is sufficiently (absolutely) large we shall then pass over the pole at the point $s = 1$, where there is a residue $y(\log y - 1)$. The integral over the semi-circle which we have introduced is clearly $O(K^{-1}x\log x)$.

Thus we have proved that as $x \to \infty$ the estimate

$$x^{-1}\sum_{n \le x} |r(n)|\log n = (1 + O(K^{-1}))F\exp(l(\sigma_0))\log x + O(K^{-\delta}\log x)$$

holds for each fixed positive number K. Hence

$$\limsup_{x\to\infty} \frac{|\sum |r(n)|\log n - Fx\log x\exp(l(\sigma_0))|}{x\log x\cdot\{1 + \exp(l(\sigma_0))\}} \le c_2 K^{-\delta},$$

and letting $K \to \infty$ we complete the proof of lemma (10.6).

In this section we prove one more lemma.

Lemma 10.7.

$$\limsup_{x\to\infty} x^{-1} \sum_{n\le x} |r(n)| > 0.$$

Proof. Assume, to the contrary, that

$$\sum_{n\le x} |r(n)| = o(x) \qquad (x \to \infty).$$

Let n_1 run through those integers which are squarefree; divisible only by those primes $p > 8p_0$, for which $|g(p)| \le 2$. Let n_2 run through the sequence of all remaining positive integers. Let us consider the integer 1 to be a member both of the sequence $\{n_1\}$ and of the sequence $\{n_2\}$.

Clearly, every integer n occurs at least once amongst the products $n_1 n_2$ with $(n_1, n_2) = 1$, and therefore

$$\sum_{n\le x} |g(n)| \le \sum_{n_2\le x} |g(n_2)| \sum_{n_1\le x/n_2} |g(n_1)|.$$

Let ε be a positive real number. For a suitably large fixed real number N, and all $x > N$, we have

$$\sum_{n_1\le x/n_2} |g(n_1)| = \sum_{n_1\le x/n_2} |r(n_1)| < \varepsilon x/n_2,$$

uniformly for $n_2 \le x/N$. Therefore

$$\sum_{n\le x} |g(n)| < \varepsilon x \sum_{n_2\le x/N} |g(n_2)| n_2^{-1} + \sum_{x/N<n_2\le x} |g(n_2)| O(xn_2^{-1}),$$

this last step by means of the hypothesis (2) and the Cauchy–Schwarz inequality. Let us suppose, for the moment, that the series

$$\sum |g(n_2)| n_2^{-1} \tag{12}$$

converges. Then dividing by x, and letting $x \to \infty$, we would obtain

$$\limsup_{x\to\infty} x^{-1} \sum_{n\le x} |g(n)| = O(\varepsilon)$$

for every $\varepsilon > 0$. Hence

$$\sum_{n\le x} |g(n)| = o(x) \qquad (x \to \infty),$$

and this would contradict the hypothesis (1) of the theorem, that the limit A is non-zero.

We shall now establish the convergence of the series (12), and thereby complete the proof of lemma (10.7).

For each prime p, hypothesis (2) of the theorem shows that $g(p^k) = O(p^{k/2})$, so that the series

$$\sum_{k=1}^{\infty} p^{-k}|g(p^k)|$$

converges. Then in terms of the number p_0 which appears in the inequality (9):

$$\sum |g(n_2)|n_2^{-1} \le \prod_{\substack{p \le 8p_0 \\ \text{or } |g(p)| > 2}} \left(1 + \sum_{k=1}^{\infty} p^{-k}|g(p^k)|\right) \prod_{p > 8p_0} \left(1 + \sum_{k=2}^{\infty} p^{-k}|g(p^k)|\right).$$

In the first of these two products the terms corresponding to primes $p \le 8p_0$ contribute $O(1)$. As for the terms with $p > p_0$, their contribution does not exceed

$$\exp\left(\sum_{|g(p)| > 2} \sum_{k=1}^{\infty} p^{-k}|g(p^k)|\right).$$

But from what we have proved so far in §2,

$$\sum_{|g(p)| > 2} p^{-1}|g(p)| \le 4 \sum p^{-1}|g(p) - 1|^2 < \infty,$$

and

$$\sum_{p > p_0} \sum_{k=2}^{\infty} p^{-k}|g(p^k)| \le \sum_{p > p_0} \sum_{k=1}^{\infty} p^{-k}[1 + |g(p^k)|^2]$$

$$\le \sum_{p > p_0} \frac{1}{p(p-1)} + \sum_{p > p_0} \sum_{k=2}^{\infty} p^{-k}|g(p^k)|^2 < \infty.$$

The second product is likewise seen to be convergent.

In this way we have completed the proof of lemma (10.7).

We now prove that if p is a prime which does not exceed $8p_0$, then the series

$$\sum_{k=2}^{\infty} p^{-k}|g(p^k)|^2$$

converges.

Let α be a positive real number, $0 < \alpha < 1$. Let p be a fixed prime $p \leq 8p_0$, and consider the sum

$$W = \sum_{\substack{p^k m \leq x \\ p \nmid m,\, k \geq 2,\, p^k \leq \alpha^{-1}}} \sum{}'' |g(p^k m)|^2 = \sum_{\substack{p^k \alpha \leq 1 \\ k \geq 2}} |g(p^k)|^2 \sum_{\substack{m \leq xp^{-k} \\ p \nmid m}}{}'' |g(m)|^2$$

where $''$ indicates that m is constrained to run through the members n_1 of the first sequence defined in the proof of lemma (10.7). For these integers $p \nmid m$ anyway, and $|g(m)| = |r(m)|$.

Let x run through a sequence of numbers x_ν, ($\nu = 1, 2, \dots$; $x_\nu \to \infty$ as $\nu \to \infty$), for which

$$\sum_{m \leq x} |g(m)| \geq c_1 x.$$

The existence of such a sequence, for a suitable positive constant c_1, is guaranteed by lemma (10.7). We now apply the result of lemma (10.6). Thus

$$\log x \sum_{x^{1/2} < n \leq x} |r(n)| \leq 2 \sum_{n \leq x} |r(n)| \log n \leq 3Fx \log x \cdot \exp(l(\sigma_0)) + o(x \log x),$$

and uniformly for all sufficiently large members $x = x_\nu$ of the special sequence,

$$\exp(l(\sigma_0)) \geq c_2 > 0.$$

We shall extend this inequality in a certain manner. Namely, if $\alpha x \leq y \leq x$, and $\sigma_2 = 1 + (\log y)^{-1}$, then for all large values of x,

$$\exp(l(\sigma_2)) \geq \tfrac{1}{2} c_2. \tag{13}$$

Here (of course) the constant $c_2/2$ will be independent of α. In fact, since

$$\sigma_2 = 1 + (\log x + O(1))^{-1} = \sigma_0 + O((\log x)^{-2}),$$

then for a suitably chosen value of $\sigma_1, \sigma_0 \leq \sigma_1 \leq \sigma_2$, the mean-value theorem of calculus asserts that

$$\begin{aligned} l(\sigma_2) - l(\sigma_0) &= -(\sigma_2 - \sigma_0) \sum_p (|r(p)| - 1) p^{-\sigma_1} \log p \\ &= O((\log x)^{-2} \sum p^{-\sigma_0} \log p) = O\left((\log x)^{-2} \cdot -\frac{\zeta'}{\zeta}(\sigma_0)\right) \\ &= O((\log x)^{-1}). \end{aligned}$$

The asserted inequality (13) is now evident for all sufficiently large values of $x = x_\nu$.

We can apply lemma (10.6) once again, to deduce that uniformly for $\alpha x \le y \le x$, ($x = x_\nu$ sufficiently large),

$$y^{-1}\sum_{m\le y}{}'' |g(m)| \ge (y \log y)^{-1} \sum_{m\le y} |r(m)| \log m$$

$$\ge (1 + o(1))F \exp(l(\sigma_2)) \ge c_3 > 0.$$

Applying the Cauchy–Schwarz inequality we deduce that

$$\sum_{m\le y}{}'' |g(m)|^2 \ge c_3^2 y.$$

On the one hand, we can use this inequality to obtain a lower bound for the sum W, thus:

$$W \ge \sum_{k\ge 2,\, p^k\alpha\le 1} |g(p^k)|^2 c_3^2 x p^{-k} \qquad (x = x_\nu).$$

On the other hand, it follows at once from hypothesis (2) of the theorem that

$$W \le \sum_{n\le x} |g(n)|^2 \le c_4 x.$$

Putting these bounds for W together, dividing by x in the resulting inequality, and letting $x_\nu \to \infty$, we deduce that

$$\sum_{p^k\le\alpha^{-k}} p^{-k}|g(p^k)|^2 \le c_4 c_3^{-2} < \infty.$$

This inequality is uniform in all values of α which satisfy $0 < \alpha < 1$, so that the series

$$\sum_{k=2}^{\infty} p^{-k}|g(p^k)|^2$$

is convergent. This is what we wished to prove.

The validity of the second and third of the three conditions in (3) is now completely established.

In the next section we show that condition (4) and the first part of condition (3) are satisfied.

Application of the Hardy–Littlewood Tauberian Theorem [§5]

Consider the Dirichlet series

$$G(s) = \sum_{n=1}^{\infty} g(n)n^{-s},$$

and its Euler product

$$G(s) = \prod_p \left(1 + \sum_{k=1}^{\infty} p^{-ks} g(p^k)\right).$$

Define, for each prime p, the function

$$\eta_p(s) = \sum_{k=1}^{\infty} p^{-ks} g(p^k).$$

Suppose that for the prime p_1 we have $\eta_{p_1}(1) = -1$. We shall show that this leads to a contradiction with the hypothesis that the limit A is non-zero. In fact, for $\sigma > 1$,

$$|G(s)| \le |1 + \eta_{p_1}(s)| \prod_{p \ne p_1} (1 + |\eta_p(s)|).$$

The infinite product does not exceed

$$\prod_p \left(1 + \sum_{k=1}^{\infty} p^{-k\sigma} |g(p^k)|\right) = \sum_{n=1}^{\infty} |g(n)| n^{-\sigma} = O\left(\frac{1}{\sigma - 1}\right),$$

this last step following from hypothesis (2). Moreover, if $\sigma > 1/2$ then by the Cauchy–Schwarz inequality and the third part of condition (3),

$$\sum_{k=1}^{\infty} p^{-k\sigma} |g(p^k)| \le \left(\sum_{k=1}^{\infty} p^{-k} |g(p^k)|^2 \cdot \sum_{k=1}^{\infty} p^{-k(2\sigma - 1)}\right)^{1/2}.$$

so that for each prime p, $\eta_p(s)$ is analytic in the half-plane $\sigma > 1/2$. In particular, as $\sigma \to 1+$ we have $\eta_{p_1}(\sigma) = O((\sigma - 1))$.

We can now deduce from our temporary assumption that in the strip $1 < \sigma \le 2$,

$$G(\sigma) = O\left((\sigma - 1) \cdot \frac{1}{\sigma - 1}\right) = O(1).$$

However, an integration by parts shows that

$$G(\sigma) = (1 + o(1)) \frac{A}{\sigma - 1} \qquad (\sigma \to 1+),$$

and this gives a contradiction. Therefore, for each prime p,

$$\sum_{k=1}^{\infty} p^{-k} g(p^k) \ne -1,$$

which is condition (4).

It follows from our remarks concerning $\eta_p(s)$ that for a suitably chosen prime p_2 the inequality $|\eta_p(s)| \le 1/2$ holds uniformly for all $p > p_2$ in the half-plane $\sigma \ge 1$. We write

$$G(s) = \prod_{p \le p_2} (1 + \eta_p(s)) \prod_{p > p_2} (1 + \eta_p(s))e^{-\eta_p(s)} \cdot \exp\left(\sum_{p > p_2} \eta_p(s) \right).$$

In the half-plane $\sigma \ge 1$,

$$\begin{aligned}\sum_{p > p_2} |\log(1 + \eta_p(s)) - \eta_p(s)| &\le \sum_{p > p_2} |\eta_p(s)|^2 \\ &\le 2 \sum_{p > p_2} \left(p^{-2}|g(p)|^2 + \sum_{k=2}^{\infty} p^{-k\sigma}|g(p^k)| \right)^2 \\ &\le 4 \sum_{p > p_2} p^{-2}(1 + |g(p) - 1|^2) + 4 \sum_{p > p_2} \sum_{k=2}^{\infty} p^{-k}|g(p^k)|^2 \sum_{k=2}^{\infty} p^{-k} < \infty,\end{aligned}$$

whilst

$$\sum_{p > p_2} |\eta_p(s) - g(p)p^{-s}| \le \sum_{p > p_2} \sum_{k=2}^{\infty} p^{-k}|g(p^k)| < \infty.$$

There is therefore a non-zero number γ_1 so that, as $\sigma \to 1+$,

$$G(\sigma) = (1 + o(1))\gamma_1 \exp\left(\sum_{p > p_2} p^{-\sigma}g(p) \right).$$

Likewise (or by lemma (6.6)) there is a non-zero constant γ_2 so that, as $\sigma \to 1+$,

$$\zeta(\sigma) = (1 + o(1))\gamma_2 \exp\left(\sum_{p > p_2} p^{-\sigma} \right).$$

Thus, as $\sigma \to 1+$,

$$\begin{aligned}\exp\left(\sum_{p > p_2} p^{-\sigma}(g(p) - 1) \right) &\sim G(\sigma)\gamma_2/(\zeta(\sigma)\gamma_1) \\ &\sim A\gamma_2(\sigma - 1)^{-1}/(\gamma_1(\sigma - 1)^{-1}) \sim A\gamma_2/\gamma_1,\end{aligned}$$

so that

$$\omega = \lim_{\sigma \to 1+} \sum (g(p) - 1)p^{-\sigma}$$

exist, and is finite.

For real numbers u define the function

$$\kappa(u) = \sum_{p \le e^u} (g(p) - 1)p^{-1}.$$

Then we can write this last result in the form

$$\int_0^\infty e^{-\beta u} d\kappa(u) \to \omega \qquad (\beta \to 0+).$$

Moreover, if $x \le y$, $x \to \infty$, $y/x \to 1$, then

$$|\kappa(y) - \kappa(x)|^2 \le \sum_{x < p \le y} p^{-1} \cdot \sum_{x < p \le y} |g(p) - 1|^2 p^{-1} = o(1).$$

We may therefore apply the Hardy–Littlewood tauberian theorem, lemma (2.18), to each of the functions Re $\kappa(u)$ and Im $\kappa(u)$ in turn, to deduce that

$$\lim_{u \to \infty} \kappa(u) = \omega.$$

This proves that the series

$$\sum \frac{g(p) - 1}{p}$$

converges.

We have now established the validity of all the conditions in (3) and (4), so completing the proof that they are necessary.

Application of a Theorem of Halász [§6]

From now until the final remarks of the present chapter we shall assume that the conditions (3) and (4) are satisfied. We prove that when this is so, then so are the conditions (1) and (5), and so (2). We shall do this by applying theorem (9.1) several times.

Let $r(n)$ be the multiplicative function which was introduced at the beginning of the proof of lemma (10.4), save that the condition $p > 8p_0$ is replaced by $p \ge 8$. It is clear that this function satisfies conditions (i), (ii) and (iii) of theorem (9.1); we shall now show that with a suitable number A it also satisfies condition (iv).

According to hypothesis (3)

$$\sum p^{-1}|r(p) - 1|^2 \le \sum p^{-1}|g(p) - 1|^2 < \infty.$$

In view of the identity

$$|\lambda - 1|^2 = |\lambda|^2 - 1 + 2(1 - \text{Re } \lambda),$$

which is valid for all complex numbers λ, the same hypothesis ensures that the series

$$\sum p^{-1}(|r(p)|^2 - 1)$$

is convergent. As a consequence, an integration by parts shows that for $\sigma > 1$

$$\begin{aligned}\sum p^{-\sigma}(|r(p)|^2 - 1) &= (\sigma - 1)\int_1^\infty y^{-\sigma}\sum_{p \le y} p^{-1}(|r(p)|^2 - 1)dy \\ &\le c_1(\sigma - 1)\int_1^\infty y^{-\sigma}dy = c_1 < \infty.\end{aligned}$$

In a manner entirely analogous to that which was used in establishing lemma (10.5), we can show that if $|\tau| \le K(\sigma - 1)$, then, as $\sigma \to 1+$,

$$\Delta(s) = \sum_{n=1}^\infty r(n)n^{-s} = (1 + o(1))\frac{L}{s - 1}\exp\left(\sum p^{-\sigma}(r(p) - 1)\right), \tag{14}$$

where the constant L has the non-zero value

$$\prod(1 + r(p)p^{-1})e^{-r(p)/p}\prod\left(1 - \frac{1}{p}\right)e^{1/p}.$$

Moreover, in the strip $1 < \sigma \le 2$ the inequality

$$|\Delta(s)| \le c_2|\zeta(s)\zeta(\sigma)|^{1/2}$$

obtains.

We now need a result due essentially to Abel. Let

$$\rho = \sum p^{-1}(r(p) - 1).$$

Consider the integral

$$J(\sigma) = \sum p^{-\sigma}(r(p) - 1) = (\sigma - 1)\int_1^\infty y^{-\sigma}\sum_{p \le y} p^{-1}(r(p) - 1)dy.$$

Let ε be a positive real number. Then for an appropriately large number y_0, and all $y > y_0$, we have

$$\left|\sum_{p \le y} p^{-1}(r(p) - 1) - \rho\right| < \varepsilon.$$

Therefore

$$\left|J(\sigma) - (\sigma - 1)\int_1^{\infty} y^{-\sigma}\rho\,dy\right| \leq (\sigma - 1)\int_{y_0}^{\infty} y^{-\sigma}\varepsilon\,dy + (\sigma - 1)\int_1^{y_0} y^{-\sigma}O(1)dy,$$

$$\limsup_{\sigma\to 1+}|J(\sigma) - \rho| \leq \varepsilon.$$

Since ε may be chosen arbitrarily small, $J(\sigma) \to \rho$ as $\sigma \to 1+$.

Accordingly, we deduce from (14) that if $\varepsilon > 0$, $|\tau| \leq K(\sigma - 1)$, and σ is sufficiently near to 1, then

$$\left|\Delta(s) - \frac{M}{s - 1}\right| \leq \frac{\varepsilon}{|s - 1|} \leq \frac{\varepsilon}{\sigma - 1},$$

with the constant

$$M = \prod\left(1 + \frac{r(p)}{p}\right)\left(1 - \frac{1}{p}\right).$$

But if $|\tau| > K(\sigma - 1)$, then

$$\left|\Delta(s) - \frac{M}{s - 1}\right| \leq c_2|\zeta(s)\zeta(\sigma)|^{1/2} + \frac{|M|}{|s - 1|} = O(K^{-1/2}(\sigma - 1)^{-1}).$$

Hence, in either case

$$\limsup_{\sigma\to 1+}\,(\sigma - 1)\left|\Delta(s) - \frac{M}{s - 1}\right| \leq \varepsilon + O(K^{-1/2}).$$

We let $\varepsilon \to 0+$, and then $K \to \infty$. This shows that the function $r(n)$ satisfies condition (iv) of theorem (9.1) with $A = M$. A direct application of theorem (9.1) gives

$$x^{-1}\sum_{n\leq x} r(n) \to M \qquad (x \to \infty).$$

Define the multiplicative function $q(n)$ by

$$\sum_{n=1}^{\infty} q(n)n^{-s}\cdot\sum_{n=1}^{\infty} r(n)n^{-s} = \sum_{n=1}^{\infty} g(n)n^{-s}.$$

Comparing Euler products shows that for each prime p an equivalent definition is that

$$1 + q(p)p^{-s} + q(p^2)p^{-2s} + \cdots = \left(1 + \sum_{k=1}^{\infty} g(p^k)p^{-ks}\right)(1 + r(p)p^{-s})^{-1}.$$

Thus

$$q(p^k) = \sum_{l=0}^{k} g(p^{k-l})(-r(p))^l,$$

from which it follows that

$$\sum p^{-1}|q(p)| = \sum_{|g(p)|>2} p^{-1}|g(p)| \le 2\sum p^{-1}|g(p) - 1|^2 < \infty,$$

and

$$\sum_p \sum_{k=2}^{\infty} p^{-k}|q(p^k)| \le \sum_p \sum_{k=2}^{\infty} \sum_{l=0}^{\infty} p^{-(k-l)}|g(p^{k-l})| \cdot p^{-l}|r(p)|^l$$

$$= \sum_p \sum_{k=2}^{\infty} p^{-m}|g(p^m)|(1 + |r(p)|p^{-1} + |r(p)|^2p^{-2} + \cdots)$$

$$\le \sum_p \left(1 - \frac{|r(p)|}{p}\right)^{-1} \sum_{k=2}^{\infty} p^{-k}(1 + |g(p^k)|^2) < \infty.$$

The series

$$\sum_{m=1}^{\infty} q(m)m^{-1}$$

is therefore absolutely convergent. We apply lemma (2.19) to obtain

$$x^{-1}\sum_{n\le x} g(n) \to M \sum_{m=1}^{\infty} q(m)m^{-1} \qquad (x \to \infty).$$

The limiting value in this result has the alternative representation

$$\prod\left(1 - \frac{1}{p}\right)\left(1 + \sum_{k=1}^{\infty} p^{-k}g(p^k)\right),$$

which condition (4) guarantees to be non-zero.

This completes our proof that condition (1) follows from (3) and (4).

Conclusion of Proof [§7]

We complete the proof of theorem (10.1) by showing that the validity of (5) follows from that of (3) and (4). We shall outline the argument, which proceeds along the lines of §6. Suffice it to say that we consider $|r(n)|^2$ in place of $r(n)$, and then make use of the following facts:

For each prime p

$$||r(p)|^2 - 1|^2 = ||r(p)| - 1|^2||r(p)| + 1|^2 \le 9|r(p) - 1|^2,$$

so that

$$\sum p^{-1}||r(p)|^2 - 1|^2 \le 9 \sum p^{-1}|g(p) - 1|^2 < \infty.$$

By means of the identity

$$|\lambda|^4 - 1 = ||\lambda|^2 - 1|^2 + 2|\lambda - 1|^2 - 4\,\mathrm{Re}(1 - \lambda)$$

and the first and second parts of hypothesis (3) of theorem (10.1), we obtain the convergence of the series

$$\sum p^{-1}(|r(p)|^4 - 1).$$

Applying theorem (9.1) we show that

$$\lim_{x \to \infty} x^{-1} \sum_{n \le x} |r(n)|^2$$

exists and is finite.

We introduce the multiplicative function $v(n)$, defined by

$$\sum v(n)n^{-s} \cdot \sum |r(n)|^2 n^{-s} = \sum |g(n)|^2 n^{-s},$$

so that

$$1 + \sum_{k=1}^{\infty} v(p^k)p^{-ks} = \left(1 + \sum_{k=1}^{\infty} |g(p^k)|^2 p^{-ks}\right)(1 + |r(p)|^2 p^{-s})^{-1}.$$

It is easy to see that

$$\sum p^{-1}|v(p)| = \sum_{|g(p)| > 2} p^{-1}|g(p)|^2 \le 2 \sum p^{-1}|g(p) - 1|^2 < \infty,$$

and

$$\sum_p \sum_{k=2}^{\infty} p^{-k}|v(p^k)| \le \sum_p \left(1 - \frac{|r(p)|^2}{p}\right)^{-1} \sum_{k=2}^{\infty} p^{-k}|g(p^k)|^2 < \infty.$$

A further application of lemma (2.19) and we deduce the finite existence of

$$\lim_{x \to \infty} x^{-1} \sum_{n \le x} |g(n)|^2.$$

This completes the proof of the theorem (10.1).

Concluding Remarks

Let $h(n)$ be a multiplicative function which, for simplicity, satisfies the conditions $h(p^m) = 0$ if $m \geq 2$, and $|h(p)| \leq p/2$. Assume further that the two conditions

$$(15) \qquad \sum p^{-1}|h(p) - 1|^2 < \infty \qquad \sup_{\sigma > 1} \sum p^{-\sigma}(|h(p)|^2 - 1) < \infty$$

are satisfied. Then along the lines of lemma (10.6) we may obtain the estimate

$$\sum_{n \leq x} h(n) = (1 + o(1))xL(\log x) + o(x),$$

where

$$L(u) = \exp\left(\sum_p (h(p) - 1)p^{-1-u^{-1}}\right).$$

It is not difficult to show that $L(u)$ is a slowly-oscillating function of u (as $u \to \infty$), but in general $|L(u)| = 1$ will not hold, even for large values of u.

As it stands this result has two imperfections: the second of the two conditions is not necessary, and the term $o(x)$ in the asymptotic estimate can be essentially removed.

Define the Dirichlet series

$$H(s) = \sum_{n=1}^{\infty} h(n)n^{-s}$$

and

$$H_0(s) = \sum_{n=1}^{\infty} |h(n)|n^{-s}.$$

Since

$$||h(p)| - 1|^2 \leq |h(p) - 1|^2$$

the estimate

$$\sum_{p \leq x} \frac{|h(p)|}{p} \log p \sim \log x \qquad (x \to \infty),$$

follows readily from the convergence of the first series in (15). By employing the trivial bound $|H(s)| \leq H_0(\sigma)$ for $\sigma > 1$, the argument given in the concluding remarks to Chapter 9 enables us to prove that if $T = (\sigma - 1)^{-D}$ and $\beta > 1$ is fixed, then

$$\int_{-T}^{T} \left| \frac{H(s)}{s} \right|^{\beta} d\tau \leq c_0(\sigma - 1)H_0(\sigma)^{\beta}$$

for all σ sufficiently near to 1.

Moreover,

$$\sum_{|h(p)| \leq 1/2} \frac{1}{p} \leq \sum \frac{4|h(p) - 1|^2}{p} < \infty$$

and a simple change in the proof of lemma (19.6) shows that for a certain positive constant δ the inequality

$$|H(s)| \leq K^{-\delta} H_0(\sigma)$$

holds uniformly for $K(\sigma - 1) < |\tau| \leq T$ and all σ sufficiently near to 1. Here δ does not depend upon K.

Hence (as in Chapter 9 and lemma (10.6)) we arrive at the estimate

$$\sum_{n \leq x} h(n) = (1 + o(1))xL(\log x) + o(xH_0(\sigma_0)\zeta(\sigma_0)^{-1}) \qquad (x \to \infty),$$

making no use of the second condition in (15).

From the identity

$$||h| - h|^2 = 2|h|(|h| - \operatorname{Re} h)$$

we see that for any positive N

$$\sum_{\substack{p \leq N \\ |h(p)| \geq 1/2}} p^{-1}(|h(p)| - \operatorname{Re} h(p)) \leq \sum_{p \leq N} 2|h(p)|(|h(p)| - \operatorname{Re} h(p))$$

$$\leq \sum_p p^{-1}||h(p)| - h(p)|^2 \leq 2 \sum p^{-1}\{||h(p)| - 1|^2 + |1 - h(p)|^2\}$$

$$\leq 4 \sum p^{-1}|h(p) - 1|^2 \leq c_1,$$

and

$$\sum_{\substack{p \leq N \\ |h(p)| \leq 1/2}} p^{-1}(|h(p)| - \operatorname{Re} h(p)) \leq \sum_{|h(p)| \leq 1/2} p^{-1} \leq 4 \sum p^{-1}|h(p) - 1|^2 \leq c_1,$$

the constant c_1 being independent of N. It is now straightforward to show that

$$H_0(\sigma_0)\zeta(\sigma_0)^{-1} \le c_2 \exp(\sum p^{-\sigma_0}(|h(p)| - 1)) \le c_3 L(\log x),$$

so that

$$\sum_{n \le x} h(n) = (1 + o(1))xL(\log x) \qquad (x \to \infty).$$

Along these lines the following result may be established (we suppress the details):

Theorem 10.8. *Let $h(n)$ be a multiplicative function for which the series*

$$\sum_p p^{-1}|h(p) - 1|^2, \sum_{p,\, k \ge 2}\sum p^{-k}|h(p^k)|$$

converge.

Then

$$\frac{\sum_{n \le x} h(n)}{xL(\log x)} \to A \qquad (x \to \infty),$$

where

$$L(u) = \exp\left(\sum_p p^{-1-u^{-1}}(h(p) - 1)\right)$$

is a slowly-oscillating function of u, and the constant A is given by

$$A = \prod_p \left(1 - \frac{1}{p}\right)\left(1 + \frac{h(p)}{p} + \frac{h(p^2)}{p^2} + \cdots\right)\exp\left(\frac{1 - h(p)}{p}\right).$$

This result may be compared with that of theorem (6.2).

If one is interested only in establishing the necessity of the conditions in theorem (10.1) then the need for lemma (10.6) can be avoided. This was pointed out by Daboussi and Delange [1]. In the same paper they show that in place of the Hardy-Littlewood tauberian theorem one may use the following argument:

With

$$\kappa(u) = \sum_{p \le e^u} (g(p) - 1)p^{-1}$$

we have

$$\int_0^\infty e^{-\beta u} d\kappa(u) \to \omega \qquad (\beta \to 0+).$$

Integrating by parts and changing the variable this may be alternatively expressed as

$$\int_0^\infty e^{-y} \kappa\left(\frac{y}{\beta}\right) dy \to \omega \qquad (\beta \to 0+).$$

We next note that for $y \geq 1, 0 < \beta \leq 1/2$,

$$\left|\kappa\left(\frac{y}{\beta}\right) - \kappa\left(\frac{1}{\beta}\right)\right|^2 \leq \sum_{p > \exp(1/\beta)} p^{-1}|g(p) - 1|^2 \cdot \sum_{\exp(1/\beta) < p \leq \exp(y/\beta)} p^{-1}$$
$$\leq c_4(\log y + 1)$$

with similar arguments when $y < 1$, so that

$$\int_0^\infty e^{-y}\left|\kappa\left(\frac{y}{\beta}\right) - \kappa\left(\frac{1}{\beta}\right)\right| dy$$
$$\leq c_5 \int_1^\infty e^{-y}(|\log y| + 1)^{1/2} dy + c_5 \int_0^1 y^{-1/2} dy \leq c_6.$$

Moreover, for each fixed positive value of y

$$\kappa\left(\frac{y}{\beta}\right) - \kappa\left(\frac{1}{\beta}\right) \to 0 \qquad (\beta \to 0+).$$

Hence, by Lebesgue's theorem on dominated convergence

$$\sum_{p \leq \exp(1/\beta)} p^{-1}(g(p) - 1) = \omega + o(1) \qquad (\beta \to 0+).$$

This establishes the convergence of the first series in (3).

Although it seems to be a difficult question to give necessary and sufficient conditions in order that a finite mean-value exist for a given multiplicative function, perhaps with a restriction to non-negative functions it would be more tractible. The corresponding problem for non-negative additive functions is simply answered.

What is the analogue of theorem (10.1) when $A = 0$ is required? Here the difficulty is to formulate appropriate *necessary* conditions.

References

van Aardenne-Ehrenfest, T., de Bruijn, N. G., Korevaar, J.
1. A note on slowly oscillating functions.
Nieuw Arch. Wiskunde (2)**23**(1949), 77–86.

Aczél, J.
1. *Lectures on Functional Equations and their Applications.*
Academic Press, New York and London, 1966.

Akilov, G. P., Kantorovich, L. V.
1. *Functional Analysis in Normed Spaces.*
Translated from the Russian. Pergamon, Macmillan, New York, 1964.

Axer, A.
1. Beitrag zur Kenntnis der zahlentheoretischen Funktionen $\mu(n)$ und $\lambda(n)$.
Prace matematyczno-fizyczne (*Warsaw*) **21**(1910), 65–95.

Babu, G. J.
1. On the distribution of additive arithmetical functions of integral polynomials.
Technical Report Math.-Stat. 19/1971, Indian Statistical Institute, 1971.
2. Probabilistic methods in the theory of arithmetical functions.
Indian Statistical Institute, Calcutta, 1973.

Bakstys, A. (See Cyrillic Index.)

Barban, M. B. (See Cyrillic Index.)
9. The 'Large Sieve' method and its applications in the Theory of Numbers.
Uspekhi Mat. Nauk **21**(1966) No. 1 (127), 51-102; =*Russian Math. Surveys* **21**(1966), No. 1, 49-103.

Barban, M. B., Vinogradov, A. I.
1. On the number theoretic basis of probabilistic number theory.
Dokl. Akad. Nauk SSSR **154**(1964), 495–496; =*Soviet Math. Doklady* **5**(1964), 96–98.

Barban, M. B., Vinogradov, A. I., Levin, B. V. (See Cyrillic Index.)

Behrend, F.
1. Three reviews; of papers by Chowla, Davenport and Erdös.
Jahrbuch über die Fortschritte der Mathematik. **60**(1935), 146–149.
2. On sequences of numbers not divisible one by another.
Journ. London Math. Soc. **10**(1935), 42–44.

Berry, A. C.
1. The accuracy of the Gaussian approximation to the sum of independent variates.
Trans. Amer. Math. Soc. **49**(1941), 122–136.

Billingsley, P.
1. *Convergence of Probability Measures.*
Wiley, New York, 1968.
2. Additive functions and Brownian motion.
Notices Amer. Math. Soc. **17**(1970), 1050, *Abstract No.* 681-A9.

Birch, B. J.
1. Multiplicative functions with non-decreasing normal order.
Journ. London Math. Soc. **42**(1967), 149–151.

Bombieri, E.
1. Maggiorazione del resto nel "Primzahlsatz" col metodo di Erdös-Selberg.
Ist. Lombardo Accad. Sci. Lett. Rend. A**96**(1962), 343–350.
2. Sulle formule di A. Selberg generalizzate per classi di funzioni aritmetiche e le applicazioni al problema de resto nel "Primzahlsatz".
Riv. Mat. Univ. Palermo (2)3(1962), 393–440.
3. On the large sieve.
Mathematika **12**(1965), 201–225.

Bombieri, E., Davenport, H.
1. "On the large sieve method".
Abh. aus Zahlentheorie und Analysis zur Erinneruing an Edmund Landau. Deut. Verlag. Wiss., Berlin, 1968, 11–22.

de Bruijn, N. G.
1. Pairs of Slowly Oscillating functions occurring in asymptotic problems concerning the Laplace transform.
Nieuw Archief voor Wiskunde (3)**7**(1959), 20–26.

de Bruijn, N. G., van Aardenne-Ehrenfest, T., Korevaar, J.
1. A note on slowly oscillating functions.
Nieuw Arch. Wiskunde (2)**23**(1949), 77–86.

Burgess, D. A.
1. The distribution of quadratic residues and non-residues.
Mathematika **4**(1957), 106–112.
2. On character sums and primitive roots.
Proc. London Math. Soc. (3)**12**(1962), 179–192.

Cauchy, A.
1. *Œuvres complètes d'Angustin Cauchy.*
Gauthier-Villars et fils, Paris, 1897.

Chowla, S.
1. On abundant numbers.
Journ. Indian Math. Soc. (2)**1**(1934), 41–44.

Chowla, S., Erdös, P.
1. A theorem on the distribution of values of L-series.
Journ. Indian Math. Soc. **15**A(1951), 11–18.

Daboussi, H., Delange, H.
1. On a theorem of P.D.T.A. Elliott on multiplicative functions.
Journ. London Math. Soc. (2)**14**(1976), 345–356.

Davenport, H.
1. Über numeri abundantes.
Sitzungsbericht Akad. Wiss. Berlin, **27**(1933), 830–837.
2. *Multiplicative Number Theory.*
Markham, Chicago, 1967.

Davenport, H., Bombieri, E.
1. "On the large sieve method."
Abh. aus Zahlentheorie und Analysis zur Erinnerung an Edmund Landau.
Deut. Verlag. Wiss., Berlin, 1968, 11–22.

Davenport, H., Halberstam, H.
1. The values of a trigonometrical polynomial at well spaced points.
Mathematika **13**(1966), 91–96. See, also, Corrigendum and addendum, *Mathematika* **14**(1967), 229–232.

Delange, H.

1. Sur la distribution des valeurs de certaines fonctions arithmétiques.
   Colloque sur la Théorie des Nombres, Bruxelles (1955), 147–161.
2. On some arithmetical functions.
   Illinois Journ. Math. **2**(1958), 81–87.
3. Un théorème sur les fonctions arithmétiques multiplicatives et ses applications.
   Ann. Scient. Éc. Norm. Sup. 3e *série* t. **78**(1961), 1–29.
4. Sur les fonctions arithmétiques multiplicatives.
   Ann. Scient. Éc. Norm. Sup. 3e *série* t. **78**(1961), 273–304.
5. Sur le nombre des diviseurs premiers de n.
   Acta Arith. **7**(1961/62), 191–215.
6. On a class of multiplicative arithmetical functions.
   Scripta Math. **26**(1963), 121–141.
7. Sur un théorème de Rényi, II.
   Acta Arithmetica **13**(1968), 339–362.
8. On the distribution modulo 1 of additive functions.
   Journ. Indian Math. Soc. **34**(1970), 215–235.
9. On finitely distributed additive functions.
   Journ. London Math. Soc. (2)**9**(1975), 483–489.

Delange, H., Daboussi, H.

1. On a theorem of P. D. T. A. Elliott on multiplicative functions.
   Journ. London Math. Soc. (2)**14**(1976), 345–356.

Diamond, H. G.

1. The distribution of values of Euler's phi function.
   Amer. Math. Soc. Symposia in Pure Math. XXIV(1973), 63–75.

Diamond, H. G., Steinig, J.

1. An elementary proof of the prime number theorem with a remainder term.
   Invent. Math. **11**(1970), 199–258.

Dirichlet, P. G. L.

1. *Werke*, Berlin 1889; see Gedächtnissrede auf Gustav Peter Lejeune Dirichlet, by E. E. Kummer, pp. 311–344, in particular p. 324.

Doeblin, W.

1. Sur les sommes d'un grand numbres de variables aléatoires indépendantes.
   Bull. Soc. math. France **53**(1939), 23–32; 35–64.

Edwards, R. E.

1. *Fourier Series. A modern introduction, Vol. 1.*
   Holt, Rinehart and Winston, New York, Toronto, London, 1967.

Elliott, P. D. T. A.

1. On sequences of integers.
   Quart. Journ. Math. (Oxford)(2)**16**(1965), 35–45.
2. On certain additive functions.
   Acta Arith. **12**(1967), 365–384.
3. On certain additive functions II.
   Acta Arith. **14**(1968), 51–64.
4. The distribution of the quadratic class number.
   Liet. mat. rinkinys = Litovsk Mat. Sbornik, **10**(1970), 189–197.
5. On the mean value of $f(p)$.
   Proc. London Math. Soc. (3)**21**(1970), 28–96.
6. The Turán-Kubilius inequality, and a limitation theorem for the large sieve.
   Amer. Journ. Math. **92**(1970), 293–300.
7. On inequalities of Large Sieve type.
   Acta Arith. **18**(1971), 405–422.
8. On the limiting distribution of additive functions (mod 1).
   Pacific Journ. of Math. **38**(1971), 49–59.

9. The least prime k-th-power residue.
Journ. London Math. Soc. (2) **3**(1971), 205–210.
10. On Distribution Functions (mod 1): Quantitative Fourier Inversion.
Journ. of Number Theory, **4**, No. 6 (1972), 509–522.
11. On the least pair of consecutive quadratic non-residues (mod p).
Proceedings of the 1972 Number Theory Conference, Boulder, 75–79.
12. On connections between the Turán-Kubilius inequality and the Large Sieve: Some applications.
Amer. Math. Soc. Proceedings of Symposia in Pure Math, Vol. **24** Providence, 1973, 77–82.
13. On additive functions whose limiting distributions possess a finite mean and variance.
Pacific Journ. Math. **48**(1973), 47–55.
14. A class of additive arithmetic functions with asymptotically normal distribution.
Mathematika **20**(1973), 144–154.
15. On the Distribution of the Values of Quadratic L-series in the Half-plane $\sigma > 1/2$.
Inventiones Math. **21**(1973), 319–338.
16. On the limiting distribution of additive arithmetic functions.
Acta. math. **132**(1974), 53–75.
17. On the limiting distribution of $f(p + 1)$ for non-negative additive functions.
Acta Arith. **25**(1974), 259–264.
18. A conjecture of Kátai.
Acta Arith. **26**(1974), 11–20.
19. On the limiting distribution of additive arithmetic functions: Logarithmic Renormalization.
Proc. London Math. Soc. (3)**31**(1975), 364–384.
20. The law of large numbers for additive arithmetic functions.
Math. Proc. Camb. Phil. Soc. **78**(1975), 33–71.
21. A mean-value theorem for multiplicative functions.
Proc. London Math. Soc. (3)**31**(1975), 418–438.
22. General asymptotic distributions for additive arithmetic functions.
Math. Proc. Camb. Phil. Soc. **79**(1976), 43–54.
23. On a conjecture of Narkiewicz about functions with non-decreasing normal order.
Colloquium Math. **36**(1976), 289–294.
24. On a problem of Hardy and Ramanujan.
Mathematika **23**(1976), 10–17.
25. On two conjectures of Kátai.
Acta Arithmetica **30**(1976), 35–59.
26. On certain asymptotic functional equations.
Aequationes Mathematicae **17**(1978) 44–52.

Elliott, P. D. T. A., Halberstam, H.
1. Some applications of Bombieri's theorem.
Mathematika **13**(1966), 196–203.

Elliott, P. D. T. A., Ryavec, C.
1. The Distribution of the Values of Additive Arithmetical Functions.
Acta math. **126**(1971), 143–164.

Erdös, P.
1. On the density of abundant numbers.
Journ. London Math. Soc. **9**(1934), 278–282.
2. On the density of some sequences of numbers I.
Journ. London Math. Soc. **10**(1935), 120–125.
3. On the normal number of prime factors of $p - 1$ and some related problems concerning Euler's φ-function.
Quart. Journ. Math. (Oxford)**6**(1935), 205–213.
4. On a problem of Chowla and some related problems.
Camb. Phil. Soc. Proc. **32**(1936), 534–540.
5. On the density of some sequences of numbers II.
Journ. London Math. Soc. **12**(1937), 7–11.

6. Note on the number of prime divisors of integers.
Journ. London Math. Soc. **12**(1937), 308–314.
7. On the density of some sequences of numbers III.
Journ. London Math. Soc. **13**(1938), 119–127.
8. On the smoothness of the asymptotic distribution of additive arithmetical functions.
Amer. Journ. Math. **61**(1939), 722–725.
9. On a theorem of Littlewood and Offord.
Bull. Amer. Math. Soc. **51**(1945), 898–902.
10. On the distribution function of additive functions.
Annals of Math. **47**(1946), 1–20.
11. Some remarks about additive and multiplicative functions.
Bull. Amer. Math. Soc. **52**(1946), 527–537.
12. Some remarks and corrections to one of my papers.
Bull. Amer. Math. Soc. **53**(1947), 761–763.
13. On a new method in elementary number theory which leads to an elementary proof of the prime number theorem.
Proc. Nat. Acad. Sci. U.S.A. **35**(1949), 374–384.
14. Remarks on number theory I and II.
Acta Arith. **5**(1959), 23–33; 171–177.
15. On the distribution of numbers of the form $\sigma(n)/n$ and some related questions.
Pacific Journ. Math. **52**(1974), 59–65.

Erdös, P., Chowla, S.
1. A theorem on the distribution of values of L-series.
Journ. Indian Math. Soc. **15**A(1951), 11–18.

Erdös, P., Kac, M.
1. On the Gaussian law of errors in the theory of additive functions.
Proc. Nat. Acad. Sci. U.S.A. **25**(1939), 206–207.
2. The Gaussian law of errors in the theory of additive number-theoretic functions.
Amer. Journ. Math. **62**(1940), 738–742.

Erdös, P., Rényi, A.
1. On the mean value of non-negative multiplicative number-theoretical functions.
Michigan Journ. Math. **12**(1965), 321–328.

Erdös, P., Ruzsa, I., Sárközi, A.
1. On the number of solutions of $f(n) = a$ for additive functions.
Acta Arith. **24**(1973), 1–9.

Erdös, P., Ryavec, C.
1. A characterization of finitely monotonic additive functions.
Journ. London Math. Soc. (2)**5**(1972), 362–367.

Erdös, P., Turán, P.
1. On a problem in the theory of uniform distribution, I, II.
Indigationes Math. **10**(1948), 370–378; 406–413.

Erdös, P., Wintner, A.
1. Additive arithmetical functions and statistical independence.
Amer. Journ. Math. **61**(1939), 713–721.

Esseen, C. G.
1. On the Liapounoff limit of error in the theory of probability.
Ark. Mat. Astr. o Fysik 28 A(1942), 1–19.
2. Fourier analysis of distribution functions. A mathematical study of the Laplace-Gaussian law.
Acta math. **77** No. 1–2, (1945), 1–125.
3. On the Concentration Function of a Sum of Independent Random Variables.
Z. Warscheinlichk. verw. Geb. **9**(1968), 290–308.

Faĭnleĭb, A. S.
1. Distribution of values of Euler's function.
Mat. Zametki **1**(1967), 645–652; =*Math. Notes* **1**(1967), 428–432.
2. A generalisation of Esseen's inequality and its application to Probabilistic Number Theory.
Izv. Akad. Nauk, ser. Mat. **32**, No. 4, (1968), 859–879; =*Mathematics of the USSR—Izvestija Vol.* **2** (1968), No. 4, 821–844.

Faĭnleĭb, A. S., Levin, B. V.
1. Distribution of values of additive arithmetic functions.
Dokl. Akad. Nauk SSSR **171**(1966), 281–284; =*Soviet Math. Doklady* **7**(1966), 1474–1477.
2. A summation method for multiplicative functions.
Izv. Akad Nauk, ser. Mat. **31**, No. 3, (1967), 697–710; =*Mathematics of the USSR—Izvestija*, Vol. **1**, No. 3, (1967), 677–690.
3. Applications of some integral equations to problems in number theory.
Uspekhi Mat. Nauk 22(3)**135**(1967), 119–198; =*Russian Math. Surveys* **22**(3), (1967), 119–204.
4. Integral Limit Theorems for Certain Classes of Additive Arithmetic Functions.
Trud. Mosk. Mat. Ob. **18**(1968), 19–54; =*Moscow Math. Soc., translated by Amer. and Lond. Math. Societies*, **18**(1968), 19–57.
5. Multiplicative functions and probabilistic number theory.
Izv. Akad. Nauk, ser. Mat. **34**(1970), 5, 1064–1109; =*Mathematics of the USSR—Izvestija* **4**(1970), No. 5, 1071–1118.

Faĭnleĭb, A. S., Toleuov, J. (See Cyrillic Index.)

Forti, M. C., Viola, C.
1. On large sieve type estimates for the Dirichlet series operator.
Amer. Math. Symposia XXIV, Amer. Math. Soc., Providence, Rhode Island (1973), 31–49.

Fréchet, M., Shohat, J.
1. A proof of the generalized central limit theorem.
Trans. Amer. Math. Soc. **33**(1931), 533–543.

Galambos, J.
1. A probabilistic approach to mean values of multiplicative functions.
Journ. London Math. Soc. (2)**2**(1970), 405–419.
2. Limit distribution of sums of (dependent) random variables with applications to arithmetical functions.
Z. Warsch. und verw. Gebiete **18**(1971), 261–270.
3. On the distribution of strongly multiplicative functions.
Bull. London Math. Soc. **3**(1971), 307–312.
4. Distribution of additive and multiplicative functions.
The theory of Arithmetic Functions. Proceedings of conf. at Western Michigan Univ., 1971. Springer Lecture Notes—Mathematics **251**, Berlin, 1972, 127–139.

Gallagher, P. X.
1. The large sieve.
Mathematika **14**(1967), 14–20.
2. Bombieri's mean value theorem.
Mathematika **15**(1968), 1–6.
3. The Large Sieve and Probabilistic Galois Theory.
Amer. Math. Soc. Symposia Proceedings XXIV, Providence, Rhode Island, (1973), 91–101.

Gnedenko, B. V., Kolmogorov, A. N.
1. *Limit distributions for sums of independent random variables.*
Translated from the Russian and annotated by K. L. Chung, Addison-Wesley, Reading, Mass., London, 1968.

Halász, G.
1. Über die Mittelwerte multiplikativer zahlentheoretischer Funktionen.
Acta Math. Acad. Sci. Hung. **19**(1968), 365–403.

2. Über die Konvergenz multiplikativer zahlentheoretischer Funktionen.
Studia Scient. Math. Hungarica **4**(1969), 171–178.
3. On the distribution of additive and the mean values of multiplicative arithmetic functions.
Studia Scient. Math. Hungarica **6**(1971), 211–233.
4. Remarks to my paper, "On the distribution of additive and mean values of multiplicative arithmetic functions."
Acta Math. Acad. Scient. Hungar. **23**(3–4), (1972), 425–432.
5. On the distribution of additive arithmetic functions.
Acta Arithmetica **27**(1975), 143–152.
6. Estimates of the concentration function of Combinatorial Number Theory and Probability.
Preprint. Lecture at Oberwolfach, November, 1975.

Halberstam, H.
1. On the distribution of additive number-theoretic functions I.
Journ. London Math. Soc. **30**(1955), 43–53.
2. Über additive zahlentheoretische Funktionen.
J. reine und angew. Math. **195**(1955), 210–214.
3. On the distribution of additive number-theoretic functions II.
Journ. London Math. Soc. **31**(1956), 1–14.
4. On the distribution of additive number-theoretic functions III.
Journ. London Math. Soc. **31**(1956), 14–27.

Halberstam, H., Davenport, H.
1. The values of a trigonometrical polynomial at well spaced points.
Mathematika **13**(1966), 91–96; See, also: Corrigendum and addendum. *Mathematika* **14**(1967), 229–232.

Halberstam, H., Elliott, P. D. T. A.
1. Some applications of Bombieri's theorem.
Mathematika **13**(1966), 196–203.

Halberstam, H., Richert, H.-E.
1. Brun's method and the Fundamental Lemma.
Acta Arithmetica. **24**(1973), 113–133.
2. *Sieve Methods.*
Academic Press, London, New York, 1974.

Halberstam, H., Roth, K. F.
1. *Sequences, Vol. I.*
Oxford, 1966.

Hall, R. R.
1. Halving an estimate obtained from Selberg's upper bound method.
Acta Arithmetica **25**(1974), 347–351.

Hamel, G.
1. Eine Basis aller Zahlen und die unstetigen Lösungen der Funktionalgleichung $f(x + y) = f(x) + f(y)$.
Math. Annalen. **60**(1905), 459–462.

Hardy, G. H.
1. *Ramanujan. Twelve lectures on subjects suggested by his life and work.*
Cambridge, 1940.
2. *Divergent Series.*
Oxford, 1949.

Hardy, G. H., Ramanujan, S.
1. The normal number of prime factors of a number n.
Quart. Journ. Math. (Oxford) **48**(1917), 76–92.
2. Asymptotic formulae in Combinatory Analysis.
Proc. London Math. Soc. (2)**17**(1918), 75–115.

Hardy, G. H., Littlewood, J. E., Pólya, G.
1. *Inequalities.*
Cambridge, 1934.

Hardy, G. H., Wright, E. M.
1. *An introduction to the Theory of Numbers.*
Oxford, Fourth Edition, 1960.

Hartman, P.
1. Infinite convolutions on locally compact Abelian groups and additive functions.
Trans. Amer. Math. Soc. **214**(1975), 215–231.

Heilbronn, H.
1. Zeta functions and L-functions. Chapter 8 of *Algebraic Number Theory*, edited by J. W. S. Cassels and A. Fröhlich. Academic Press, London, New York, 1967.

Hengartner, W., Theodorescu, R.
1. *Concentration Functions.*
Academic Press, New York, London, 1973.

Hooley, C.
1. On the representation of a number as the sum of two squares and a prime.
Acta Math. **97**(1957), 189–210.
2. On the greatest prime factor of a quadratic polynomial.
Acta Math. **117**(1967), 281–299.

Ibragimov, I. A., Linnik, Yu. V.
1. *Independent and stationary sequences of random variables.*
Edited by J. F. C. Kingman, translated from the Russian, Wolters-Noordhoff, Groningen, The Netherlands, 1971.

Jessen, B., Wintner, A.
1. Distribution functions and the Riemann Zeta Function.
Trans. Amer. Math. Soc. **38**(1935), 48–88.

Jutila, M.
1. On Character Sums and Class Numbers.
Journ. of Number Theory. **5**(1973), 203–214.

Kac, M.
1. Note on the distribution of values of the arithmetic function $d(n)$.
Bull. Amer. Math. Soc. **47**(1941), 815–817.
2. Probability Methods in some problems of Analysis and Number Theory.
Bull. Amer. Math. Soc. **55**(1949), 641–665.

Kac, M., Erdös, P.
1. On the Gaussian law of errors in the theory of additive functions.
Proc. Nat. Acad. Sci. U.S.A. **25**(1939), 206–207.
2. The Gaussian law of errors in the theory of additive functions.
Amer. Journ. Math. **62**(1940), 738–742.

Kantorovich, L. V., Akilov, G. P.
1. *Functional Analysis in Normed Spaces.*
Translated from the Russian, Pergamon, Macmillan, New York, 1964.

Karamata, J.
1. Sur un mode de croissance régulière des fonctions.
Mathematica (Cluj)**4**(1930), 38–53.
2. Neuer Beweis und Verallgemeinerung der Tauberschen Sätze welche die Laplacesche und Stieltjesche Transformation betreffen.
J. reine und angew. Math. **164**(1931), 27–39.

Kátai, I.
1. On distribution of arithmetical functions on the set of prime plus one.
Compositio Math. **19**(1968), 278–289.

2. Some remarks on additive arithmetical functions.
Liet. mat. rinkinys. = *Litovsk Mat. Sb.* **9**(1969), 515–518.

Kesten, H.
1. A sharper form of the Doeblin-Lévy-Kolmogorov-Rogozin inequality for concentration functions.
Math. Scand. **25**(1969), 133–144.
2. Sums of Independent Random Variables—Without moment conditions.
Annals of Math. Stat. **43**(1972), 701–732, p. 702.

Khinchine, A.
1. Zur additiven Zahlentheorie.
Mat. Sb. (N.S.)**39**(1932), 27–34.

Khinchine, A. Ya., Lévy, P.
1. Sur les lois stables.
C.R. Acad. Sci. Paris. **202**(1936), 374–376.

Kobayashi, I.
1. Remarks on the large sieve method.
Seminar on Modern Methods in Number Theory; Proceedings of a meeting held in Tokyo, at the Institute of Statistical Mathematics, 1971, three pages.
2. A note on the Selberg sieve and the large sieve.
Proc. Japan Acad. **49**(1973), 1–5.

Kolmogorov, A. N.
1. *Grundbegriffe der Warscheinlichkeitsfelder und –räume.*
Springer, Heidelberg, 1933.
2. Two uniform limit theorems for sums of independent random variables.
Teor. Ver. Prim. **1**(1956), 426–436; = *Theory of Prob. and its Appl.* **1**(1956), 384–394.
3. Sur les propriétés des fonctions de concentration de M. P. Lévy.
Université de Paris, Ann. de l'Inst. Henri Poincaré **16**, 1, (1958), 27–34.

Kolmogorov, A. N., Gnedenko, B. V.
1. *Limit distributions for sums of independent random variables.*
Translated from the Russian and annotated by K. L. Chung, Addison-Wesley, Reading, Mass., London, 1968.

Korevaar, J., van Aardenne-Ehrenfest, T., deBruijn, N. G.
1. A note on slowly oscillating functions.
Nieuw Arch. Wiskunde (2)**23**(1949), 77–86.

Kubik, L.
1. The limiting distributions of cumulative sums of independent two valued random variables.
Studia Mathematica **18**(1959), 295–309; See also remarks on this paper, *Studia Math.* **19**(1960), 249.

Kubilius, J. (See Cyrillic Index.)
1. Probabilistic Methods in the Theory of Numbers.
Uspekhi Mat. Nauk (N.S.) **11**(1956), 2(68), 31–66; = Amer. Math. Soc. Translations, Vol. **19**(1962), 47–85.
5. *Probabilistic Methods in the Theory of Numbers.*
Amer. Math. Soc. Translations of Math. Monographs, No. 11, Providence, 1964.
6. On the distribution of number-theoretic functions.
Univ. de Paris, Sem. Delange-Pisot-Poitou, Théorie des nombres, 1969/1970, No. 23, eleven pages.
7. On an inequality for additive arithmetic functions.
Acta Arithmetica **28**(1975), 371–383.

Kuipers, L., Niederreiter, H.
1. *Uniform distribution of sequences.*
Wiley, New York, 1974.

Landau, E.
1. Neuer Beweis des Primzahlsatzes und Beweis des Primidealsatzes.
Math. Annalen **56**(1903), 645–670.
2. Über die Äquivalenz zweier Hauptsätze der analytischen Zahlentheorie.
Sitzungsberichte der Kaiserlichen Akad. der Wiss. in Wien. Math. Naturw. Klasse **1202**[a]. (1911), 1–16.
3. *Vorlesungen über Zahlentheorie.* Three volumes.
Hirzel, Leipzig, 1927.
4. *Elementary Number Theory.*
Chelsea, New York, 1958.

Landau, E., Walfisz, A.
1. Über die Nichtfortsetzbarkeit einiger durch Dirichletsche Reihen definierter Funktionen.
Rend. di Palermo **44**(1919), 82–86.

Lang, S.
1. *Algebraic Numbers*
Addison-Wesley, Reading, Palo Alto, London, 1964.

Lavrik, A. F.
1. A method for estimating double sums with a real quadratic character, and applications.
Izv. Akad. Nauk SSSR, ser. Mat. **35**(1971), No. 6, 1189–1207; =*Mathematics of the USSR—Izvestija* **5**(1971), No. 6, 1195–1214.

LeVeque, W. J.
1. On the size of certain number-theoretic functions.
Trans. Amer. Math. Soc. **66**(1949), 440–463.
2. An inequality connected with Weyl's criterion for Uniform Distribution.
Amer. Math. Soc. Proceedings of Symposia in Pure Math., VIII, Providence, 1965, 22–30.

Levin, B. V., Barban, M. B., Vinogradov, A. I. (See Cyrillic Index.)

Levin, B. V., Faĭnleĭb, A. S.
1. Distribution of values of additive arithmetic functions.
Dokl. Akad. Nauk SSSR **171**(1966), 281–284; =*Soviet Math. Dokl.* **7**(1966), 1474–1477.
2. A summation method for multiplicative functions.
Izv. Akad. Nauk ser. Mat. **31**(1967), No. 3, 697–710; =*Mathematics of the USSR—Izvestija* **1**, No. 3(1967), 677–690.
3. Applications of some integral equations to problems in number theory.
Uspekhi Mat. Nauk. 22(3)**135**(1967), 119–198; =*Russian Math. Surveys* **22**(3)(1967), 119–204.
4. Integral Limit Theorems for Certain Classes of Additive Arithmetic Functions.
Trud. Mosk. Mat. Ob. **18**(1968), 19–54; =*Trans. of the Moscow Math. Soc., translated by American and London Math. Soc.*, **18**(1968), 19–57.
5. Multiplicative functions and probabilistic number theory.
Izv. Akad. Nauk ser. Mat. **34**(1970), 5, 1064–1109; =*Mathematics of the USSR—Izvestija* **4**(1970), No. 5, 1071–1118.

Levin, B. V., Timofeev, N. M. (See Cyrillic Index.)
3. On the distribution of values of additive functions.
Acta Arithmetica **26**(1974/75), No. 4, 333–364.

Levin, B. V., Timofeev, N. M., Tuliagonov, S. T. (See Cyrillic Index.)

Lévy, P.
1. Sur les séries dont les termes sont des variables éventuelles indépendants.
Studia Math. **3**(1931), 119–155.
2. *Théorie de l'addition des variables aléatoires.*
Gauthier-Villars, Paris (1937), deux, edit. (1954).
3. *Quelques Aspects de la Pensée d'un Mathématicien.*
Blanchard, Paris, 1970.

Lévy, P., Khinchine, A. Ya.
1. Sur les lois stables.
C.R. Acad. Sci., Paris **202**(1936), 374–376.

Linnik, Yu. V.
1. "The Large Sieve".
Dokl. Akad. Nauk SSSR **30**(1941), 292–294.
2. A remark on the least quadratic non-residue.
Dokl. Akad. Nauk. SSSR(N.S.) **36**(1942), 119–120.
3. On Erdös's theorem on the addition of numerical sequences.
Mat. Sb. (N.S.) **10**(52), (1942), 67–78.
4. *Decomposition of Probability Distributions.*
Oliver and Boyd, Dover, N.Y. (1964), Russian original Moscow (1960).
5. *The Dispersion Method in Binary Additive Problems.*
Translated from the Russian, Amer. Math. Soc. Translations of Math. Monographs, No. 4, Providence, Rhode Island, 1963.

Linnik, Yu. V., Ibragimov, I. A.
1. *Independent and stationary sequences of random variables.*
Edited by J. F. C. Kingman; translated from the Russian, Wolters-Noordhoff, Groningen, The Netherlands, 1971.

Littlewood, J. E., Offord, A. C.
1. On the number of real roots of a random algebraic equation III.
Mat. Sbornik **54**(1943), 277–286.

Littlewood, J. E., Pólya, G., Hardy, G. H.
1. *Inequalities.*
Cambridge, 1934.

Loève, M.
1. *Probability Theory.*
Third Edit., D. Van Nostrand, Princeton, N.J., 1962.

Lubell, D.
1. A short proof of Sperner's lemma.
J. Combinatorial Theory **1**(1966), p. 299.

Lukacs, E.
1. *Characteristic functions.*
Second Edit., Griffin, London, 1970.

Manstavičius, E. (See Cyrillic Index.)

Marcinkiewicz, J.
1. Sur un propriété de la loi de Gauss.
Math. Zeit. **44**(1938), 622–638.

Marek, G. R.
1. *Gentle Genius—The story of Felix Mendelssohn.*
Thomas Y. Crowell Company, New York.

Matthews, K. R.
1. On an inequality of Davenport and Halberstam.
Journ. London Math. Soc. (2)**4**(1972), 638–642.
2. On a bilinear form associated with the Large Sieve.
Journ. London Math. Soc. (2)**5**(1972), 567–570.
3. Hermitian Forms and the Large and Small Sieves.
Journ. Number Theory **5**(1973), 16–23.

Mertens, F.
1. Ein Betrag zur analytischen Zahlentheorie.
Journ. für die reine und angew. Math. **78**(1874), 46–62.

Mirsky, L.
1. *An Introduction to Linear Algebra.*
Oxford, 1955.

Montgomery, H. L.
1. A note on the large sieve.
Journ. London Math. Soc. **43**(1968), 93–98.
2. Mean and large values of Dirichlet polynomials.
Invent. Math. **8**(1969), 334–345.
3. *Topics in Multiplicative Number Theory.*
Springer lecture notes, No. 227, Berlin, 1971.
4. Hilbert's inequality and the Large Sieve.
Proceedings of the Number Theory Conference, Boulder, Colorado (1972), 156–161.

Montgomery, H. L., Vaughan, R. C.
1. Hilbert's inequality.
Journ. London Math. Soc. (2)**8**(1974), 73–82.

Munroe, M. E.
1. *Measure and Integration.*
Addison-Wesley, Reading Mass., London, 1970.

Narkiewicz, W.
1. On additive functions with a non-decreasing normal order.
Colloquium Math. **32**(1974), 137–142.

Niederreiter, H., Kuipers, L.
1. *Uniform distribution of sequences.*
Wiley, New York, 1974.

Norton, K.
1. On the number of restricted prime factors of an integer, I.
Illinois J. Math. **20**(1976), 681–705.
2. On the number of restricted prime factors of an integer, II.
Acta math. **143**(1979), 9–38.

Offord, A. C., Littlewood, J. E.
1. On the number of real roots of a random algebraic equation, III.
Mat. Sbornik **54**(1943), 277–286.

Pan, Chen-tong (Chen-tung)
1. A new application of the Ju. V. Linnik large sieve method.
Acta Mat. Sinica **14**(1964), 597–606; = *Chinese Math.-Acta*, **5**(1964), 642–652.

Paul, E. M.
1. Some properties of additive arithmetic functions.
Sankyā, Series A **29**(1967), 279–282.

Philipp, W.
1. *Mixing sequences of random variables and Probabilistic Number Theory.*
American Math. Soc. Memoirs, No. 114, Providence, Rhode Island, 1971.
2. Arithmetic functions and Brownian motion.
Amer. Math. Soc. Proceedings of Symposia in Pure Math., Vol. XXIV, Analytic Number Theory, 233–246; Providence, Rhode Island, 1973.

Pólya, G., Hardy, G. H., Littlewood, J. E.
1. *Inequalities.*
Cambridge, 1934.

Postnikov, A. G. (See Cyrillic Index.)

Prachar, K.
1. *Primzahlverteilung.*
Springer, Berlin, 1957.

Raikov, D. A. (See Cyrillic Index.)

Ramanujan, S.
1. *Collected Papers.*
Cambridge, 1927.

Ramanujan, S., Hardy, G. H.
1. The normal number of prime factors of a number n.
Quart. Journ. Math. (Oxford)**48**(1917), 76–92.
2. Asymptotic formulae in Combinatory Analysis.
Proc. London Math. Soc. (2)**17**(1918), 75–115.

Rényi, A. (See Cyrillic Index.)
2. On the representation of an even number as the sum of a prime and an almost prime.
Izv. Akad. Nauk SSSR Ser. Mat. **12**(1948), 57–78; =*Amer. Math. Soc. Translations* (2)19(1962), 299–321.
3. On the large sieve of Ju. V. Linnik.
Compositio Math. **8**(1950), 68–75.
4. On the density of certain sequences of integers.
Publ. Inst. Math. Acad. Serbe. Sci. **8**(1955), 157–162.
5. New Version of the Probabilistic Generalization of the Large Sieve.
Acta Math. Hung. **10**(1959), 217–226.
6. A new proof of the theorem of Delange.
Publ. Math. Debrecen **12**(1965), 323–330.
7. *Foundations of Probability.*
Holden-Day, San Francisco, 1970.

Rényi, A., Erdös, P.
1. On the mean value of non-negative multiplicative number-theoretical functions.
Michigan J. Math. **12**(1965), 321–328.

Rényi, A., Turán, P.
1. On a theorem of Erdös-Kac.
Acta Arithmetica **4**(1958), 71–84.

Richert, H.-E., Halberstam, H.
1. Brun's method and the Fundamental Lemma.
Acta Arithmetica **24**(1973), 113–133.
2. *Sieve Methods.*
Academic Press, London, New York, 1974.

Rodosskii, K. A. (See Cyrillic Index.)

Rogozin, B. A.
1. An estimate for concentration functions.
Teor. Ver. i.e. Prim. **6**(1961), No. 1, 103–105; =*Theory of Prob. and its applications* **6**(1961), 94–96.
2. On the increase of dispersion of sums of independent random variables.
Teor. Ver. i.e. Prim. **6**(1961), 106–108; =*Theory of Prob. and its applications* **6**(1961), 97–99.

Roth, K. F.
1. On the Large Sieves of Linnik and Rényi.
Mathematika **12**(1965), 1–9.

Roth, K. F., Halberstam, H.
1. *Sequences, Vol. I.*
Oxford, 1966.

Ruzsa, I., Sárközi, A., Erdös, P.
1. On the number of solutions of $f(n) = a$ for additive functions.
Acta Arithmetica **24**(1973), 1–9.

Ryavec, C.
1. A characterization of finitely distributed additive functions.
Journ. Number. Theory **2**(1970), 393–403.

Ryavec, C., Elliott, P. D. T. A.
1. The Distribution of the values of Additive Arithmetical Functions.
Acta math. **126**(1971), 143–164.

Ryavec, C., Erdös, P.
1. A characterization of finitely monotonic additive functions.
Journ. London Math. Soc. (2)**5**(1972), 362–367.

Sárközi, A., Erdös, P., Ruzsa, I.
1. On the number of solutions of $f(n) = a$ for additive functions.
Acta Arithmetica **24**(1973), 1–9.

Sathe, L. G.
1. On a problem of Hardy on the distribution of integers having a given number of prime factors, I–IV.
J. Indian Math. Soc. **17**(1953), 63–82, 83–141, and **18**(1954), 27–42, 43–81.

Schmidt, R.
1. Über divergente Folgen and lineare Mittelbildungen.
Math. Zeit. **22**(1925), 89–152.

Schoenberg, I. J.
1. Über die asymptotische Verteilung reeller Zahlen (mod 1).
Math. Zeit. **28**(1928), 171–199.
2. Über total monotone Folgen mit stetiger Belegungsfunktionen.
Math. Zeit. **30**(1929), 761–767.
3. On asymptotic distributions of arithmetical functions.
Trans. Amer. Math. Soc. **39**(1936), 315–330.

Segal, S. L.
1. A note on normal order and the Euler φ-function.
Journ. London Math. Soc. **39**(1964), 400–404.
2. On non-decreasing normal order.
Journ. London Math. Soc. **40**(1965), 459–466.

Selberg, A.
1. On an elementary method in the theory of primes.
Norske Vid. Selsk. Forh. Trondhjem **19**(1947), No. 18, 64–67.
2. An elementary proof of the prime number theorem.
Ann. of Math. (2)**50**(1949), 305–313.
3. Note on a paper by L. G. Sathe.
J. Indian Math. Soc. **18**(1954), 83–87.

Shapiro, H. N.
1. Distribution functions of additive arithmetic functions.
Proc. Nat. Acad. Sci. U.S.A. **42**(1956), 426–430.

Shohat, J., Fréchet, M.
1. A proof of the generalized central limit theorem.
Trans. Amer. Math. Soc. **33**(1931), 533–543.

Sierpinski, W.
1. Sur l'équation fonctionelle $f(x + y) = f(x) + f(y)$.
Fund. Math. **1**(1920), 116–122.

Steinhaus, H.
1. Sur les distances des points des ensembles de mesure positive.
Fund. Math. **1**(1920), 93–104.

Steinig, J., Diamond, H.
1. An elementary proof of the prime number theorem with a remainder term.
Invent. Math. **11**(1970), 199–258.

Stepanov, S. A.
1. On the number of points of a hyperelliptic curve over a finite field.
Izv. Akad. Nauk **33**(1969), No. 5, 1171–1181; =*Mathematics of the USSR—Izvestija* **3**(1969), No. 5, 1103–1114.

Szüsz, P.
1. Remark to a theorem of P. Erdös.
Acta Arithmetica **26**(1974), 97–100.

Tchudakoff, N. G.
1. Theory of the characters of number semigroups.
J. Indian Math. Soc. N.S. **20**(1956), 11–15.

Theodorescu, R., Hengartner, W.
1. *Concentration Functions.*
Academic Press, New York, London, 1973.

Timofeev, N. M.
1. Estimation of the Remainder Term in one-dimensional Asymptotic Laws.
Dokl. Akad. Nauk **200**(1971), No. 2, 298–302; =*Soviet Math. Dokl.* **12**(1971), No. 5, 1401–1405.

Timofeev, N. M., Levin, B. V. (See Cyrillic Index.)
3. On the distribution of values of additive functions.
Acta Arithmetica **26**(1974/1975), No. 4, 333–364.

Timofeev, N. M., Tuliagonov, S. T., Levin, B. V. (See Cyrillic Index.)

Titchmarsh, E. C.
1. A divisor problem.
Rend. del. Circ. Mat. di Palermo **54**(1930), 414–419; **57**(1933), 478–479.
2. *The Theory of Functions.*
Oxford, second edit., 1939.
3. *Introduction to the theory of Fourier integrals.*
Oxford, second edit., 1948.
4. *The Theory of the Riemann Zeta-Function.*
Oxford, 1951.

Tjan, M. M. (See Cyrillic Index.)

Toleuov, J., Faĭnleĭb, A. S. (See Cyrillic Index.)

Tuliagonov, S. T., Levin, B. V., Timofeev, N. M. (See Cyrillic Index.)

Turán, P.
1. Az egész számok prímosztóinak számáról.
Mat. Lapok. **41**(1934), 103–130.
2. On a theorem of Hardy and Ramanujan.
Journ. London Math. Soc. **9**(1934), 274–276.
3. Über einige Verallgemeinerungen eines Satzes von Hardy und Ramanujan.
Journ. London Math. Soc. **11**(1936), 125–133.

Turán, P., Erdös, P.
1. On a problem in the theory of uniform distribution, I, II.
Indig. Math. **10**(1948), 370–378; 406–413.

Turán, P., Rényi, A.
1. On a theorem of Erdös–Kac.
Acta Arithmetica **4**(1958), 71–84.

Uzdavinis, R. V. (See Cyrillic Index.)

Vaughan, R. C., Montgomery, H. L.
1. Hilbert's Inequality.
Journ. London Math. Soc. (2)**8**(1974), 73–82.

Vinogradov, A. I. (See Cyrillic Index.)

Vinogradov, A. I., Barban, M. B.
1. On the number theoretic basis of probabilistic number theory.
Dokl. Akad. Nauk SSSR **154**(1964), 495–496; =*Soviet Math. Doklady* **5**(1964), 96–98.

Vinogradov, A. I., Barban, M. B., Levin, B. V. (See Cyrillic Index.)

Vinogradov, I. M.
1. *The Method of Trigonometrical Sums in the Theory of Numbers.*
Translated from the Russian, revised and annotated, by Anne Davenport and K. F. Roth. Interscience, London, New York.

Viola, C., Forti, M. C.
1. On large sieve type estimates for the Dirichlet series operator.
Amer. Math. Soc. Symposia in Pure Math. XXIV, Providence, Rhode Island (1973), 31–49.

Walfisz, A., Landau, E.
1. Über die Nichtfortsetzbarkeit einiger durch Dirichletsche Reihen definierter Funktionen.
Rend. di Palermo **44**(1919), 82–86.

Weil, A.
1. Sur les courbes algébriques et les variétés qui s'en deduisent.
Actualités math. sci., No. 1041, *Paris*, 1945, *Deuxième Partie*, §IV.

Weyl, H.
1. Über die Gleichverteilung von Zahlen mod Eins.
Math. Annalen **77**(1916), 313–352.

Wintner, A.
1. Über den Konvergenzbegriff der matematischen Statistik.
Math. Zeit. **28**(1928), 476–480.
2. Statistics and Prime Numbers
Letter to *Nature* **147**(1941), 208–209.
3. *Eratosthenian Averages.*
Waverly Press, Baltimore, 1943.
4. *The Theory of Measure in Arithmetical Semigroups.*
Waverly Press, Baltimore, 1944.

Wintner, A., Erdös, P.
1. Additive arithmetical functions and statistical independence.
Amer. Journ. Math. **61**(1939), 713–721.

Wintner, A., Jessen, B.
1. Distribution functions and the Riemann zeta function.
Trans. Amer. Math. Soc. **38**(1935), 48–88.

Wirsing, E.
1. Das asymptotische Verhalten von Summen über multiplikative Funktionen.
Math. Annalen **143**(1961), 75–102.
2. Elementare Beweise des Primzahlsatzes mit Restglied, I.
J. reine und angew. Math. **211**(1962), 205–214.
3. Elementare Beweise des Primzahlsatzes mit Restglied, II.
J. reine und angew. Math. **214/215**(1964), 1–18.
4. Das asymptotische Verhalten von Summen über multiplikative Funktionen, II.
Acta Math. Acad. Sci. Hung. **18**(1967), 411–467.

5. A characterization of log *n* as an additive function.
Symposia Mathematica IV(1970), 45–57. Academic Press, London, New York.

Wolke, D.
1. Moments of the Number of Classes of Primitive Quadratic Forms with Negative Discriminant.
Journ. Number Theory **1**(1969), 502–511.
2. Das Selbergsche Sieb für zahlentheoretische Funktionen, I.
Archiv. der Mathematik, **24**(1973), 632–639.

Wright, E. M., Hardy, G. H.
1. *An introduction to the Theory of Numbers.*
Oxford, fourth edit., 1960.

Yong, Chi-hsing
1. *Asymptotic behaviour of Trigonometric Series with modified monotone coefficients.*
The Chinese University of Hong Kong, Monograph (1974), Lib. of Congress Number 73-78286.

Yosida, Kôsaku
1. *Functional Analysis.*
Springer, Berlin, third edit., 1971.

Zolotarev, V. M. (See Cyrillic Index.)
1. On a general theory of multiplication of independent random variables.
Izv. Akad. Nauk **142**(1962), No. 4, 788–791; =*Soviet Math.* (*Amer. Math. Soc. Translation*) **3**(1962), 166–170.

Литература

In this short cyrillic index I give a number of references which are difficult or at the moment impossible to obtain in English translation. A number of papers by some of these authors, and which are available in English, are listed in the previous collection of references.

Бакштис, А.

1. О предельных законах распределения мультипликативных арифметических функций.
   Liet. mat. rinkinys. = Лит. Мат. Сб. 8 (1968), (1), 5–20.

Барбан, М.Б.

1. Новые применения "Большого Решета" Ю.В. Линника.
   Акад. Наук Узбекской ССР, Труд. Инст. мат. В.И. Романовского теор. вероят. мат. стат. 22 (1961), 1–20.
2. Арифметические функции на "редких" множествах.
   Акад. Наук Узбекской ССР, Труд. Инст. мат. В.И. Романовского теор. вероят. мат. стат. 22 (1961), 21–35.
3. Нормальный порядок аддитивных арифметических функций на множестве "сдвинутых" простих чисел.
   Acta. Math. Acad. Sci. Hungar. 12 (1961), 409–415.
4. Аналог закона Больших Чисел для аддитивных арифметических функций заданных на множестве "сдвинутых" простих чисел. Д.А.Н. УзССР 12 (1961), 8–12.
5. Об одной теореме И.П. Кубилюса.
   Изв. А.Н. УзССР Сер. физ. мат. н. (1961), (5), 3–9. See also the corrigendum in (1963), (1), 82–83.
6. Арифметические функции на "редких" множествах.
   УзССР фанлар Акад. докладнари, Д.А.Н. УзССР (1961), (8), 10–12.
7. "Большое Решето" Ю.В. Линника и предельная теорема для числа классов идеалов мнимого квадратического поля.
   Изв. А.Н. СССР, Сер. мат. 26 (1962), (4), 573–580.
8. Замечание к работе автора "Новые применения", "Большого Решета", "Ю.В. Линника".
   Акад. Наук Узбекской ССР, Труд. Инст. мат. В.И. Романовского теор. вероят. мат. стат. (1964), (1), 130–133.

Барбан, М.Б., Виноградов, А.И., Левин, Б.В.

1. Пределные законы для функций класс H И.П. Кубилюса, заданных на множестве "сдвинутых" простих чисел.
   Liet. mat. rinkinys = Лит. мат. Сб. 5 (1965), 5–8.

Виноградов, А.И.

1. О плотностной гипотезе для L — рядов Дирихле.
   И.А.Н. СССР. Сер. мат. 29 (1965), 903–934.
2. Исправления к работе А.И. Виноградова "О плотностной гипотезе для L — рядов Дирихле".
   И.А.Н. СССР. Сер. мат. 30 (1966), 719–720.

Виноградов, А.И., Левин, Б.В., Барбан, М.Б.
1. Пределные законы для функций класса H И.П. Кубилюса, заданных на множестве "сдвинутых" простих чисел.
Liet. mat. rinkinys = Лит. мат. Сб. 5 (1965), 5–8.

Золотарев, В.М.
2. Безранично делимых законов класса L.
Liet. mat. rinkinys = Лит. мат. Сб. 3 (1963), 123–140.

Кубилюс, И.П.
2. Об одном классе аддитивных арифметических функций, распределенных асимптотических по нормальному закону.
Fiz.-techn. inst. darbai. Liet. TSR mokslu akad., Тр. Физ.-техн. ин.-та. А.Н. Лит. ССР 2 (1956), 5–15.
3. Асимптотическое разложение законов распределения некоторых арифметических функций.
Liet. mat. rinkinys = Лит. мат. Сб. 2 (1962), (1), 61–73.
4. О некоторых задачах вероятностной теории чисел.
"Тр. VI Всес. совещания по теории вероятностей и матем. стат. 1960", Вилнюс Гос. изд.-во полит. и научн. лит. ЛитССР, (1962), 57–68.

Левин, Б.В., Барбан, М.Б., Виноградов, А.И.
1. Пределные законы для функций класса H И.П. Кубилюса, заданных на множестве "сдвинутых" простих чисел.
Liet. mat. rinkinys = Лит. мат. Сб. 5 (1965), 5–8.

Левин, Б.В., Тимофеев, Н.М.
1. Аналитический метод в Вероятностной Теории Чисел.
Уч. зап. Владимирского гос. пед. ин.-та. мат. 57, (2), (1971), 57–150.
2. Распределеные значений аддитивных функций.
Успехи мат. наук 28 (1), (169), (1973), 243–244.

Левин, Б.В., Тимофеев, Н.М., Туляганов, С.Т.
1. Распределение значений мультипликативных функций.
Liet. mat. rinkinys = Лит. Мат. Сб. 13 (1), (1973), 87–100.

Манставичюс, Э.
1. О распределении аддитивных арифметических функции (mod 1). Liet. mat. rinkinys = Лит. Мат. Сб. 13 (1973), 101–108.

Постников, А.Г.
1. Введние в Аналитическую Теорию Чисел.
Изд. "Наука" Глав. пед. физ.-мат. лит., Москва, 1971.

Райков, Д.А.
1. О разложении законов Пуассона.
И.А.Н. СССР 14 (1937), (1), 9–12.
2. О разложении законов Гаусса и Пуассона.
И.А.Н. СССР 2 (1938), 91–124.

Реньи, А.
1. О представлении четных чисел в виде сумны простого и почти простого числа.
Д.А.Н. СССР 56 (1947), 455–458.

Родосский, К.А.
1. О степенных невычетах и нулях L — функций.
И.А.Н. Сер. Мат. 20 (1956), 303–306.

Тимофеев, Н.М., Левин, Б.В.
1. Аналитический метод в Вероятностной Теории Чисел.
Уч. зап. Владимирского гос. пед. ин.-та. мат. 57, (2), (1971), 57–150.
2. Рарпределение значений аддитивных функций.
Успехи мат. наук 28, 1, (169), (1973), 243–244.

Тимофеев, Н.М., Туляганов, С.Т., Левин, Б.В.
1. Распределение значений мультипликативных функций.
   Liet. mat. rinkinys = Лит. Мат. Сб. 13 (1), (1973), 87–100.

Толеуов, Ж., Файнлейб, А.С.
1. О распределении значений арифметических функций на некоторых подмножествах натурального ряда.
   УзССР Фанлар Акад. ахбороти. физ. — матем. фанлари сер., Изв. А.Н. УзССР, Сер. физ. матем. Н., 13 (1969), (5), 23–27.

Туляганов, С.Т., Левин, Б.В., Тимофеев, Н.М.
1. Распределение значений мультипликативных функций.
   Liet. mat. rinkinys = Лит. Мат. Сб. 13 (1), (1973), 87–100.

Тян, М.М.
1. К вопросу о распределении значений функций Эйдера Ф(n).
   Liet. mat. rinkinys = Лит. Мат. Сб. 6 (1), (1966), 105–119.

Уждавинис, Р.В.
1. О распределении значений аддитивных арифметических функций от целочисленных полиномов.
   Liet. TSR Mokslu Akad darbai, Ser. B 2 (18), (1959) = Труд. Акад. Наук ЛитССР, Сер. Б. 2 (18), (1959), 9–29.
2. Кандидатская диссертация, Вилнюс, 1961.
3. Некоторые предельные теоремы для аддитивных арифметических функций.
   Liet. mat. rinkinys = Лит. Мат. Сб. (1961), (1), (1–2), 355–364.
4. Аналог теоремы Эрдёша — Винтнера для последовательности значений целочисленного полинома.
   Liet. mat. rinkinys = Лит. Мат. Сб. 7 (1967), (2), 329–338.

Файнлейб, А.С., Толеуов, Ж.
1. О распредлении значений арифметических функций на некоторых подмножествах натурального ряда.
   УзССР Фанлар Акад. ахбороти Физ. — матем. фанлари сер., Изв. А.Н. УзССР, Сер. физ. матем. Н., 13 (1969), (5), 23–27.

Author Index

van Ardenne–Ehrenfest, de Bruijn and Korevaar 18, 77
Abel 352
Aczél 76
Akilov and Kantorovich 181

Bakstys 280
Banach–Hahn 181
Barban 9, 12, 80, 92, 93, 136, 184
Barban and Vinogradov 122, 126
Barban, Vinogradov and Levin 174
Behrend 4, 189, 214
Bernoulli 36
Berry 74, 78
Berry–Esseen 74, 101, 117, 217
Bertrand 3
Bessel–Hagen 3, 4, 189
Billingsley 30
Birch 14
Bombieri 12, 90, 92, 93, 184, 185, 186, 317
Bombieri and Davenport 185
Borel 3
Borel–Cantelli 45, 46
Borel–Carathéodory 58
de Bruijn 77
de Bruijn, van Aardenne–Ehrenfest and Korevaar 18, 77
Brun 4, 9, 80, 134, 184, 210, 213
Brun–Titchmarsh 90, 120, 135, 160, 161, 213
Burgess 154, 157

Cantelli–Borel 45, 46
Carathéodory–Borel 58
Cartwright 94
Cauchy 7, 14, 17, 20, 46, 76, 231, 239, 283, 289, 290, 329
Cauchy–Schwarz 42, 68, 135, 149, 150, 151, 160, 161, 164, 168, 174, 196, 197, 201, 202, 234, 243, 246, 271, 300, 304, 317, 338, 339, 345, 348, 349
Chowla, S. 4, 189
Courant–Fisher 163

Daboussi and Delange 358
Davenport 4, 111, 189, 214
Davenport and Bombieri 185
Davenport and Halberstam 12, 185
Dedekind 114
Delange 10, 11, 218, 219, 225, 226, 254, 256, 258, 283, 285, 286, 301, 305, 333
Delange and Daboussi 358
Desargues 12
Diamond 219
Dirichlet 10, 12, 14, 15, 79, 92, 94, 96, 98, 100, 101, 108, 109, 110, 111, 112, 114, 183, 184, 186, 222, 224, 225, 235, 301, 308, 311, 317, 322, 326, 331, 335, 337, 341, 342, 348, 356
Doeblin 217

Edwards 68
Egoroff 18
Einstein 3
Elliott 12, 13, 75, 76, 144, 153, 155, 185, 217, 218, 219, 269, 283, 286, 292, 295, 333
Elliott and Ryavec 10, 257, 258, 265, 268, 283, 306
Eratosthenes 221, 254, 266
Erdös 3, 4, 5, 10, 11, 90, 91, 118, 187, 189, 203, 207, 210, 211, 212, 213, 214, 218, 219, 220, 254, 258, 265, 283, 285, 301, 302
Erdös and Kac 4, 5, 14, 146, 182, 214
Erdös–Kac–Kubilius 9
Erdös and Ryavec 268
Erdös and Selberg 10, 256
Erdös and Turán 75
Erdös and Wintner 4, 10, 187, 214, 254, 280
Esseen 74, 78, 203, 218
Esseen–Berry 74, 101, 117, 217
Euclid 4
Euler 89, 95, 96, 97, 114, 134, 188, 203, 230, 326, 338, 341, 349, 353

Faĭnleĭb 74, 75, 203, 219
Faĭnleĭb and Levin 11, 286
Fejér 3, 70, 78, 101, 229
Feller–Lindeberg 56
Fisher–Courant 163
Forti and Viola 12, 185, 186, 317
Fourier–Stieltjes 61, 305
Fréchet–Shohat 59, 78

Galambos 218, 280, 281
Gallagher 12, 93, 185
Gauss 1, 51, 94
Gershgorin 165, 316
Gnedenko 53, 55, 56
Gnedenko and Kolmogorov 24, 26, 28, 29, 49, 50, 51, 53, 54, 55, 56, 57

Hadamard 1, 29
Hahn–Banach 181
Halász 10, 12, 21, 77, 114, 224, 225, 226, 233, 252, 255, 256, 286, 308, 312, 317, 330, 331
Halberstam 6
Halberstam and Davenport 12, 185
Halberstam and Richert 80, 185
Halberstam and Roth 77
Hamel 76
Hardy 1, 3, 102
Hardy–Littlewood (Hardy and Littlewood) 2, 77, 102, 254, 348, 351, 358
Hardy, Littlewood and Pólya 186
Hardy and Ramanujan 2, 3, 5, 13, 14
Hardy and Wright 85, 89, 108, 112, 118, 133, 134, 224
Hartman 292
Heilbronn 94, 114
Hengartner and Theodorescu 218
Hermite 12, 162, 166
Hilbert 181, 185
Hölder 179, 186, 236, 313, 329

Ibragimov and Linnik 78
Ikehara–Wiener 100, 101, 102

Jessen and Wintner 46, 78, 193

Kac 4, 5, 146
Kac–Erdös (Kac and Erdös) 4, 5, 14, 146, 182, 214
Kac–Kubilius–Erdös 9
Kantorovich and Akilov 181
Karamata 18, 77
Kesten 218

Khinchine 55, 77
Khinchine and Lévy 49, 57
Kobayashi 12, 185
Kolmogorov 3, 32, 44, 45, 51, 52, 53, 56, 77, 217
Kolmogorov and Gnedenko 24, 26, 28, 29, 49, 50, 51, 53, 54, 55, 56, 57
Korevaar, van Aardenne–Ehrenfest and de Bruijn 18, 77
Kronecker 110, 111
Kubilius 5, 6, 11, 80, 115, 119, 122, 123, 125, 126, 128, 129, 138, 139, 140, 142, 144, 145, 146, 147, 148, 180, 181, 258, 286
Kubilius–Erdös–Kac 9
Kubilius–Turán 6, 13, 147, 152, 158, 173, 182, 185, 192, 199, 218
Kuipers and Niederreiter 78
Kummer 111

Lambert 219
Landau 1, 94, 111, 254
Landau and Walfisz 100
Lang 101
Lebesgue 16, 19, 21, 22, 23, 48, 59, 67, 107, 222, 228, 258, 260, 359
Legendre 110, 154
LeVeque 6, 75, 76
Levin, Barban and Vinogradov 174
Levin and Faĭnleĭb 11, 286
Levin and Timofeev 10, 12, 257, 258
Levin, Timofeev and Tuliaganov 274, 275, 280, 281
Lévy 3, 14, 24, 25, 31, 46, 50, 51, 52, 53, 57, 78, 193, 195, 217, 219
Lévy and Khinchine 49, 57
Liapounoff 218
Lindeberg–Feller 56
Linnik 8, 9, 58, 78, 93, 112, 113, 183, 184, 218, 317
Linnik and Ibragimov 78
Littlewood–Hardy (Littlewood and Hardy) 2, 77, 102, 254, 348, 351, 358
Littlewood and Offord 217
Littlewood, Pólya and Hardy 186
Loève 127
Lubell 78
Lukacs 78

von Mangoldt 97, 311
Mann 77, 293
Manstavičius 286, 305
Marcinkiewicz 58
Marek 111
Matthews 12, 185
Mellin 61, 78, 279, 307, 322
Mellin–Stieltjes 61, 141
Mendelssohn, Fanny 111
Mendelssohn, Felix 111
Mendelssohn, Moses 112
Mendelssohn, Rebecka 111, 112
Mertens 89
Mirsky 162, 166
Möbius 85, 251, 254, 282, 301
Montgomery 93, 185, 229, 235, 317
Montgomery and Vaughan 185
Mordell 213
Munroe 18

Niederreiter and Kuipers 78

Offord and Littlewood 217

Pan 184
Parseval 8, 23, 228, 235, 236
Perron 94, 95, 322
Philipp 127
Plancherel 22, 23
Poincaré 3
Poisson 52
Pólya, Hardy and Littlewood 186

Pólya–Vinogradov 154
Postnikov 11, 74, 78
Prachar 90, 91, 92, 95

Rado 214
Raikov 58, 112, 113
Ramanujan 1
Ramanujan and Hardy 2, 3, 5, 13, 14
Rényi 3, 9, 12, 24, 47, 92, 93, 183, 184, 218, 317
Rényi and Turán 6
Richert and Halberstam 80, 185
Riemann 96, 97, 154, 338, 341
Rogozin 32, 78, 218
Roth 12, 184
Roth and Halberstam 77
Ryavec 283, 289
Ryavec and Elliott 10, 257, 258, 265, 268, 283, 306
Ryavec and Erdös 268

Schmidt, R 18, 77
Schnirelmann 77, 292, 293, 294, 296
Schoenberg (Schönberg) 4, 189, 213, 214
Schur 214
Schwarz–Cauchy 42, 68, 135, 149, 150, 151, 160, 161, 164, 168, 174, 196, 197, 201, 202, 234, 243, 246, 271, 300, 304, 317, 338, 339, 345, 348, 349
Selberg 79, 80, 84, 119, 127, 129, 142, 145, 176, 182, 185, 213
Selberg and Erdös 10, 256
Shohat–Fréchet 59, 78
Siegel–Walfisz 91
Sierpinski 76
Sperner 32, 78, 214, 216, 218
Steinhaus 16, 76, 261, 283
Stepanov 155
Stieltjes–Fourier 61, 305
Stieltjes–Mellin 61, 141
Stirling 34
Stolz 338
Stone–Weierstrass 307
Szüsz 224

Tauber 77, 100, 101, 102, 254, 351, 358
Taylor 104
Tchebycheff 3, 36, 44, 192, 235
Theodorescu and Hengartner 218
Timofeev and Levin 10, 12, 257, 258
Timofeev, Tuliaganov and Levin 274, 275, 280, 281
Titchmarsh 10, 22, 58, 59, 70, 90, 94, 95, 97
Titchmarsh–Brun 90, 120, 135, 160, 161, 213
Tjan 219
Tuliaganov, Timofeev and Levin 274, 275, 280, 281
Turán 3, 4, 5, 6, 147, 180, 181, 182, 185
Turán and Erdös 75
Turán–Kubilius 6, 13, 147, 152, 158, 173, 182, 185, 192, 199, 218
Turán and Rényi 6

Uzdavinis 134

de la Vallée–Poussin 1, 29
Vaughan and Montgomery 185
Vinogradov, A. I. 92, 93, 184
Vinogradov, A. I. and Barban 122, 126
Vinogradov, A. I., Barban and Levin 174
Vinogradov, I. M. 2, 7, 8, 154, 155
Vinogradov–Pólya 154
Viola and Forti 12, 185, 186, 317

Walfisz and Landau 100
Walfisz–Siegel 91
Weierstrass–Stone 307

Weil 155
Weyl 69, 75, 284
Wiener 68, 100
Wiener–Ikehara 100, 101, 102
Wintner 3, 10, 59, 78, 254, 285
Wintner and Erdös 4, 10, 187, 214, 254, 280
Wintner and Jessen 46, 78, 193
Wirsing 10, 11, 90, 144, 225, 226, 227, 254, 255, 256, 273, 331
Wolke 183
Wright and Hardy 85, 89, 108, 112, 118, 133, 134, 224

Yong, Chi–hsing 77
Yosida 181

Zolotarev 61, 141, 274, 279, 307

Subject Index

Abundant numbers 3–4, 189
Additive function, finitely distributed 11, (definition), 258, 259, 260, 267, 270, 275, 276, 283
Adjoint of operator 181–182
Algebra of sets 29–30, 115–146
Analytic characteristic functions 57, 112–113
Asymptotic density, *See* Notation
  lower 295, 297
Asymptotic relations 19–21

Basis 294
Berry–Esseen theorem 74, 117
Borel–Cantelli lemma, applications 45–46
Borel–Carathéodory lemma 58
Brownian motion 3
Brun's sieve 4, 9, 80, 129

Cauchy law 14
Cauchy's functional equation 17, 20, 76, 283, 289
Central Limit Theorem 218
Character, Dirichlet 110–111
  primitive 110
Characteristic function 9, (definition) 27
  analytic 57, 112–113
  component of 113–114
  convergence of 28

Circle method 2, 6–9
Classification 13
Class Number, quadratic 14, 110–111, 117
Coefficients, Fourier 66
Compactness lemma 25
Component
  of characteristic function 113–114
  of distribution function 113
Concentration function 31, 217–218
Conditional probability 31, 35
Conjugate 12, 181–182
Continuity criterion, Lévy's 46, 78
Continuity of distribution function 48–49
Control, mathematical 13
Courant–Fisher theorem 163–164
Convergence
  characteristic functions 28
  Fourier coefficients 67–68
  Mellin–Stieltjes transforms 63
  modified–weak 63, (definition) 273–274, 280
  weak 24–30
Convolution 30, 254–255
Convolutions, infinite 37
Cyclotomic field 114

Decomposition 13
Dedekind–Dirichlet series 114
Density
  asymptotic, *see* Notation

Density [*cont.*]
lower = density, lower asymptotic 295, 297
Schnirelmann 77, 293
Desargues' theorem 12
Differences 16
Differential equation, approximate 13
Dirichlet
L-series 14
marriage of 111–112
multiplication, convolution 98, 109
Dirichlet series 79, 94
component 114
operator 186
Discriminant
fundamental 111
of quadratic field 110–111
Dispersion method 93
Distribution function 24
continuity of 48–49
convergence 24, 25
improper 25
proper 25
Distribution functions (mod 1) 65
continuity of 67
convergence 65–68
discontinuous, quantitative Fourier inversion 75–76
Distribution law 24
Dual 12, 13
of operator 12, 181–182
space 181–182
of Turán–Kubilius inequality 13, 147–186, 194, 335
Duality principle 150, 162, 185–186, 316–317

Erdös and Turán inequality 75
Erdös' sample paper 207–210
commentary on 210–213
Erdös–Wintner theorem 187–224
Esseen–Berry theorem 74, 117
Esseen's inequality 74
Euler product 95, 97, 114, 230, 326, 338, 341, 349, 353
Euler's constant 89
Euler's function, distribution of 188–189, 213, 214

Fejér kernel 78, 101, 229
Feller–Lindeberg condition 56
Finite probability space 5, 115–146
Finitely distributed additive function 11, (definition) 258, 259, 260, 267, 270, 275, 276, 283
Finitely monotonic additive function 268–269
Fisher–Courant theorem 163–164
Fourier coefficients 66
Fourier inversion (mod 1), quantitative 74–76
Fourier inversion, quantitative 69
Fourier–Stieltjes transform 61 *See also* Characteristic Function
Fréchet–Shohat–Wintner 59–60, 78
Functional equation, Cauchy's 17, 20, 76, 283, 289
Functions, slowly oscillating 18
Fundamental lemma 80

Gaussian component 51, 94
Geometry, Plane Projective 12
Gershgorin discs 165, 316
Group of substitutions 13

Hardy–Littlewood circle method 2, 6–9
Hardy–Littlewood tauberian theorem 77, (statement) 102, 254, 348, 351, 358
Hermitian matrix, operator 162
Hermitian operator 12
spectral radius 12, 162
Highest common divisor of a sequence 295
Hilbert's inequality 185

Improper distribution function 25
Independence and divisibility by primes 146
Independence in probabilistic number theory 4, 146
Independent random variables 30
Infinite convolutions 37
Infinitely divisible law 49
  characteristic function according to Kolmogorov 51
  characteristic function according to Lévy–Khinchine 49–50
Infinitely divisible laws, convergence of 53
Infinitesimal variable 54
Integral equation, approximate 11
Inversion formula 28

Khinchine–Lévy representation 49–50, 57
Kolmogorov's inequality 44
Kronecker symbol 110–111
  primitive character 110
Kubilius' model 115–146
  construction of 119
Kubilius–Turán inequality 6, 13, 147–186, 192, 199
  dual of 13, 147–186, 194, 335

L^2-norm 12, 23
L-series, Dirichlet 14, 110–111, 114
Lambert series 219
Large deviation inequality for random variables 127
Large Sieve 9, 10, 12, 13, 93, 165, 183–186, 317
Law
  Cauchy 14
  normal 11, (definition) 52
  Poisson 52–53
  stable 11
Laws on a finite interval 58
Lebesgue L^2-class 22
Legendre symbol 110, 154
LeVeque's conjecture 6
Lévy metric (distance) 14, 24
Lévy representation (formula) 50
Lévy representation (modified) 51
Lévy–Khinchine representation 49–50, 57
Lévy's continuity criterion 46, 78
Limit Law 24
Limiting distribution of a strongly additive function 4
Lindeberg–Feller condition 56

Major Arcs, intervals 7
von Mangoldt's function 97, 311
Mann's theorem 77, 293
Mean of a random variable 30
Measure 16, probability 29, 30, 115–146
Mellin transform, M-transform 61, 141, 279, 307, *See also,* 94–95, 233, 322–326
Mellin–Stieltjes transform, 61, 141, 279, 307
Metric, Lévy 14, 24
Minor Arcs, intervals 7, 8, 9
Möbius inversion 85
Model for multiplicative functions 140
Models for strongly additive functions 115–146
Modified–Lévy representation 51
Moments
  determination by 60–61
  method of 59

Natural boundary 100
Norm
  algebraic 114
  L^2 12, 23
  operator 181, 186
Normal law 11, (definition) 52
Normal number of prime factors 2
Normal order of an arithmetic function 2, 14

Operator
adjoint of 181–182
Dirichlet-series type 186
dual of 12, 181–182
norm 181, 186

Parsevals' relation (*See also* Plancherel's identity) 8, 23, 228, 235, 236
Partitions 2
Perfect numbers 4
Perron's theorem 94–95, 322–326
Plancherel's identity (*See also* Parseval's relation) 23, 228, 235, 236
Plancherel's theory 22
Poisson law 52–53
Pólya–Vinogradov inequality 154
Prime Ideal Theorem 94
Prime Number Theorem 10, 90, 145, 254, 283
elementary proof 10, 90
Prime numbers
in Arithmetic Progressions 90–92
distribution 89
Primitive root, least positive 158
Probability measure 29, 30, 115–146
Probability space 29, 115–146
Products of independent random variables 141, 142, 144
Proper distribution function 25
Purity of type 46, 78, 292

Quadratic Class Number 14, 110–111, 117
Quadratic residues, least pair of 153

Random variable 29
infinitely divisible 49
Relations asymptotic 19–21
Riemann Hypothesis 97
Riemann zeta function 96
functional equation 97

Schnirelmann density 77, 293
Schnirelmann sum 293
Selberg's sieve method 79–89, 120–121, 129, 142, 145, 176, 185, 213
'Shifted primes' 9
Sieve
Brun 4, 9, 80, 129
large 9, 10, 12, 13, 93, 165, 183–186, 317
Selberg 79–89, 120–121, 129, 142, 145, 176, 185, 213
Skew–Hermitian form 166
Slowly decreasing functions 102
Slowly oscillating functions 18
Spectral radius 12, 162
of Hermitian matrix, operator 12, 162
Sperner's lemma 32, 78, 214, 216, 218
Stable law 11
Substitutions, group of 13
Sums of independent random variables, limit theorems 54
Surrealistic Continuity Theorem 265, 269

Tauberian theorem
Hardy and Littlewood 77, (statement) 102, 254, 348, 351, 358
Wiener–Ikehara 100–101
Tchebycheff's inequality 3, 192
Three Series Theorem, Kolmogorov 37–38, 77
Total event 29
Truncated additive functions 5
Turán and Erdös inequality 75

Turán–Kubilius inequality 6, 13, 147–186, 192, 199
dual of 13, 147–186, 194, 335
Type 26

Uniform distribution (mod 1) 66, 69
Uniform law (mod 1) 66

Variance, random variable, σ^2, D^2 30–31
Vinogradov–Pólya inequality 154

Weak convergence
of distribution functions 24, 25
of distribution functions (mod 1) 65–68
of measures 24, 30
Weyl's criterion 69
quantitative 75
Wiener–Ikehara tauberian theorem 100–101
Wintner–Erdös theorem, 187–224

$\mathbb{Z}$-module 289
Zeta function (Riemann) 97